# Microwave Discharges: Theory & Applications

## International Workshop
## Zvenigorod, Russia

Authors Listed at Back

Edited By

Yu. A. Lebedev

## Plasma Physics Series

*Wexford Press*
*2008*

# PREFACE

This book is a result of the IV International Workshop on "Microwave Discharges: Fundamentals and Applications" which was held in September 18-22, 2000 in Zvenigorod, Russia. The main purposes were to discuss recent achievements in the study of microwave plasma, identify directions for future researches, and promote close relationship between scientists from different countries.

Most of the authors of the book are well known specialists in different areas of microwave plasma physics, technique and plasma processing. Topics cover all problems of theory, experiments and applications of microwave discharges and yield the state-of-the-art and trends in:

- discharge theory, modelling, and diagnostics,
- methods of microwave plasma generation,
- high and low pressure microwave discharges,
- continuos wave and pulsed microwave discharges,
- interaction of microwaves with a plasma,
- applications of microwave plasma (surface treatment, etching, film deposition, growth of structures, light sources, analytical chemistry, etc).

We hope that this book will be useful for all specialists who work in low temperature plasma physics and processing.

The Organizing Committee wishes to thank the following for their contribution for success of this Workshop:

- The Russian Business
- European Office of Aerospace Research and Development, Air Force Office of Scientific Research, United States Air Force Research laboratory
- The Russian Foundation for Basic Research.

We would like to acknowledge the people who helped to organize the Workshop. Our thanks to the Organizing Committee with a special thanks to Dr. V.A. Ivanov, Dr. I.V. Malyukova and T.E. Krivonosova for their invaluable contribution in the held of the Workshop. We would like also to acknowledge all participants for interesting reports and discussions which made this scientific meeting successful.

Yuri A. Lebedev

# CONTENTS

# THEORY
# AND MODELLING

# INFLUENCE OF THE WAVE FREQUENCY ON THE HEAVY PARTICLES AXIAL BEHAVIOR IN ARGON SURFACE-WAVE SUSTAINED PLASMA

## E. Benova, Ts. Petrova*

Chair of Physics, Department for Language Learning (Institute for Foreign Students), Sofia University, 27 Kosta Loulchev Str., BG-1111 Sofia, Bulgaria
*Faculty of Physics, Sofia University, 5 James Bourchier Blvd., BG-1126 Sofia, Bulgaria

**Abstact.** The axial behavior of the excited atoms, atomic and molecular ions in argon surface-wave sustained plasma columns is investigated theoretically at various wave frequencies and plasma radii in the pressure range 0.1-10 Torr. The numerical model is based on self-consistent solving of a full set of electrodynamic and kinetic equations: the local dispersion equation, wave energy balance equation, electron Boltzmann equation, electron energy balance equation, neutral gas thermal equation and the balance equations for the electrons, excited atoms, atomic and molecular ions. It is found out that, depending on the conditions, the populations of $3p^5 4s$ levels decrease or increase along the column, while the populations of $3p^5 4p$-block of levels and the densities of the charged particles always decrease from the wave launcher to the column end. The dynamics of the populations of the excited states and partial contributions of different elementary processes in the particle balances is studied. The results are compared with available experimental data and other models.

## 1. INTRODUCTION

Gas discharges sustained by high-frequency (HF) surface waves have been systematically investigated both experimentally and theoretically over the past two decades. The interest in such discharges is due to the wide range of operating conditions: gas pressure from 1 mTorr to a few atmospheres, discharge tube radius from about 0.5 mm to above 10 cm, wave frequency from a few hundred of kHz to ~10 GHz. Since the plasma is sustained through wave propagation, the wave launcher (surfatron [1], surfaguide [2], waveguide-surfatron [3], Ro-box [4]) can be very small compared to the plasma column length. Figure 1 shows a typical experimental situation.

Surfatron

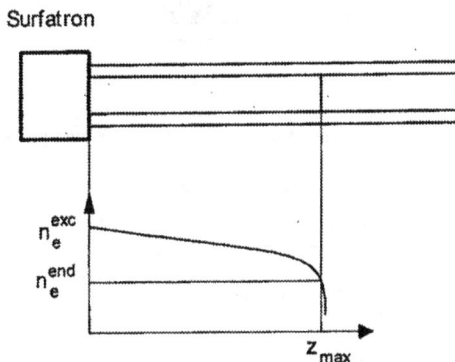

Figure 1. Typical experimental situation

The wave electric field heats the electrons, thus the electrons absorb the wave energy and it decreases along the column. The electrons expend the obtained energy for ionization and excitation of the neutral atoms creating and sustaining in this way the plasma. From the other side, the plasma is a part of the waveguide structure for wave propagation. Because the wave power decreases along the column, the electron number density decreases too and the plasma column is axially inhomogeneous. At low and intermediate pressures the plasma is strongly nonequilibrium one – the temperature of the neutral component ($T_g \approx 300$ K) is less than the electron temperature ($T_e \sim 1$–$2$ eV). The fact that the wave creates the medium for its propagation by itself makes the modelling a rather complex problem. A physically acceptable model must include two aspects: the electromagnetic one (describing the wave propagation) and the gas-discharge description (the balances of all the particles creation and loss) [5]. There exist models studying only the wave propagation and not concerning the particle balances [6–9] and kinetic models which deal with the balance mechanisms in the plasma but ignore the wave aspect of the problem [10–13]. All these models do not allowed a complete description of the surface-wave sustained plasmas and the efforts are directed in self-consistent modelling where both aspects are taken into account and coupled together [5,14–18].

In fact, the plasma column is both radially and axially inhomogeneous and the complete self-consistent model should be two-dimensional. Except of a few works studying only the radial particles distribution [14,19], all the above mentioned models describe the axial distributions of the wave or/and plasma characteristics assuming radially averaged plasma densities.

In this paper, an axial self-consistent model of an argon plasma column sustained by a travelling azimuthally symmetric electromagnetic wave [18] is applied to investigate the role of the wave frequency on the plasma characteristics in the pressure range 0.1–10 Torr. The model is based on simultaneous solving of the local dispersion relation, the wave energy balance equation, the electron energy and electron Boltzmann equations, a set of particle balance equations for electrons, excited atoms, atomic and molecular ions point by point along the column. In fact, the model consists of two parts: electrodynamic and kinetic ones. The electron energy balance equation plays the role of the self-consistent link between these parts – the wave power dissipated by the electrons (calculated in the electrodynamic part) is expanded for elastic and inelastic collisions with heavy particles (which can be derived from the kinetic part).

## 2. THEORETICAL FORMULATION

The propagation of an azimuthally symmetric TM surface wave of a constant frequency $\omega$ is described by Maxwell's equations with appropriate boundary conditions. The HF plasma permittivity $\varepsilon_p$ is derived from the simplest model of a cold weakly collisional electron plasma with an effective electron–neutral collision frequency for momentum transfer $\nu_{eff} < \omega$

$$\varepsilon_p = 1 - \frac{\omega_p^2}{\omega(\omega + i\nu_{eff})}$$

where $\omega_p = \sqrt{4\pi e^2 n_e / m}$ is the electron plasma angular frequency and $n_e$ is the radially averaged electron number density. The basic equations of the electrodynamic part of the model obtained under these assumptions are the local wave dispersion relation and the wave energy balance equation [18]. The dispersion equation

$$D(\omega, k, \varepsilon_p, \varepsilon_d, R, R_d, R_m) = 0 \qquad (1)$$

yields a relation between the electron number density $n_e$ and the wave number $k$ at a given axial position (phase diagram) depending on the experimental conditions (the wave frequency $\omega$; the permittivity $\varepsilon_d$, the internal and external radii of the dielectric tube – $R$ and $R_d$ respectively; the metal screen radius $R_m$).

The wave energy balance equation obtained from Poynting's theorem can be written in the form

$$\frac{dS}{dz} = -Q, \qquad (2)$$

where $S$ is the wave energy flux being sum of the axial components of Poynting's vector averaged over the wave period $2\pi/\omega$ and integrated over the plane normal to the plasma column, from the axis to the metal screen radius, at the given axial position $z$:

$$S = 2\pi \int_0^R r S_z^p dr + 2\pi \int_R^{R_d} r S_z^d dr + 2\pi \int_{R_d}^{R_m} r S_z^v dr. \qquad (3)$$

Here $p$, $d$, and $v$ denote plasma, dielectric, and vacuum, respectively. For an azimuthally symmetric wave the axial component of Poynting's vector is

$$S_z = \frac{c}{8\pi} \mathrm{Re}(E_r^* B_\varphi)$$

In the wave energy balance equation (2) $Q$ is the wave power per unit column length absorbed by electrons,

$$Q = 2\pi \int_0^R r \langle \mathbf{j} \cdot \mathbf{E} \rangle dr \propto n_e E_0^2, \tag{4}$$

where the angular brackets mean averaging over the wave period and

$$E_0^2 \frac{2}{R^2} \int_0^R r |\mathbf{E}|^2 dr \tag{5}$$

is the radially averaged squared wave electric field.

At steady state conditions the absorbed by the electrons wave power $Q$ should equal the power that the electrons expend in elastic and inelastic collisions with the heavy particles. The equation for electron energy balance takes the form

$$Q = \pi R^2 n_e \theta, \tag{6}$$

where $\theta(n_e)$ is the mean power loss per electron. It is expressed by collision and diffusion frequencies (for details see Eq. (2) in [18]).and can be calculated in the kinetic part of the model. This requires the knowledge of the electron energy distribution function (EEDF), which determines the electron transport parameters, the rates of elementary processes, and several other key quantities. The EEDF, obtained by solving the electron Boltzmann equation, is assumed to be time independent because the wave frequency $\omega$ is larger than the energy relaxation frequency and the condition for stationarity [20] is fulfilled. The heating of the electrons by the HF wave electric field is taken into account in the Boltzmann equation through the wide used approximation of the effective electric field

$$E_{\mathrm{eff}} = \frac{E_0}{\sqrt{2}} \frac{v_c(u)}{\sqrt{v_c^2(u) + \omega^2}} \tag{7}$$

Here $E_0$ is given by (5), $v_c$ is the electron–neutral collision frequency for momentum transfer, and $u$ is the electron energy.

It is of major importance to choose appropriate excited states of the atom and charged particles involved in the discharge. Figure 2 shows the energy levels' diagram of the argon atom used in the model. The $Ar(3p^5 4s)$ four levels (two metastable and two resonant) are treated separately and the $Ar(3p^5 4p)$ ten levels as one lumped block of levels. This choice enables an accurate description of both the electron and heavy particle kinetics and is applicable at gas pressure up to few Torr at degree of ionization less than $10^{-2}$.

The electron creation is due to several ionization processes: ionization from the ground state, stepwise, Penning, and associative ionization. It is assumed that after an ionization event the primary and the ejected electrons share the remaining energy. The loss of electrons is through dissociative recombination and ambipolar diffusion. The treatment of the homogeneous electron Boltzmann equation is based on the conventional two-term Legendre polynomial expansion assuming a quasistationary approximation of the anisotropic part of the distribution. The EEDF has been calculated using conventional numerical technique.

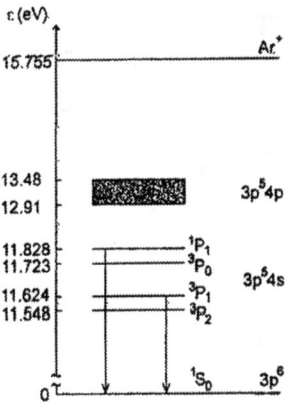

Figure 2. Energy levels' diagram of the argon atom

The heavy particle kinetics includes a set of particle balance equations for all excited atoms considered as well as the balance equations for both atomic and molecular ions. The elementary processes included are diffusion to the wall, excitation from the ground state, electron impact excitation and deexcitation between excited states, molecular ion formation, and radiative transitions with accounting for the trapping of radiation in all allowed radiative processes under consideration [18].

The heavy particle kinetics is coupled with the electron kinetics in self-consistent way in order to satisfy all balance equations. The obtained mean power loss per electron $\theta$ is used in solving the electron energy balance equation (6) together with the wave energy balance equation (2) and the local wave dispersion relation (1). The basic concept of the numerical calculations is presented schematically in Fig. 3. As a result of the calculations the axial distributions of all plasma and wave characteristics are derived.

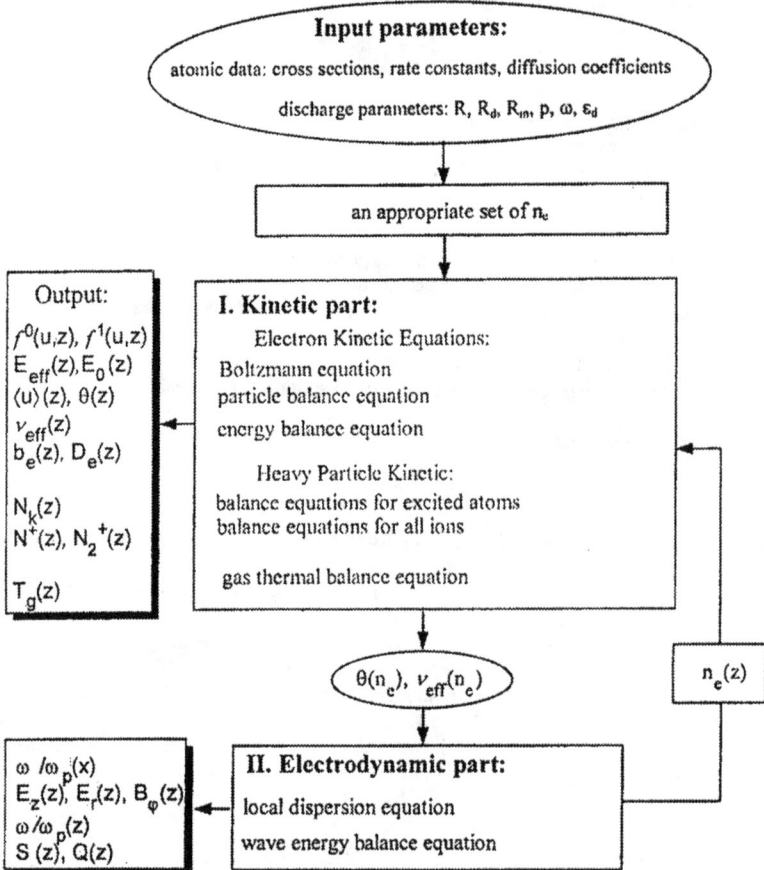

**Figure 3.** Basic concept of the numerical calculations

## 3. RESULTS AND DISCUSSION

The axial behavior of the excited atoms' densities can be drastically different depending on the discharge conditions (see Fig. 7 in Ref. 18). It has been found both experimentally and theoretically [21] that the population densities of $3p^5 4s$ levels increase from the wave launcher to the column end while in another work [22] they decrease with plasma density decrease, respectively along the column. One can see also that the behavior and the values of the two metastable and two resonant $4s$ levels' populations are very different in these two papers: In Ref. 21 the most populated is the $^3P_2$ metastable level, the least populated is $^3P_0$ – the other metastable level, the two resonant levels have nearly the same populations lying between those of the metastable states and all four levels have the same axial behavior. In Ref. 22 at low plasma density the populations of the metastable states are higher than the populations of resonant states; at higher plasma densities the populations of the four $4s$ levels are arranged in order of their statistical weights as in an equilibrium situation; with plasma density increasing the two metastable levels have similar behavior which is different from that of the two resonant states. In order to investigate the role of the experimental conditions on the excited atoms axial distribution the calculations have been done at various gas pressures, plasma radii and wave frequencies.

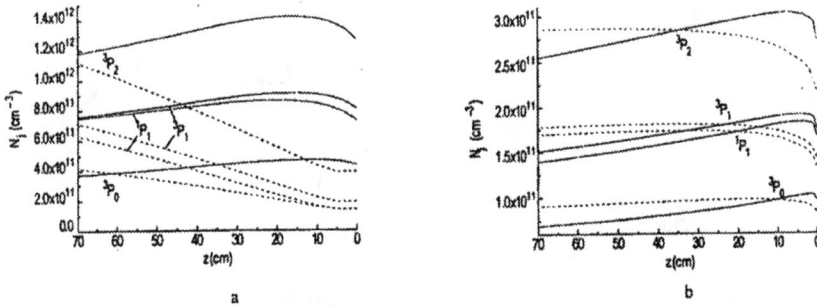

a                                                         b

Figure 4. Axial distributions of 4s levels' population at plasma radius 0.45 cm, wave frequency 2.45 GHz (solid lines) and 210 MHz (dashed lines) and gas pressure 0.150 Torr (a) and 10 Torr (b)

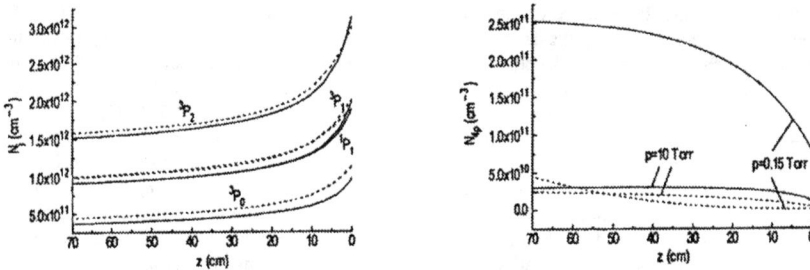

Figure 5. Axial distributions of $3p^5 4s$ levels' population at plasma radius 0.15 cm, wave frequency 2.45 GHz (solid lines) and 210 MHz (dashed lines) and gas pressure 1 Torr

Figure 6. Axial distributions of $3p^5 4p$ block of levels population at plasma radius 0.45 cm, wave frequency 2.45 GHz (solid lines) and 210 MHz (dashed lines) and gas pressure 0.150 Torr and 10 Torr.

The axial profiles of $3p^5 4s$ levels population at gas pressure 0.150 Torr, plasma radius 0.45 cm and wave frequencies 210 MHz and 2.45 GHz are plotted in Fig. 4a.The results at gas pressure 10 Torr are presented in Fig. 4b. For convenience the end of the plasma column is at $z = 0$. It can be seen that at higher wave frequency the number density of the excited atoms mainly increases in $z$ direction while at lower frequency the tendency is just the opposite. This is not true at plasma radius 0.15 cm – in this case the axial behavior of the $4s$ levels population does not depend very much on the wave frequency in all the pressure range. Figure 5 illustrate this at gas pressure of 1 Torr.

The axial profiles of $3p^5 4p$ block of levels are plotted in Fig. 6 at the same experimental conditions as in Fig. 4. At low gas pressure both the population values and behavior depend strongly on the wave frequency while at higher pressure frequency dependence is almost negligible. And in all considered discharge conditions the $4p$-block population decreases along the plasma column (being almost constant in some cases).

Excited atoms population depends on the EEDF (via the elementary processes' rate coefficients) and on the electron number density. It has been shown [18] that in wide range of experimental conditions the EEDF has a non-Maxwellian character and changes along the plasma column. The EEDF at two axial positions – near the wave exciter ($z = 70$ cm) and close to the column end ($z = 10$ cm) at low and high frequencies and gas pressures of 150 mTorr and 10 Torr is presented in Fig. 7. It is seen that the distribution function is more sensitive to the wave frequency at low gas pressures than at higher (compare Fig. 7 a and b).

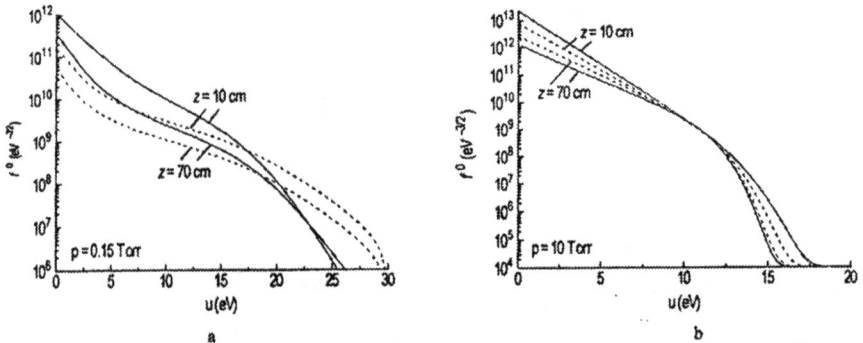

a        b

Figure 7. Frequency dependence of the EEDF at two $z$ positions. The discharge conditions are the same as in Fig. 4

Figure 8 illustrates the dependence of the excited atom populations on $n_e$ in wide range of electron number densities. The population of the $4s$ levels can increase or decrease with $n_e$ decreasing (away from the launcher) while the population of $4p$ block always decreases. The plasma density of a real column is actually in a much narrow interval, which depends on the discharge conditions. At low wave frequency the electron number density of the plasma column is $n_e \sim 10^{10}$ cm$^{-3}$ [22] while at 2.45 GHz it is in the range $10^{11}$–$10^{13}$ cm$^{-3}$ [21]. One can see from Fig. 8b that in this range the population of $4s$ levels increases when $n_e$ decreases away from the launcher and in some cases only at low plasma densities near the column end this behavior changes to the opposite one. At plasma densities $n_e > 10^{12}$ cm$^{-3}$. the population of $4p$ block is high enough and this block plays an important role in the elementary processes. The main population mechanism for all $4s$ levels except $^1P_1$ is the excitation from the ground state. The population of $^1P_1$ level is owing to the radiative and electron impact transitions from the other excited states and especially from the $4s$ block – the latest makes clear why at these discharge conditions $^1P_1$ is more populated than the corresponding metastable state $^3P_0$. The main process at these conditions is the electron impact exchange of excitation between excited states, which determine the same axial behavior of all $4s$ levels. At $n_e \sim 10^{10}$ cm$^{-3}$ the population of $4p$ block is negligibly small (Fig. 8a) and all $4s$ levels are populated by the excitation from the ground state. As a result $^1P_1$ is less populated than $^3P_0$ and the densities of all the excited atoms decrease with the $n_e$ decreasing. The different main loss mechanisms at low $n_e$ – radiative decay of the resonant states and diffusion to the wall of the metastables is the reason for the different behavior that these levels have.

It also has to be mentioned that if the EEDF is assumed to be Maxwellian, the calculated 4s levels' population always decreases along the column [13,17], which is not in agreement with the experimental results.

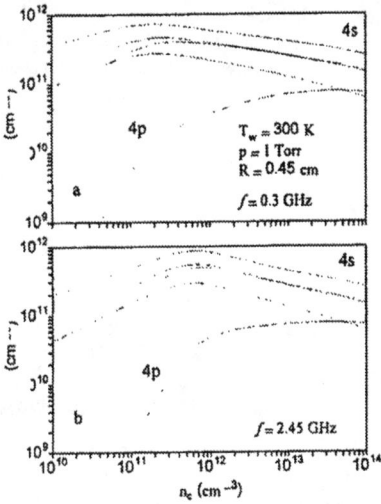

Figure 8. Excited 4s and 4p atoms populations as function of the plasma density at low and high wave frequency

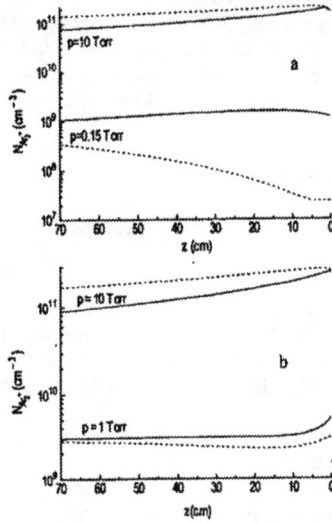

Figure 9. Molecular ions axial distribution at plasma radius 0.45 cm (a) and 0.15 cm (b), wave frequency 2.45 GHz (solid lines) and 210 MHz (dashed lines) and various gas pressure

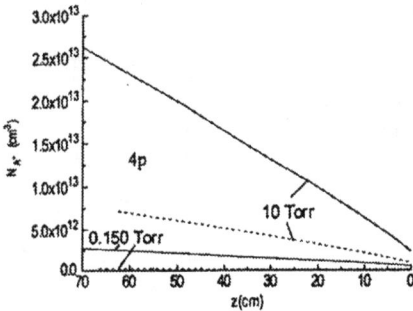

Figure 10. Axial distribution of atomic ions. The discharge conditions are the same as in Fig. 6

It also has to be mentioned that if the EEDF is assumed to be Maxwellian, the calculated 4s levels' population always decreases along the column [13,17], which is not in agreement with the experimental results.

The Ar molecular ions number densities also increase or decrease from the launcher to the column end depending on the discharge conditions (Fig. 9) while the atomic ions axial distribution follows that of the electrons (Fig. 10).

14

## 4. CONCLUSION

A self-consistent model previously reported has been used in this investigation. The model is based on the numerical solving of the full set of kinetic and electrodynamics equations describing both the wave propagation and the gas discharge processes. It has been shown that the frequency of the electromagnetic wave sustaining the plasma column has an important influence on the axial distributions of the ions and excited argon atoms. There are discharge conditions at which the $3p^5 4s$ levels' population and molecular ion density increase along the column and others at which the behavior is just the opposite. The population of $3p^5 4p$ block of levels always decreases in $z$ direction. Important role in this behavior plays the EEDF, which is far from Maxwellian one and is changing with the wave frequency and along the plasma column.

## Acknowledgments

This work was supported by the National Fund for Scientific Research under Grant No. F-821/98.

## References

1. Moisan M., Zakrzewski Z., and Pantel R. J. Phys. D: Appl. Phys., 1979, **12**, 219.
2. Moisan M., Zakrzewski Z., Pantel R., and Leprince P. IEEE Trans. Plasma Sci., 1984, **PS-12**, 203.
3. Moisan M., Chaker M., Zakrzewski Z., and Paraszczak J. J. Phys. E: Sci. Instrum., 1987, **20**, 1356.
4. Moisan M., Zakrzewski Z. Rev. Sci. Instrum., 1987, **58**, 1895
5. Sá A. B., Ferreira C. M., Pasquiers S., Boisse-Laporte C., Leprince P., and Marec J J. Appl. Phys., 1991, **70**, 4147.
6. Glaude V. M. M., Moisan M., Pantel R., Leprince P., and Marec J. J. Appl. Phys., 1980, **51**, 5693.
7. Zhelyazkov I., Benova E., and Atanassov V. J. Appl. Phys., 1986, **59**, 1466.
8. Boisse-Laporte C., Granier A., Dervisevic E., Leprince P., and Marec J. J. Phys. D: Appl. Phys.,1987, **20**, 197.
9. Aliev Yu. M., Ivanova K. M., Moisan M., and Shivarova A. Plasma Sources Sci. Technol., 1993, **2**, 145.
10. Ferreira C. M. and Loureiro J. J. Phys. D: Appl. Phys., 1984, **17**, 1175.
11. Ferreira C. M., Alves L. L., Pinheiro M., and Sá A. B. IEEE Trans. Plasma Sci., 1991, **19**, 229.
12. Kortshagen U. J. Phys. D: Appl. Phys., 1993, **26**, 1691.
13. Cotrino J. and Gordillo-Vázquez F. J. J. Phys. D: Appl. Phys., 1995, **28**, 1888.
14. Ferreira C. M. J. Phys. D: Appl. Phys., 1981, **14**, 1811.
15. Ferreira C. M. J. Phys. D: Appl. Phys., 1983, **16**, 1673.
16. Sá A. B. Microwave Discharges: Fundamentals and Applications. Ed. C. M. Ferreira and M. Moisan. New York: Plenum, 1993, 75.
17. Benova E., Petrova Ts., Blagoev A., and Zhelyazkov I. J. Appl. Phys., 1998, **84**, 147.
18. Petrova Ts., Benova E., Petrov G., and Zhelyazkov I. Phys. Rev. E, 1999, **60**, 875.
19. Darchicourt R., Pasquiers S., Boisse-Laporte C., Leprince P., and Marec J. J. Phys. D: Appl. Phys., 1988, **21**, 293.
20. Winkler R., Capitelli M., Dilorando M., Gorse C., and Wilhelm J. Plasma Chem. Plasma Process., 1986, **6**, 437.
21. Lao C., Gamero A., Sola A., Petrova Ts., Benova E., Pterov G. M., Zhelyazkov I. J. Appl. Phys., 2000, **87**, 7652.
22. Böhle A., Kortshagen U. Plasma Sources Sci.& Technol., 1994, **3**,80.

# MOLECULAR DISCHARGES SUSTAINED BY A TRAVELLING WAVE

C.M. Ferreira, F. M. Dias, V. Guerra and E. Tatarova

Centro de Fisica de Plasmas, Instituto Superior Técnico, 1049-001 Lisboa, Portugal

**Abstract.** Theoretical models have been developed for surface wave sustained $N_2$ and $H_2$ discharges which account in a self-consistent way for the main plasma balances governing the discharge production, including bulk and surface processes. The approach used describes self-consistently the axial discharge structure, i.e., the axial distributions of charged particle concentrations, population densities of excited species and neutrals, taking into account inhomogeneous gas heating along the plasma column as well as plasma-wall interactions. Spatially resolved experimental investigations confirm the main trends of the model predictions.

## 1. INTRODUCTION

Technological applications involving molecular plasmas usually call for a complex investigation of the transport and reactions of numerous neutral and charged species and of the energy exchange channels in the plasma source, in order to predict correctly the general trends of discharge behaviour and relevant species concentration. As far as microwave plasma generation is concerned, the investigation of discharges sustained by travelling surface waves (SWs) continues to provide major breakthroughs, due to their flexible operation and easy access to a variety of diagnostics. This enables detailed comparisons between modelling and experiment which provide pronounced impact on the understanding of HF discharges in general. However the operation of travelling wave sustained discharges in molecular gases is extremely complex due to non-linear coupling between different quantities concerning the discharge, propagating wave and interface conditions. For this reason, a plausible description of such systems can only be achieved by coupling the particle balance equations for all relevant charged and neutral species to the electron Boltzmann equation and the equations describing the wave propagation, the gas thermal balance and plasma-wall interactions.

This work presents a number of important aspects and discusses current issues in the modelling of surface wave sustained discharges operating in $H_2$ and $N_2$. The self-consistent approach used describes the axial discharge structure, i.e. the axial distribution of charged particle concentrations, population densities of the excited species and neutrals taking into account inhomogeneous gas heating and plasma-wall interactions. To illustrate that self-consistent modelling is a powerful tool to understand discharge operation important problems such as molecular dissociation, inhomogeneous gas heating along the discharge length, atomic reassociation at the wall, $H^-$ negative ions creation and losses are discussed in terms of the obtained theoretical and experimental results. In particular, the important aspect of plasma-wall interaction is addressed. The correlation between wall temperature and degree of $H_2$ dissociation and the effects on the $H^-$ concentration in $H_2$ discharges are discussed on the basis of obtained experimental data.

## 2. SELF-CONSISTENT MODELLING OF TRAVELLING WAVE DISCHARGES

In this section we discuss the self-consistent modelling of microwave discharges featuring an extended zone of operation (outside the field applicator) which are sustained by a propagating azimuthally symmetric ($m=0$) surface wave at a frequency $f = \omega/2\pi$. The plasma column is generated in a cylindrical dielectric tube with permitivity $\varepsilon_d$. The z coordinate is directed along the tube axis and the wave propagates along it with wave vector $k = \alpha + i\beta$ ($\beta$ and $\alpha$ are the axial wave number and attenuation coefficient, respectively). As the wave propagates and creates its own propagating structure, the wave power flow decreases along the wave path since the power is progressively absorbed by the plasma electrons which dissipate this power in collisions with the gas particles i.e. the gas is excited, ionised and heated. There is a self-consistent "interplay" between wave dynamics, generated plasma and conditions at the interface (wall). This interplay defines the physical basis of discharge sustaining and to express it, all the quantities concerning the generated plasma column, wall conditions and wave propagation should be self-consistently coupled. Two important interrelated aspects of discharge physics must be investigated self-consistently, namely:

 i)   axial variation of different steady-state discharge balances – electron and heavy particle balances, electron energy balance, gas thermal balance, etc.;

 ii)  characteristics of the surface wave propagation along the generated inhomogeneous plasma column, i.e. axial variation of wave dispersion characteristics and wave power balance.

For a self-consistent modelling, the input parameters are the usual externally controlled ones: gas pressure, wave frequency, tube radius, and the power delivered to the launcher or the electron density at the position of the launcher. In order to derive the axial variation of all quantities of interest, the model must be based on a set of coupled equations for the plasma bulk describing the kinetics of free electrons, the vibrational kinetics of electronic ground state molecules, the kinetics of excited electronic states of molecules and atoms, the chemical kinetics of neutrals and ions, the gas thermal balance and the charge particle balance determining the electric field sutaining the discharge. Further, a set of equations for the wave dispersion properties and power balance has to be coupled in order to close the formulation.

Self-consistent models as described above have recently been developed for travelling wave sustained discharges in nitrogen [1-6] and hydrogen [7,8]. These models determine the axial structure of the discharge, i.e. the following properties: electron energy distribution function (EEDF) and related energy averaged quantities (mean electron energy, electron rate coefficients etc.); concentration of H and N atoms in the electronic ground state; population of electronically excited states $H_2(c^3\Pi_u), N_2(A^3\Sigma_u^+, a'^1\Sigma_u^-, B^3\Pi_g, C^3\Pi_u, a^1\Pi_g, w^1\Delta_u)$ of molecules and atoms $H(n = 2\text{-}8)$; concentration of ions $(N_2^+, N_4^+, H^+, H_2^+, H_3^+, H^-)$; population of vibrational levels of molecules in the electronic ground state $H_2(X^1\Sigma_g^+), N_2(X^1\Sigma_g^+)$; electric field maintaining the discharge $(E)$; radially averaged gas temperature $(T_g)$; wave propagation characteristics.

It is beyond the scope of this paper to discuss all the processes occurring in the molecular plasmas considered. In this section we will only address a number of important issues for the understanding of basic discharge workings and we will briefly describe the set of equations used.

## 2.1. Particle kinetics

### 2.1.1. Kinetics of free electrons.

The electron energy distribution function (EEDF) is determined by solving the homogeneous quasistationary electron Boltzman equation. Assuming the anisotropy caused by spatial inhomogeneities and the field to be sufficiently small, the electron velocity distribution is approximated by the usual two-term expansion in spherical harmonics. Electron collisions of the first and the second kind and electron-electron collisions are accounted for. Since vibrationally excited molecules in nitrogen discharges constitute an appreciable fraction of the total molecular population, the EEDF is generally a function of the vibrational distribution function (VDF). The electron transport parameters and the rate coefficients, as calculated from the EEDF, are functions of the reduced electric field $E/N$ ($N$ is the total density of the neutrals) and, for nitrogen, they depend also on VDF. For simplicity a radially homogeneous field is assumed in the kinetic models. The electron cross sections used in the collision terms of the Boltzmann equation are the same as in [7,8] for $H_2$ and as in [2,3,6,11,12] for $N_2$. The reader should refer to these works for details.

### 2.1.2. Heavy particle kinetics.

The population of $H_2(X^1\Sigma_g^+, v)$ and $N_2(X^1\Sigma_g^+, v)$ vibrational levels have been determined from the usual coupled system of master equations taking into account (i) vibrational excitation and deexcitation by electron collisions (e–V processes); (ii) vibration-vibration (V–V) exchanges; (iii) vibration-translation (V–T) exchanges in collisions of $N_2(X^1\Sigma_g^+,v)$ with $N_2$, N and of $H_2(X^1\Sigma_g^+,v)$ with $H_2$, H; (iv) one-quantum energy exchange in collisions of $N_2(X^1\Sigma_g^+,v)$ and $H_2(X^1\Sigma_g^+,v)$ with the wall. Note that only single quantum transition, which are the most likely ones, are usually considered in the (V–V) and (V–T) collisional exchange processes. The sole exception concerns V–T exchanges in $N_2$–N and $H_2$–H collisions for which multiquantum transitions are known to be important. For hydrogen there exists an efficient vibrational excitation mechanism (especially for upper vibrational levels $v > 3$) through electron impact excitation of $H_2(B^1\Sigma_u^+)$ and $H_2(C^1\Pi_u)$ states followed by fast decay to excited vibrational levels of the electronic ground state $H_2(X^1\Sigma_g^+,v)$. A detailed list of processes and corresponding rate coefficients is given in [2,3,6,11,12]. A large number of collisional-radiative processes involving electronically excited states of $N_2$ and $H_2$ molecules was taken into consideration in the models developed. A list of the major processes with the corresponding rate coefficient was given in recent works [6,7,9].

### 2.1.3. Dissociation kinetics.

Dissociation kinetics is an important issue due to the numerous volume and wall processes involving atoms that strongly influence the discharge operation. In a microwave nitrogen discharge, the main dissociation process is direct electron impact dissociation.

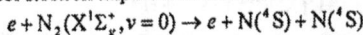

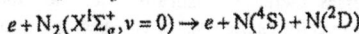

$$e + N_2(X^1\Sigma_g^+, v = 0) \rightarrow e + N(^4S) + N(^4S)$$

$$e + N_2(X^1\Sigma_g^+, v = 0) \rightarrow e + N(^4S) + N(^2D)$$

and the main loss channels of ground state atoms are the following

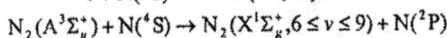

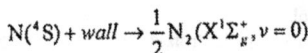

$$e + N(^4S) \rightarrow e + N(^2D,^2P)$$

$$N_2(A^3\Sigma_u^+) + N(^4S) \rightarrow N_2(X^1\Sigma_g^+, 6 \leq v \leq 9) + N(^2P)$$

$$N(^4S) + wall \rightarrow \frac{1}{2}N_2(X^1\Sigma_g^+, v = 0)$$

For hydrogen, the fast radiative decay of the group of states $c^3\Pi_u, a^3\Sigma_g^+, e^3\Sigma_u^+$ into a repulsive state $b^3\Sigma_u^+$ results in dissociation of $H_2$. Therefore, the sum of the rate coefficients for excitation of these states has been taken as the rate coefficient for dissociation of $H_2$ by electrons. Dissociation by electron impact yielding H(1s)+H(n=2); H(1s)+H(n=3), and H(1s)+H(n=4) has also been taken into account [7]. The wall reassociation is the major loss channel for H atoms.

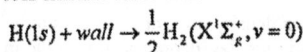

$$H(1s) + wall \rightarrow \frac{1}{2}H_2(X^1\Sigma_g^+, v = 0)$$

The concentration of hydrogen atoms resulting from the above processes is strongly influenced by the value of the wall loss probability $\gamma_H$ which in turn depends heavily on the wall conditions, in particular on the wall temperature $T_w$. According to the results in Ref. 10, the following temperature dependence has been used: $\gamma_H = \gamma_0 \exp\left(-\dfrac{840}{T_w}\right)$. Here, $\gamma_0 = 1 \times 10^{-2}$ or $6 \times 10^{-2}$ depending on the wall condition. In fact, it has been found that these values of $\gamma_0$ and the above wall temperature dependence law can describe rather well different experiments. The reason for different values of $\gamma_0$ may be connected with the different technologies used to produce Pyrex glass.

### 2.1.4. Charged Particle Balance and Maintaining Field.

For a self-consistent determination of the electric field maintaining the discharge a balance between the rates of charged particle production and loss has to be obeyed together with the quasineutrality condition. For example, the field strength necessary for the steady-state operation of the nitrogen discharge is obtained from the balance between the total rate of ionization, including direct, associative [involving collisions between the metastable species $N_2(A^3\Sigma_u^+)$ and $N_2(a'^1\Sigma_u^-)$] and step-wise [from

$N_2(A^3\Sigma_u^+)$, $N_2(a'^1\Sigma_u^-)$, $N_2(w^1\Delta_u)$, $N_2(B^3\Pi_g)$ $N_2(a^1\Pi_g)$] ionization, and the total rate of electronic losses due to diffusion to the wall (in the presence of $N_2^+$ and $N_4^+$ ions [2,3]) and electron–ion bulk recombination. In nitrogen discharges an important source of charged particles is associative ionization from the metastable states $N_2(A^3\Sigma_u^+)$ and $N_2(a'^1\Sigma_u^-)$ via the processes

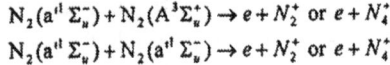

$$N_2(a''\Sigma_u^-) + N_2(A^3\Sigma_u^+) \to e + N_2^+ \text{ or } e + N_4^+$$
$$N_2(a''\Sigma_u^-) + N_2(a''\Sigma_u^-) \to e + N_2^+ \text{ or } e + N_4^+$$

Note that the electron continuity equation used to determine electric field is strongly nonlinear in $E$ due to the strong dependence of the electron ionization rate coefficients on the electric field. Small changes in the field result in large changes in the tail of the EEDF, and thus in the electron rate coefficients for excitation and ionization.

## 2.1.5. H⁻ negative ions.

High frequency discharges in hydrogen have mainly been studied as sources of H atoms, and not of negative ions, since H atoms are known to impede the H⁻ production [7,9,13].

It is now well accepted that negative hydrogen ions are mainly produced by dissociative attachment from vibrationally excited $H_2(X^1\Sigma_g^+, v)$ molecules.

$$e + H_2(X^1\Sigma_g^+, v) \to H + H^-$$

The cross-sections for these processes increase drastically with the vibrational quantum number involved. High populations in electronically excited states is a peculiarity of discharges sustained by surface waves. For this reason, dissociative attachment of metastable molecules $H_2(C^3\Pi_u, v)$ by collisions with electrons, that is, the reactions

$$e + H_2(C^3\Pi_u, v) \to H + H^-$$

can also be an important source channel.

Another important source mechanism is electron attachment to high-lying electronically excited Rydberg states due to the significant concentration of energetic electrons [7]. Detachment by dissociated atoms, according to the reaction $H^- + H(1s) \to e + H_2$. constitutes an important loss channel for negative ions.

## 2.2. Gas and wall temperature

The gas temperature determines the gas density and influences srongly the rate coefficient values for numerous bulk reactions. Its calculation must be incorporated into the models by using the gas thermal balance equation together with the particle kinetic equations. Under nearly isobaric conditions, neglecting the axial transport and assuming a parabolic temperature radial profile, the stationary gas thermal balance equation can be expressed in the form [4,7,9]:

$$\frac{8\lambda(T_g)}{R^2}(T_g - T_w) = Q_{in} \tag{1}$$

Here, $T_g$ is the radially averaged gas temperature and $T_w$ is the wall temperature. Thermal conduction $\lambda(T_g)$ to the tube wall usually is the main gas cooling mechanism. The wall temperature is introduced as an input parameter, either determined experimentally [4,5,6] or by using a semi-empirical formula [7,8]. It was found that the measured axial distribution of the wall temperature can be well fitted by the semi-empirical formula $T_w = T_0 + C(WR)^\beta$, where $T_0$ is the room temperature in K, $W$ (in Watt/cm³) is the specific discharge power, $R$ is the tube radius in cm, and $C$ and $\beta$ are fitting parameters. $Q_{in}$ accounts for the total net power transferred per unit volume to the translational modes from volume and wall sources of heat. The bulk excitation processes of vibrational states of $H_2$ and $N_2$ followed by vibrational relaxation via $N_2$-$N_2$, $N_2$-N, $H_2$-H, $H_2$- $H_2$ collisions constitute important gas heating sources. At high degrees of ionization, exothermic pooling reactions should also be taken into account [4,5]. Moreover, interactions of gas phase species with the surface such as deexcitation of electronically and vibrationally excited states on the wall and wall reassociation of atoms also contribute to gas heating.

## 2.3. Wave–to–plasma power coupling

The self-consistent interplay between the wave electrodynamics and the discharge balances results in an inhomogeneous power dissipation along the wave path, and consequently in an axial variation of the discharge parameters which is non-linearly coupled to the absorbed power. The theoretical treatment of the wave-to-plasma coupling in the models is based on a simultaneous solution of the wave power and the electron power balance equations [3,5,6,7]. The mechanism of power transfer can be expressed quantitatively by introducing $\theta$ – the mean power needed to maintain an electron ion pair [2,3]. The absorbed power in a plasma slice of thickness $\Delta z$ at position $z$ is $\Delta P = \theta(z) S n_e(z) \Delta z$, $S$ denoting the plasma cross-section. Under steady state conditions, the spatial rate of power change $dP/dz$ is due to the power absorbed by the electrons per unit discharge length. The local power balance equation is therefore

$$\theta(z) S n_e(z) = -\frac{dP}{dz} = 2\alpha(z) P(z_0) \exp\left(-\int_{z_0}^{z} 2\alpha(x)dx\right) \tag{2}$$

This equation linking the wave power flux with the electron power losses in the discharge provides the axial description of the discharge structure. As usual, the attenuation coefficient $\alpha$ is derived by solving the local wave dispersion equation [1,3].

## 3. RESULTS AND DISCUSSION.

This section presents an application of the models to different cases of SW discharge operation in nitrogen and hydrogen. Figures 1 and 2 illustrate the strong coupling between the EEDF and the VDF in a nitrogen discharge operating at 500 MHz. The "vibrational barrier" effects on the EEDF are well pronounced close to the column end but they are partly attenuated by superelastic collisions towards the SW launcher due to the higher degrees of ionization and, correspondingly, higher vibrational excitation of the nitrogen molecules in this region [2,11,12]. The degree of ionization decreases from about $10^{-5}$ near the SW launcher ($z/L_t \to 1$) to about $10^{-6}$ near to the end ($z/L_t \to 0$). In order to unify all the results, the distance towards the end $z$ is normalized to the total discharge length $L_t$. The higher degree of ionization at the beginning of the column produces a higher vibrational temperature and consequently a more enhanced high energy tail of the EEDF. As one moves to the column end, the decrease in degree of ionization causes a drop in the vibrational temperature and a faster depletion in the EEDF tail due to the vibrational barrier. The axial variation of the vibrational distribution of the electronic ground state $N_2(X^1\Sigma_g^+, v)$ is depicted in Fig.2. The VDF is mainly shaped by the following mechanisms: the lower vibrational levels are directly pumped by electron collisions; V–V exchanges propagate vibrational quanta upwards and are responsible for the formation of a plateau in the VDF for intermediate levels; finally, the simultaneous effects of vibrational dissociation and V–T exchanges (especially with atoms) influences the population in the highest levels of the VDF. As can be seen, the region corresponding to the above mentioned mechanisms can be clearly identified. An increase in the populations of the high vibrational levels ($v > 35$) towards the column end should be noted. The simultaneous effects of the decrease in the degree of ionization ($n_e/N$) and in the gas temperature (see below) cause a decrease in the population of the lower vibrational levels, i.e in vibrational temperature towards the end [3]. At the same time, the decrease in the nitrogen atom density (see below) and in $T_g$ causes an increased population of the higher vibrational levels ($v > 30$) close to the discharge end as a result of the decreasing influence of V–T depopulation mechanisms by $N_2$–N collisions [4].

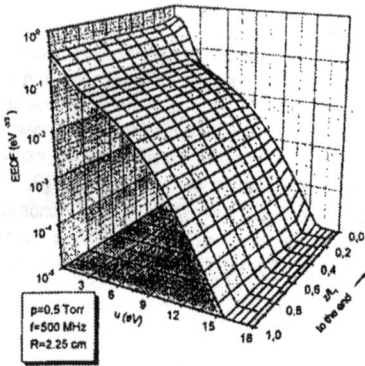

Figure 1.

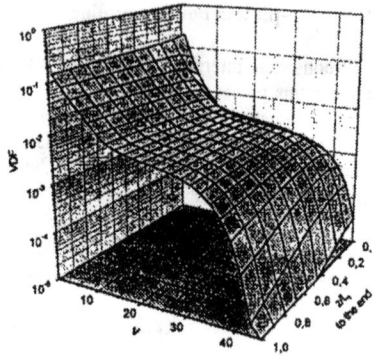

Figure 2.

Calculated and measured axial distributions of the gas temperature in $N_2$ discharges are shown in Figures 3 and 4. Also shown on these figures for completeness are the axial distributions of the measured wall temperatures. An infrared sensitive non-perturbative measurement, using an electro-optical thermometer, provides experimental results for the axial variation of the wall temperature during discharge operation. Note that the wall was cooled by natural convection only. The experimental gas temperatures were determined, as usual for nitrogen, by measuring the rotational distribution of the second positive system $N_2(C^3\Pi_u, v') \rightarrow N_2(B^3\Pi_g, v'')$ in the 375.5 – 379 nm wavelength range (i.e. 0–0 vibrational transition) assuming that the rotational and translational modes are in equilibrium [3]. Similar results for the gas temperature in the $H_2$ discharge have been obtained by applying a Doppler broadening technique [7]. As observed, the gas temperature decreases non-linearly from the position of the SW launcher towards the plasma column end, which reflects the non-uniform wave power absorption along the wave path. The experimental results clearly demonstrate that, to a first approximation, the gas temperature at a given axial position is a function of the power absorbed per unit length by the plasma electrons at that position [6,7]. It is usually considered that gas heating in $N_2$ comes mainly from V–V and V–T relaxation processes of vibrationally excited molecules. This is confirmed by the results presented in Fig. 3 where the excitation of vibrational levels by electron impact followed by V–T relaxation in $N_2$–N multiquantum processes is shown to be the dominant gas heating mechanism for the conditions considered. This result is a consequence, as previously discussed, of the high V–T depopulation rates of the VDF by $N_2$–N collisions. It should be noted that including the contribution of exothermic reactions at the wall involving long lived species, such as $N_2$ metastable molecules, provides better agreement between calculated and measured temperatures for a discharge operating at 2.45 GHz. At higher degrees of ionization, the wall deactivation of metastables can be an important source of gas heating as shown in Fig. 4. Here, $\xi$ denotes the part of the energy which is assumed to return to the gas phase and to dissipate into heat due to this process. The most important gas heating sources in the hydrogen discharge are V-T relaxation of $H_2(X)$ molecules, dissociation of $H_2$ by electron impact, with production of "hot" H atoms, and wall recombination of H atoms, with production of "hot" $H_2$ molecules [7].

Molecular dissociation is an important issue for $N_2$ discharges. At the pressures of interest here, the main source channel for $N(^4S)$ atoms is electron impact dissociation. It is assumed that $N(^2D, ^2P)$ metastable atoms are deexcited to the ground state $N(^4S)$ through collisions at the wall with an effective probability $\gamma = 0.8 - 0.9$ [11,12]. As expected, the higher electron densities obtained in the discharge operating at 2.45 GHz lead to higher dissociation degrees than in the case of discharge at 500 MHz (Fig. 5). As far as the main production mechanism of $N(^4S)$ atoms is electron impact dissociation, the decrease of both $n_e$ and $E/N$ along the discharge results in a decrease in the $N(^4S)$ atom number density. The competitive interplay of the loss channels results in different predominant loss mechanisms for nitrogen atoms respectively at low and high degrees of dissociation and ionization, Losses associated with quenching of $N_2(A^3\Sigma_u^+)$ by $N(^4S)$ are the dominant loss channel at low ionization degrees ($10^{-6}$ – $10^{-5}$) while at high electron densities, (2.45 GHz, $n_e/N = 10^{-4} - 10^{-5}$) destruction by electron reexcitation to higher atomic levels dominates. At this point it should be stressed that the present results should be regarded as merely indicative due to large uncertainties in the input data. This concerns, in particular, the values of the probabilities for wall reassociation of ground state atoms and wall destruction of metastable atoms.

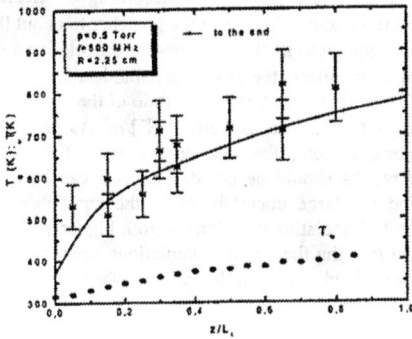

Figure 3.

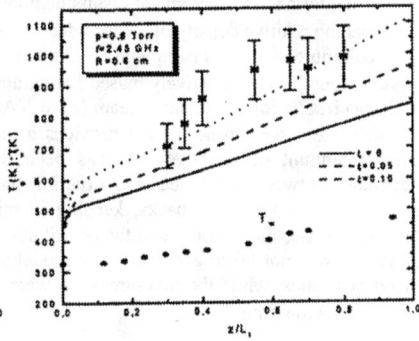

Figure 4.

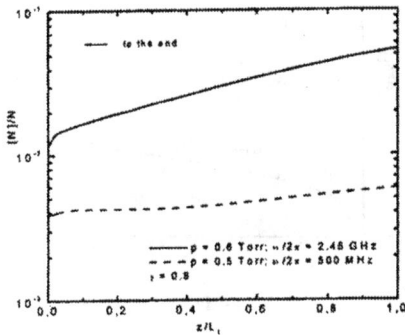

Figure 5.

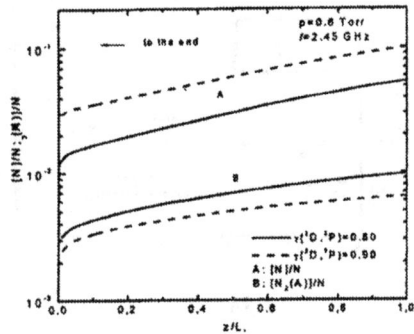

Figure 6.

The strong coupling between the $N(^4S)$ and $N_2(A^3\Sigma_u^+)$ populations and their dependence on plasma-wall interactions are illustrated in Fig. 6. Changing the probability $\gamma$ for wall deactivation of $N(^2D,^2P)$ atoms from 0.9 to 0.8 results in a decrease of the $N(^4S)$ density by about a factor of 2 and in an increase in the $N_2(A^3\Sigma_u^+)$ population by a similar factor. These results clearly show that an accurate determination of the $N(^4S)$ density in the discharge requires an accurate treatment of the kinetics of plasma wall interactions, including surface and bulk $N(^2D, ^2P)$ metastable atom kinetics, in correlation with the gas thermal balance.

The dissociation kinetics in hydrogen is somewhat simpler because it is mainly determined by the processes of direct electron impact dissociation and reassociation at the wall as previously discussed. Figure 7 demonstrates that the H atom concentration depends drastically on the probability of wall reassociation. The decrease in electron density close to the column end causes a decrease in the dissociation rate, and thus in the relative number of H atoms. The theoretical results are in closer agreement with experiment when a higher reasociation probability is assumed. An emission spectroscopy method based on the determination of the ratio of the $H_{\alpha}$ (656 nm) atomic line intensity to the integrated intensity of the Fulcher $\alpha$ system ($d^3\Pi_u^-, v \to a^3\Sigma_g^+, v$) of molecular hydrogen has been applied for the measurement of the degree of dissociation.

Discharges sustained by travelling surface waves can also be interesting negative ion sources as both the experimental and the theoretical results demonstrate. As seen from Fig. 8, according to theoretical calculations the relative density of $H^-$ ions decreases from about 1% close to the launcher to about 0.1 % near the column end [7]. A laser photodetachment technique was applied to measure the relative density of negative ions [14]. A positively biased Langmuir probe collects the electrons photodetached from $H^-$ after plasma irradiation by the laser beam (a Nd-YAG laser has been used). The ratio of the current pulse amplitude to the d.c. probe current provides a means to measure the ratio $[H^-]/n_e$. As observed, a significant amount of negative ions has been detected along the discharge length. The existing discrepancies between theoretical and experimental results should be noted. This fact can be mainly attributed to theoretical drawbacks, keeping in mind the large uncertainties in the input date used. Furthermore, it should be noted that the contribution of dissociative detachment from higher vibrational levels ($v > 3$) was not taken account in the model and that the theoretical calculations concern radially averaged quantities, while the measurements were performed at the discharge axis where the number density of $H^-$ is maximum.

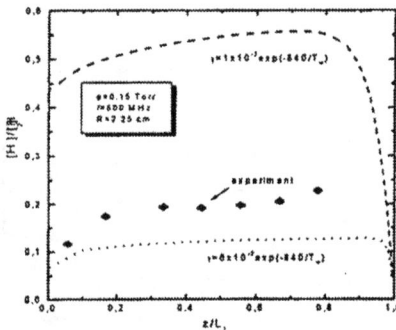 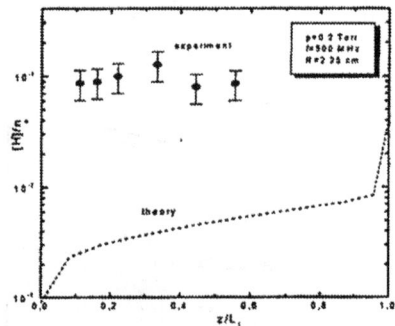

Figure 7.                                        Figure 8.

## 3. WALL EFFECTS

A difficult and still poorly understood aspect of molecular discharges is the strong coupling between volume kinetics and interface (wall) conditions. As previously discussed, the loss of H atoms is mainly controlled by wall reassociation, which strongly depends on the material and the temperature of the wall. Wall reassociation is usually associated with chemisorbed atoms, i.e. irreversibly adsorbed atoms at active sites. The surface is covered with a small fraction of such active sites. Chemisorbed atoms can recombine with atoms arriving at the active sites either directly from the gas phase (Eley-Rideal mechanism) or, for physisorbed (reversibly adsorbed) atoms, by diffusion along the surface (Langmuir-Hinshelwood mechanism [10,13,15]). The losses of gas phase atoms on a Pyrex surface in the range of wall temperatures $T_W < 500$ k may be attributed to the Langmuir-Hinshelwood mechanism of reassociation. This fact can be recognised on Figs. 9 and 11 where the dependence of the degree of dissociation on the wall temperature is depicted. For these experiments, an externally forced "local"(at a fixed axial position) cooling/heating of the tube wall was performed. Clearly the number density of H atoms decreases when $T_w$, increases up to about 400 K. At the same time, a local increase in $T_w$ results in an increase of negative ions (see Figs.10 and 12). The measured $[H^-]/n_e$ ratio exhibits an increase of about of factor of two (Fig. 10) when the "local" wall temperature raises from about 300 to 400 K. These results bring up a clear demonstration of the effects of atoms on the $H^-$ density. This influence occurs via two channels, namely: (i) H atoms effectively depopulate high vibrational levels of the electronic ground state (through V-T multiquantum relaxation processes) which have a major contribution to the formation of $H^-$ through dissociative attachment; and (ii) detachment of $H^-$ in direct colisions with H atoms. The obtained results demonstrate that changes in the wall temperature (in a certain temperature range) can be used as an effective mechanism to achieve dynamic control of the concentrations of species of interest.

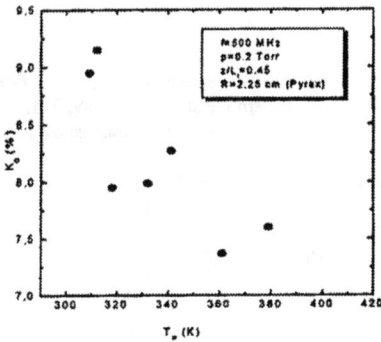

Figure 9.

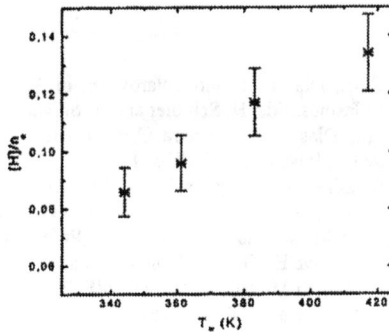

Figure 10.

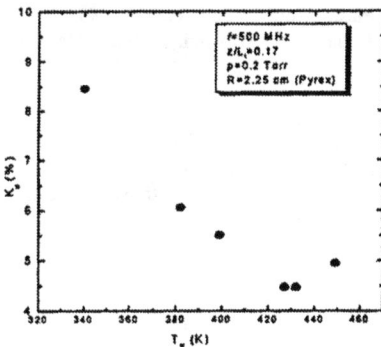

Figure 11.

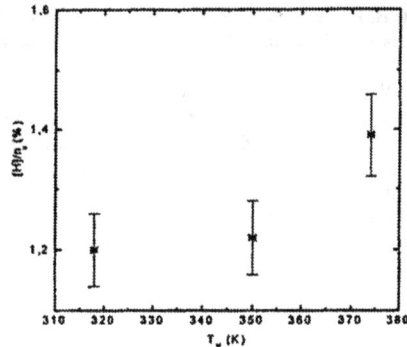

Figure 12.

# 4. CONCLUSIONS

The modelling of travelling wave discharges in molecular gases is a very complex task since it must be based on a self-consistent description of electron and heavy particles kinetics, wave electrodynamics, gas thermal balance, and plasma-wall interactions. Self-consistent modelling is an essential tool to understand the discharge behaviour and to optimise its operation with respect to the production of relevant species. A difficult and still poorly understood aspect is the strong coupling which exists between volume kinetics and wall conditions. It is clear that the predictions of such models rely heavily on the data used for the different processes. Unfortunately, numerous data concerning plasma-wall interactions are still unknown. For this reason, a deeper physical understanding of travelling wave sustained molecular discharges requires intensive experimental and theoretical work. In particular, there is a strong need for reliable measurements and modelling concerning surface kinetics and plasma-wall interactions. Nevertheless, considerable progress has been achieved in recent years as this paper illustrates.

## Acknowledgments

The authors express their gratitude to Prof. B. Gordiets, Dr. A. Ricard and Dr. M. Pinheiro for their collaboration.

## References

1. Ferreira C.M., Dias F.M. and Tatarova E. in Advanced Technologies Based on Wave and Beam Generated Plasmas, Eds. H. Schluter and A. Shivarova, Kluwer Academic Publishers, 1999, 311.
2. Tatarova E., Dias F.M., Ferreira C.M., Guerra V., Loureiro J., Stoykova E., Ghanashev I. and Zhelyazkov I. J. Phys D: Appl. Phys, 1997, **30**, 2663.
3. Tatarova E., Dias F.M., Ferreira C.M. and Ricard A. J. Appl. Phys., 1999, **85**, 49.
4. Guerra V., Tatarova E., Dias F.M. and Ferreira C.M. Proc. 14th Int. Symp. Plasma Chemistry, Eds. M. Hrabovsky, M. Konrad and V. Kopesky ,Prague, 1999, 156
5. Guerra V., Tatarova E., Dias F.M., and Ferreira C.M. Proc. XVth ESCAMPIG, Eds. Z. Donko, L. Jenik, and J. Szigeti Miscolc. 2000, vol. **24F**, 218.
6. Ferreira C.M., Guerra V., Tatarova E. and Dias F.M. (to be published).
7. Gordiets B., Pinheiro M., Tatarova E., Dias F. M., Ferreira C. M. and Ricard A. Plasma Sources Sci. Technol., 2000, **9**, 295.
8. Dias F. M., Tatarova E., Gordiets B., Pinheiro M., Sampaio E. and Ferreira C. M. Proc. XIVth ESCAMPIG, Eds. D. Riley, C. M. O. Mahony, W. G. Graham Malahide, 1998, vol. **22H**, 332.
9. Gordiets B. F., Ferreira C. M., Pinheiro M. J., Ricard A. Plasma Sources Sci. Technol., 1998, 7, 363.
10. Gordiets B F, Ferreira C M, Pinheiro M J and Ricard A Plasma Sources Sci. Technol., 1998, 7, 379.
11. Guerra V. and Loureiro J. J. Phys. D: Appl. Phys., 1995, **28**, 1903.
12. Guerra V. and LoureiroJ. Plasma Sources Sci. Technol., 1997, **6**, 361.
13. Ferreira C.M., Gordiets B. and Tatarova E. Plasma Phys. Controlled Fusion (in press).
14. Dias F.M., Tatarova E., Crespo H. and Ferreira C.M. Rev. Sci. Instr.(in press).
15. Capitelli M., Ferreira C. M., Gordiets B.F. and Osipov A.I. Plasma Kinetics in Atmospheric Gases Heidelberg: Springer – Verlag, 2000.

# ELECTRON GAS HEATING IN LOW-PRESSURE, UNMAGNETIZED MICROWAVE PLASMAS USED FOR MATERIALS PROCESSING

Timothy Grotjohn

Michigan State University, Electrical and Computer Eng., East Lansing, MI 48824 USA

Abstract. A study has been done using experimental measurements and two-dimensional simulations to understand the heating of unmagnetized microwave discharges at low pressures of 4 -60 mTorr. The microwave electric field strength and the plasma discharge parameters are measured using microcoax sampling probes and Langmuir probes, respectively. Simulations of the microwave fields are performed using a two-dimensional finite-difference time-domain model which includes the plasma currents and power absorption. Comparisons of the simulated and measured data show that at pressures of 20 mTorr and greater, collisional heating due to electron-neutral collisions (joule heating) is the dominant heating mechanism. However at lower pressures (4-20 mTorr), joule heating alone is inadequate to describe the measured results and other non-collisional heating mechanism(s) are important. The role of anomalous (stochastic) heating in the regions of microwave electric field resonance in the discharge is assessed as an important heating mechanism at low pressures.

## 1. INTRODUCTION

Microwave discharges have been developed and applied to a number of applications ranging from electron cyclotron resonance (ECR) plasma and ion sources for semiconductor etching and deposition to plasma-assisted CVD sources for diamond films. Low pressure microwave plasma sources used for materials processing generally operate as overdense plasmas with the plasma density greater than the critical density. These sources can be operated with a static magnetic field that provides for ECR heating or without a static magnetic field via ohmic, resonance and other stochastic heating mechanisms. Even when ECR strength magnetic fields are present these other heating mechanisms that occur in unmagnetized plasmas can be important[1] and may even dominate. This paper examines via a two-dimensional, microwave field and plasma model the heating of low pressure, overdense, unmagnetized plasma discharges. Further the model is constructed to closely match an experimental system that has been extensively characterized. This experimental system is a 2.45 GHz resonant cavity plasma source[2-8] that has been studied while running argon discharges at pressures of 4-60 mTorr using Langmuir probes to determine the plasma density and electron temperature and using microwave field probes to measure the microwave field strength [9-12].

Low-pressure (collision frequency << excitation frequency) microwave plasma simulations that model the spatial variation of the microwave heating fields and plasma discharge are difficult to use in the local regions where the plasma frequency matches or is near the excitation frequency. In these regions resonance effects occur and the microwave electric field can become large. Because of the localized nature of the resonance (high microwave field strength) region, stochastic heating (anamolous collisions in thin electric field regions) can occur as the electrons are accelerated/heated in this region and/or transverse through this resonance region via their initial momentum. Simple electron gas descriptions based on a plasma conductivity/dielectric value or a cold plasma momentum transport equation may not reasonably predict the plasma heating phenomenon/rate or the local microwave field strength. Rather, an electron gas model that considers the random motion of the electrons via a warm plasma or electron energy distribution function description is needed.

This paper describes a quantitative theoretical/experimental study of low pressure, unmagnetized microwave plasma discharge heating and the associated microwave energy coupling in overdense microwave discharges. The result is that at pressure ranges above approximately 20 mTorr for the discharge reactor studied the plasma heating occurs primarily by collisional heating due to electron-neutral collisions, but at lower pressures than 20 mTorr the heating of the plasma occurs by other mechanisms in addition to standard collisional heating. The non-collisional heating mechanism considered in this work is stochastic heating by anomalous collisions in electric field layers[13-17]. This paper describes the experimental measurements and the modeling studies performed that yielded this result.

## 2. EXPERIMENT/THEORY

### 2.1. Experiment

Fig. 1 displays a cross-section of the microwave plasma source studied. A detailed description of this source has been discussed in earlier publications[2-8]. The cylindrical applicator is terminated on the top by an adjustable sliding short and on the bottom by the quartz discharge dome. The z=0 plane and the cylindrical quartz dome define the discharge volume. Microwave radiation is coupled into the applicator, as shown in Fig. 1, through an adjustable end feed coupling probe. The microwave discharge is created inside the 12.5 cm diameter quartz dome by adjusting the sliding short length, $L_S$, and the coupling probe position $L_p$, to allow the cavity applicator (17.8 cm diameter) to resonate in a preselected electromagnetic mode. The discharge created for this study was an argon discharge with a flow rate of 30 sccm.

Microwave electric field measurements[9-12] were done using microcoaxial probes acting as small antennas inserted into the side of the coupling structure through small holes as shown in Figure 2. Calibrations of the probes permit absolute microwave electric field measurements. These measurements on an array of holes in the applicator walls permits the determination of the electromagnetic mode in the applicator structure.

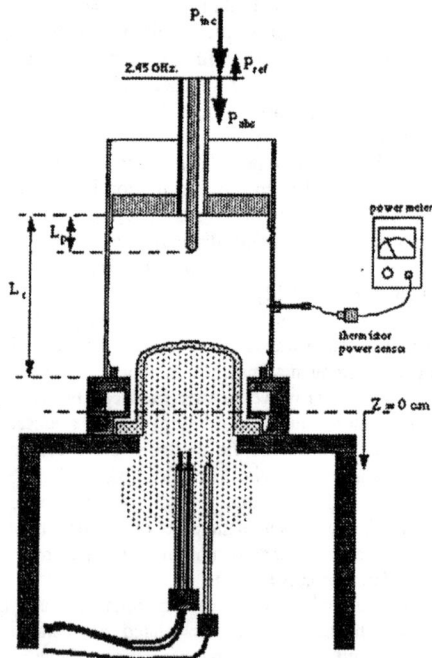

Figure 1. Cross-section of the microwave plasma reactor with Langmuir probe and microwave electric diagnostics shown.

Plasma density measurements were performed using a double Langmuir probe[5,9]. The Langmuir probe was also used to determine the electron temperature. The probes were placed into the discharge from below as shown in Fig. 1. Lastly, the incident and reflected microwave power, $P_{inc}$ and $P_{ref}$, were measured using directional couplers and microwave power meters.

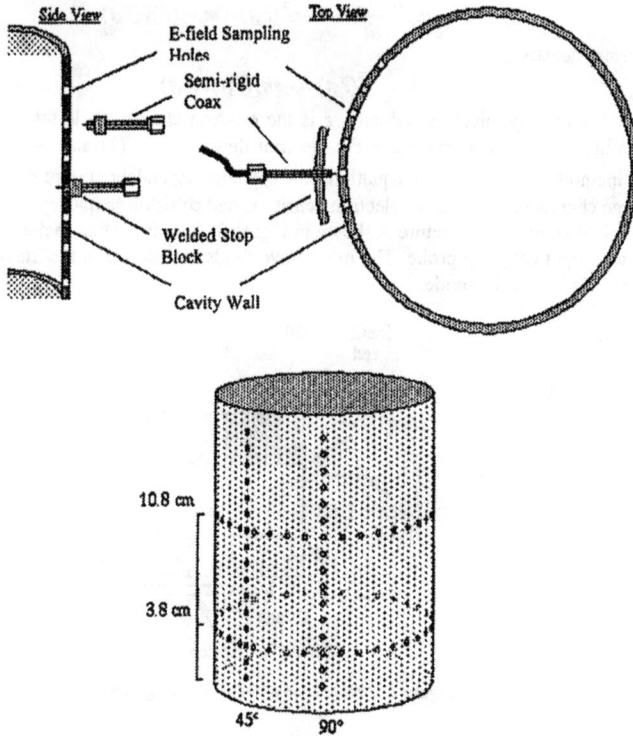

Figure 2. A microwave electric field probe is inserted into holes in the resonant cavity to determine the electric field strength and mode.

## 2.2. Modeling - Microwave Fields and Ohmic Heating

The numerical model [18,19,20,21] used in this study includes a FDTD (finite-difference time-domain) electromagnetic field model. The electromagnetic fields inside the cavity can be described by the time-dependent Maxwell's equations, which are written as

$$\nabla \times \vec{E} = -\mu \frac{\partial \vec{H}}{\partial t} \quad \text{and} \quad \nabla \times \vec{H} = -\varepsilon \frac{\partial \vec{E}}{\partial t} + \vec{J} \tag{1}$$

where $\vec{E}$ and $\vec{H}$ are the electric and magnetic fields, $\mu$ is the permeability, $\varepsilon$ is the permittivity and $\vec{J}$ is the current density. The FDTD method is formulated by discretizing Eq. (1) with a centered difference approximation in both the time and space domain. In this study, Eq. (1) is discretized in two-dimensional (r and z) cylindrical coordinates.

One of the major motivations to solve Maxwell's equations is to investigate the power dissipation inside the microwave plasma source due to the presence of plasma discharges. The power dissipation density, $P_{abs}(\vec{r},t)$ , with a power absorbing load (such as a discharge) present is

$$P_{abs}(\vec{r},t) = \vec{E}(\vec{r},t) \bullet \vec{J}(\vec{r},t) \qquad (2)$$

The current density, $\vec{J}$, which is induced by the microwave fields, can be determined by solving the momentum transport equation of electrons. The momentum transport equation of electrons can be written as

$$m_e \frac{d\vec{v}(\vec{r},t)}{dt} = -e\vec{E}(\vec{r},t) - m_e \nu_e(\vec{r})\vec{v}(\vec{r},t) \qquad (3)$$

and the current density as

$$\vec{J}(\vec{r},t) = -e\bar{n}_e(\vec{r},t)\vec{v}(\vec{r},t) \qquad (4)$$

where $\vec{v}$ is the average electron velocity, $e$ is the electron charge, $m_e$ is the electron mass, $\nu_e$ is the electron collision frequency, and $n_e$ is the electron density. Eqs. (3) and (4) are solved by the finite difference method. From the above equations, the spatially-dependent current density $\vec{J}$ is determined by the discharge characteristics such as electron density $n_e$ and collision frequency $\nu_e$.

The detailed simulation structure is shown in Figure 3. The simulation includes the cavity, the plasma discharge and input coupling probe. The microwave fields are excited in the input coaxial feed structure using an ideal TEM coaxial mode.

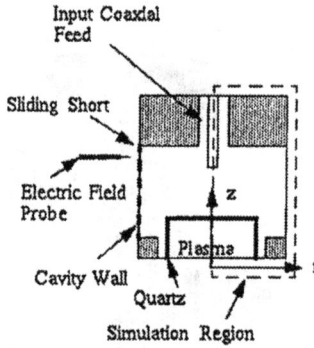

Figure 3. Cross-section of the simulation geometry.

## 2.3. Stochastic Heating

The electron gas in unmagnetized microwave plasma discharges is heated primarily by collisional (ohmic/joule heating) processes at moderate and higher pressures, but at low pressures (less than a few 10's mTorr) other non-ohmic heating mechanisms can become important. This is especially true in overdense discharges where resonance phenomenon occur. In particular, in the regions where the plasma electron frequency, $\omega_{pe}$, coincides with the microwave excitation frequency, $\omega$, a resonance occurs that can produce large electric fields. At this resonance condition, $\omega_{pe} = \omega$, a cold plasma description of the discharge using the plasma dielectric constant, $\varepsilon_p$, given by

$$\varepsilon_p = \varepsilon_o \left[ 1 - \frac{\omega_{pe}^2}{\omega(\omega - j\nu_e)} \right] \qquad (5)$$

approaches zero, $\varepsilon_p \to 0$, when $\nu_e$ is very small as occurs in low pressure plasmas.

This resonance region has a microwave electric field strength that varies versus position. Stochastic heating occurs as the electrons move (due to their thermal motion) through the region of spatially varying microwave electric fields. The initial electron motion is due to the electrons energy distribution. The stochastic heating, which the electrons experience, results in a net change in their energy as they move through this resonance region. In contrast, electrons in regions of spatially constant microwave electric fields experience no net change in their energy versus time unless a collision occurs. This type of electron heating has been treated in the theory of anomalous collisions in a thin electric field layer [13,14]. Recently this approach has been used to model the collisionless heating in low-pressure inductive discharges [15,16].

Following the development of Lieberman and Lichtenberg [17], the electric field is assumed to vary versus position in an exponential fashion given by $exp(-|r|/\delta)$ with a characteristic length $\delta$. If a Maxwellian electron distribution function with an average speed of $\bar{v}_e$ is assumed for the electron gas, the effect of this anomalous heating can be modeled with a stochastic collision frequency $v_{stoc}$. This stochastic collision frequency is found to be approximated by

$$v_{stoc} \approx \frac{\bar{v}_e^3}{\delta^3 \omega^2} \tag{6}$$

for the case of $\omega \ll v_{stoc}$, which is expected to occur for microwave frequencies. The effective collision frequency, $v_{eff}$, for the electron gas is the ohmic collision frequency, $v_e$, plus the stochastic collision frequency giving

$$v_{eff} = v_e(\bar{r}) + v_{stoc}(\bar{r}) . \tag{7}$$

The implementation of the anomalous collision term $v_{stoc}(\bar{r})$ into the plasma source simulation needs to consider that the term $\delta$ in $v_{stoc}$ depends on the microwave electric field spatial variation and that the microwave electric field depends on $v_{eff}$. To include the stochastic heating effect the method shown in Fig. 4 is used. In the first iteration the FDTD microwave field solution is done using $v_{eff} = v_e$ without considering the stochastic heating effect. Eq. (3) for the electron motion is modified to use an effective collision frequency $v_{eff}$

$$m_e \frac{d\bar{v}(\bar{r},t)}{dt} = -e\bar{E}(\bar{r},t) - m_e v_{eff}(\bar{r},t)\bar{v}(\bar{r},t) \tag{8}$$

The microwave fields calculated shows regions of resonance where the electric field varies versus position. The characteristic length of the microwave field spatial variation in the discharge is determined as

$$\delta = \left| \frac{\Delta r}{\ln\left(\frac{E(r_i)}{E(r_{i+1})}\right)} \right| \tag{9}$$

where $\Delta r$ is the grid spacing and $E(r_i)$ and $E(r_{i+1})$ are the microwave electric field at adjacent grid locations. The microwave field is then calculated again in an iterative fashion using Eq. (1). The iteration sequence is continued until $v_{eff}(\bar{r})$ and $E(\bar{r})$ converge.

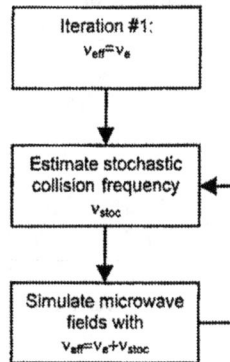

Figure 4. Iteration technique for simulation of stochastic heating.

## 3. RESULTS

### 3.1. Experimental Results

A series of plasma source measurements were taken across a pressure range of 4 to 60 mTorr. The results of this measurement set is given in Table 1. Listed in this table is the pressure, microwave incident and reflected power, plasma ion density and electron temperature as measured with a double Langmuir probe, and the microwave resonant cavity tuning adjustments including the sliding short height and probe length.

**Table 1.** Experimental Measurement Results

| Pressure (mTorr) | Incident Power (Watts) | Reflected Power (Watts) | Plasma Density (cm$^{-3}$) | Electron Temp. (K) | Short Position (cm) | Probe Position (cm) |
|---|---|---|---|---|---|---|
| 4 | 255 | 45 | $1.5 \times 10^{11}$ | 43,000 | 13.8 | 3.0 |
| 7 | 272 | 28 | $2.5 \times 10^{11}$ | 40,400 | 15.0 | 3.0 |
| 10 | 279 | 21 | $3.2 \times 10^{11}$ | 35,000 | 15.6 | 3.0 |
| 15 | 272 | 28 | $4.8 \times 10^{11}$ | 34,000 | 15.9 | 3.0 |
| 25 | 252 | 48 | $7.0 \times 10^{11}$ | 24,000 | 16.6 | 3.0 |
| 45 | 241 | 59 | $8.0 \times 10^{11}$ | 15,500 | 16.7 | 3.0 |
| 60 | 230 | 70 | $6.8 \times 10^{11}$ | 15,500 | 17.1 | 3.0 |

The table shows that the microwave plasma is a high density plasma with a plasma density in the $10^{11}$-$10^{12}$ cm$^{-3}$ range. The electron density increases with increasing pressure and the electron temperature decreases with increasing pressure.

The microwave electric field (radial component) was measured for each pressure versus vertical position as shown in Fig. 5. The measured data points are shown with the "+" symbol in units of kV/m. The measured data in Fig. 5 is for a pressure of 7 mTorr. Two peaks in the electric field are seen at about 4.5 cm and 11 cm in vertical position. Similar measurements around the circumference of the

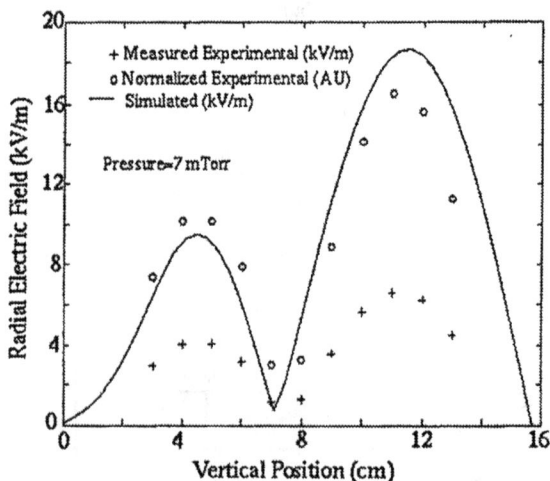

**Figure 5.** Radial microwave electric field strength versus vertical position along the outer wall of the resonant cavity.

resonant cavity at a fixed vertical position shows that the electric field is uniform along the circumferential direction. This electric field pattern is identified as a $TM_{012}$ resonant mode. This electric field measurement technique was repeated for each of the pressures studied and a curve similar to that shown in Fig. 5 was generated. At each of these pressures the electric field mode pattern remained in this $TM_{012}$ mode. Next, the peak electric field versus vertical position along the cavity wall was identified for each of the pressures. These peak measured electric field values are shown versus pressure in Fig. 6 with the open circles. The microwave electric field is seen to increase from about 3 kV/m to 7.5 kV/m as the pressure is reduced from 60 to 4 mTorr.

## 3.2. Simulation Results

Three different simulation studies are considered here to assess the important of resonance effects and stochastic heating. The first simulation set was done with a constant plasma density equal to that shown in Table 1. The second set was done using a plasma density with a spatial variation in the plasma density as produced by diffusion. In this second set of simulations localized regions exist in the discharge where resonance occurs. Finally, the important of including the stochastic heating effect in the resonance regions is simulated.

### 3.2.1. Simulation Set #1: Joule Heating Only with Constant Plasma Density

The first part of the low pressure discharge simulation investigation of non-collisional heating is to model the microwave plasma reactor using just collisional (joule) heating theory and see if the observed experimental measurements are described by this heating mechanism. The model as described above used the measured absorbed microwave energy as the power deposited into the plasma for the simulation. The simulates were done in two-dimensions (r-z) with the plasma density treated as a uniform plasma with the density and electron temperature being the measured values given in Table 1. The model just considered collisional heating as given by electron-neutral collision processes. The electron-neutral collision frequency is described by

$$v_e = n_g K(T_e) \tag{10}$$

where $n_g$ is the neutral gas density and $K$ is the collisional rate constant[17], which is dependent on the electron temperature $T_e$.

The FDTD solution of the microwave fields in the resonant cavity including the plasma discharge region yields data on the microwave electric field strength throughout the entire simulation region. The simulated radial component of the electric field data is plot versus vertical position along the outer cavity wall in Fig. 5 for comparison to measured data. The simulated data shows a similar electromagnetic field mode dependence as the measured data. Specifically, the simulated data also shows the $TM_{012}$ mode. For comparison purposes, the experimental data has also been normalized in amplitude in Fig. 5 (see the circles).

As shown in Fig. 5 the simulations predict a higher electric field for an absorbed power of 250-270 Watts as compared to the measured data at the pressure of 7 mTorr. As described above for the measured data, the simulations were also performed at pressures from 4 to 60 mTorr and the peak electric field along the outer wall of the resonant microwave cavity was determined. These simulated values are shown in Fig. 6 (see the "+" data points). For these simulated electric field values, several simulations at each pressure were performed. Each simulation considered variations of selected input parameters to determine the sensitivity of the electric field values to the input parameters. Specific variations considered included plus and minus variations of the electron density and the cavity height. The envelope formed by these variations is shown in Fig. 6.

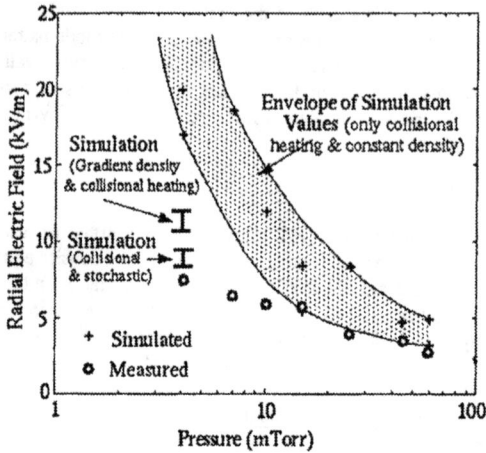

**Figure 6.** Radial component of the microwave electric field along the outer wall of the resonant cavity versus pressure.

The comparison of the simulated (shown in Fig. 6 as "only collisional heating and constant density" data) and measured electric field values in Fig. 6 reveals that the simulated data agrees approximately with measured data at the higher pressures of 20-60 mTorr, but at the lower pressures the simulated electric fields are substantially higher than the measured values. This indicates that for the simulations at lower pressures, in order to absorb the 250-270 Watts of microwave power into the plasma, a higher electric field is needed as compared to that needed in the actual experiment.

The conclusion drawn from Fig. 6 is that at the higher pressures of 20-60 mTorr, the power absorption is primarily by collisional heating from electron-neutral collisions. But at low pressures, standard collisional heating does not account for all the power transfer from the microwave fields to the electrons. Hence another power transfer mechanism besides joule heating becomes important at lower pressures.

### 3.2.2. Simulation Set #2: Inclusion of Resonant Regions with Only Joule Heating

When the plasma density is no longer constant, i.e. a diffusional profile is present, a region or regions exists where the plasma density is at critical density and resonance effects occur. To simulate this effect in this study, two-dimensional simulations of the microwave fields with a plasma density that has resonance regions was used. Fig. 7 shows a plot of the radial electric field simulation at a pressure of 4 mTorr. At the resonance regions in the plasma discharge, which occur for the radial microwave field component near the outer edge of the discharge where the excitation frequency matches the plasma frequency, a sharp increase occurs in the microwave field strength as seen in Fig. 7. Fig. 6 shows for the pressure of 4 mTorr the electric field intensity along the cavity wall needed to get the same power absorbed in the simulations as was absorbed in the experiments given in Table 1. This value (labeled as "gradient density and collisional heating" in Fig. 6) is less than that simulated assuming a constant plasma density, which is the case with no resonant regions. However, the simulation electric field at the low pressure of 4 mTorr is still larger than that measured experimentally.

33

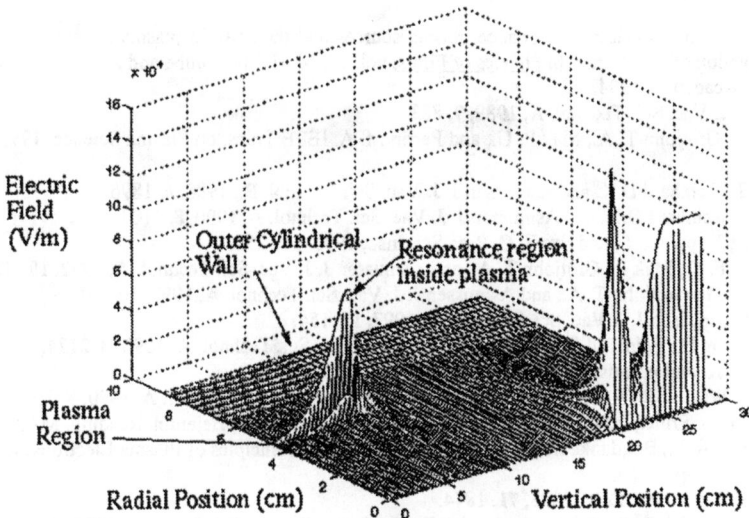

Figure 7. Radial component of the microwave electric field inside the resonant cavity including the discharge region. The plasma region extends from r=0 to r=6 cm and z=0 to z=8 cm.

### 3.2.3. Simulation Set #3: Inclusion of Stochastic Heating

The last set of simulations completed is the inclusion of the stochastic heating by anomalous collisions in the resonant regions. The technique used is that described in section 2.3 above. The simulation result at the low pressure of 4 mTorr is shown in Fig. 6 and is labeled as "collisional and stochastic". The inclusion of the stochastic collision frequency reduced the size of the microwave electric field in the resonant region and also reduced the field strength along the cavity wall to a magnitude closely matching the experimentally measured values. Hence, the simulation model, which includes joule heating, a realistic diffusion-driven plasma density profile with resonant regions and stochastic heating, predicts the microwave electric field magnitude needed to absorb a set amount of microwave power with close agreement to the experimental data.

## 4.0. SUMMARY

A study has been done using experimental measurements and two-dimensional simulations to understand the heating of unmagnetized microwave discharges at low pressures. The variable compared for the simulated and measured discharge was the microwave field strength required in the resonant cavity to absorb the microwave power. At the higher pressures greater than 20 mTorr the heating of the discharge can be modeled by including just joule heating. At the lower pressures the collision frequency is smaller and the electric field in the simulations using just joule heating is substantial higher than the measured value. This occurs for both a constant density plasma versus position and for plasma density variations having resonant regions. To get good agreement of the simulations values with the measured values at the low pressures, the inclusion of stochastic heating produced by anomalous collisions in the resonant microwave electric field regions was necessary.

Acknowledgment

This project is funded by the National Science Foundation - Grant number: DMI 9713298

34

**References**

1. Popov O., "Electron cyclotron resonance plasma sources and their use in plasma-assisted chemical vapor deposition of thin films," in Physics of Films, vol. 17, M. H. Francombe and J. L. Vossen, Eds., New York: Academic, 1994.
2. Asmussen J. J. Vac. Sci. Technol. A, 1989, **7**, 883.
3. Asmussen J., Grotjohn T. A., Mak P. U., and Perrin M. A. IEEE Trans. on Plasma Science, 1997, **25**, 1196
4. Hopwood J., Reinhard D.K. and Asmussen J. J. Vac. Sci. Technol. B., 1988, **6**, 1896.
5. Hopwood J., Reinhard D.K. and Asmussen J. J. Vac. Sci. Technol. A, 1990, **8**, 3103.
6. Asmussen J., Hopwood J., and Sze F. C. Rev. Sci. Inst., 1990, **61**, 250.
7. King G., Sze F. C., Mak P., Grotjohn T. A., and Asmussen J. J. Vac. Sci. Technol. A, 1992, **10**, 1265.
8. Mak P., King G., Grotjohn T. A., and Asmussen J. J. Vac. Sci. Technol. A, 1992, 10, 1281.
9. Mak P. and Asmussen J. J. Vac. Sci. Technol. A, 1997, **15**, 154.
10. Zhang J., Huang B., Reinhard D. K., and Asmussen J. J. Vac. Sci. Technol. A, 1990, **8**, 2124.
11. Srivastava A. K. and Asmussen J. Rev. Sci. Instrum., 1995, **66**, 1028
12. Hopwood J., Wagner R., Reinhard D. K. and Asmussen J. J. Vac. Sci. Technol. A, 1990, **8**, 2904.
13. Ichimaru I. Basic Principles of Plasma Physics: A Statistical Approach, Bejamin, Reading, MA, 1973.
14. Alexandrovich A. F., Bogdankevich L. S. and Rukhadze A. A., Principles of Plasma Electrodynamics, Springer, New York, 1984.
15. Turner M. M. Phys. Rev. Lett., 1993, **71**, 1844.
16. Godyak V. A., Piejak R. B. and Alexandrovich B. M. Plasma Sources Sci. Technol., 1995, **4**, 332.
17. Lieberman M. A. and Lichtenberg A. J. Principles of Plasma Discharges and Materials Processing, Wiley, New York, 1994, p. 80.
18. Tan W. and Grotjohn T. A. J. of Vacuum Sci. and Technol. A, 1994, **12**, 1216.
19. Gopinath V. P. and Grotjohn T. A. IEEE Trans. on Plasma Science, 1995, **23**, 602.
20. Tan W. and Grotjohn T. A., Diamond and Related Materials, 1995, **4**, 1145.
21. Hassouni K., Grotjohn T. A., and Gicquel A. J. Appl. Phys., 1999, **86**, 134.

# THEORY OF THE MICROWAVE HIGH-PRESSURE DISCHARGE

**K. Khodataev**

Moscow Radiotechnical Institute RAS, Moscow, Russia

**Abstract.** The work is devoted to theoretic study of a single streamer discharge which can be observed in the focus of microwave radiation in the open resonator. The initial stage of the discharge development from avalanche to the streamer is investigated. It is shown what the ambipolar field do not limit the electron diffusion on the top of the streamer so the streamer head velocity is defined by the free electron diffusion and local value of the ionization frequency on the streamer top. The universal value of the radius of the streamer head is found. The designed theoretic model takes into account 1)comparability of the streamer length with length of a wave of radiation, 2) limitation of energy saved in the resonator, 3)the plasma-gas dynamics and 4)the plasma-chemical processes. The computations confirm, in general, ideas that were developed earlier. The high-pressure microwave discharge is opening the wide field of applications in aeronautics and other technologies.

## 1. INTRODUCTION

The microwave high-pressure discharge is opening the wide field of applications in aeronautics and other technologies [1,2,3,4]. The electrodeless microwave discharges in a gas of high pressure demand appropriately high level of the electromagnetic field. For example the air breakdown of normal pressure needs specific energy flow of the microwave radiation 1 MW/cm$^2$.The high amplitude microwave field can be achieved in a focus of a radiation beam. Even in the case when the focusing system has a short focus so that the area of the focus cross section equals approximately wave length squared the needed generator power for 10 cm wave length is 100 MW with pulse duration more than few microseconds. It is difficult to provide so high power during some microsecond in a laboratory. But the field level needed for breakdown of a gas with high pressure (more than several atmosphere) can be achieved in a resonator because the resonator quality can be very high.

The open resonator created by two spherical copper mirrors was used in our experiments [1]. The resonator allows to create a discharge in air and other gases with pressure which does exceed one atmosphere. A high-pressure discharge in the focused traveling-wave radiation beam represents a net of the thin hot strings (connected among themselves), consistently appeared one from another (of course if the radiation intensity in the focus and pulse duration are enough). But the discharge in the resonator represents the single hot string. The string is a single because the stored energy in resonator is finite. Usually the all stored energy is adsorbed by one resonant streamer. The length of developed streamer is close to half of wave length. It is electrodynamic resonance. Appropriately the high current is inducted in the streamer. The inducted current heats the streamer plasma up to high temperature because the streamer diameter is very small. The specific heating up power related to the gas density is proportional to gas pressure because the breakdown electric field (and inducted current) is proportional to one. If the gas density is quite enough the magnetic pressure force of inducted current is able to compress the plasma in streamer (microwave pinch effect). We will show the key experimental and simulation results.

It need be noticed that it is necessary to distinguish the microwave streamer discharges in subcritical field and supercritical field. Their properties and parameters strongly differ one from another. We will describe the supercritical microwave streamer discharge below.

## 2. EXPERIMENTAL DATA

The scheme of experimental installation is shown on Fig.1. The open resonator 4 consists from two spherical copper mirrors displaced along common axis z. The curve radius $a_s$, diameter $2a_m$ and distance between them $L_0$ equal 35 cm, 55 cm and 51.7 cm correspondingly. The quartz pipe 5 filled by gas with given pressure, is placed in the centre of the resonator. The pipe represents the quarts long tube with the inner diameter equaled to 8 cm and has the optical windows on the ends. The gas pressure in the pipe can be varied from small value up to several atmosphere. The magnetron generator feeds the resonator through wave guide with circulator 2 and reactive attenuator 3 with the feeding coefficient ~$10^{-3}$. The output power of generator is 10 MW with pulse duration 40 μs. Wave length of radiation λ is 8.9 cm. The repetition frequency is less than 1 Hz. The amplitude envelop of field in resonator is monitored by the probe 6. The measurements show that the used resonator has the own time τ = 5 μs. It means that in our case the resonator allows to increase the field in the focus in 50 times. The most part of experiment was performed in air and hydrogen.

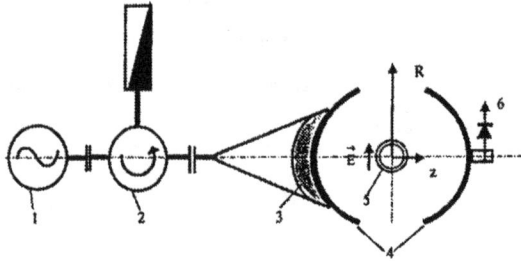

Figure 1. The scheme of the experimental installation. 1 – magnetron generator, 2 – circulator, 3 – lens, 4 – spherical mirrors of the open resonator, 5-quarts pipe, 6 – field prob.

The main results of experimental study of microwave discharge in a gas of high pressure (p>0.2 atm) in the open resonator can be easy formulated seeing on the typical example of this kind of discharge, shown on the Fig.2 [1,2,3]. The electron avalanche starts from one electron and represents the immovable in average spherical electron cloud with increasing electron number and expanding radius. When the number of electrons in avalanche is quite enough the electron cloud starts to transform to a string which is oriented along vector of electric field (Fig.2a and Fig.2b). It is a microwave streamer. The speed of the streamer arising can exceed $10^8$ cm/s if the pressure is quite enough [1]. When the streamer length achieves the resonant value (near half wave length) the field in resonator breaks down and farther exists on insignificantly small level. The streamer length development is stopped. After breakdown the streamer filament explodes and the shock wave runs out from the streamer (see Fig.2c) [1]. The measured shock wave parameters show that the streamer absorbs almost all energy stored by resonator and explodes.

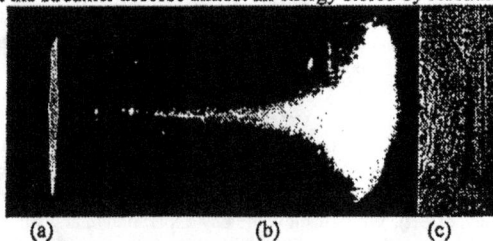

(a)　　　　　　　(b)　　　　　　　(c)

Figure 2. The streamer development. (a) – the photo of the streamer by the exposed lens, (b) – the scanned image of the streamer (the total time durance is 70 ns), (c) – the shadow of the shock wave, generated by the exploded streamer. The vertical size of the plot is 5 cm.

If gas pressure equals or more than 0.5 atm the bright core is observed in central part of the resonant streamer (see Fig2a and Fig.3). In hydrogen some times two bright cores can arise near center of the streamer. One can suppose that the most part of stored energy is absorbed in the core. It is important to note that the clear boundary on gas pressure exists between the state with the core and without one. The light intensity measurement by the densimeter of the streamer photo-image (Fig.3) shows that the streamer filament radius outside the core is very small, less than 0.05 cm. In general case the radius depends on the gas pressure. This dependence is shown in the Fig.4.

## 2. THE PHYSICAL BASIS FOR DEVELOPING OF THE STREAMER DISCHARGE THEORY

### 2.1. The initial stage - electron avalanche

At quite high gas pressure the initial electron density is very small. On this reason the discharge starts from one electron and represents the origin electron avalanche if electric field is supercritical. In difference from usual avalanche in direct current electric field the electrons of microwave avalanche have not the directed average velocity. The electron movement is oscillatory only. Initially the electron density in avalanche is insignificant for influence on the microwave field, so in average the development of avalanche in the microwave field with constant amplitude |E| can be described by known solution

$$n = \frac{\exp\left(v \cdot t - \frac{r^2}{4 \cdot D_e \cdot t}\right)}{\left(4 \cdot \pi \cdot D_e \cdot t\right)^{3/2}}, \tag{1}$$

where n – electron density, $D_e$ - electron free diffusion coefficient, $v(|E|)$ – sum of the ionization and attachment frequencies (an avalanche is developed if $v(|E|)>0$), r – radial coordinate ,t – time from the start of process. It is known that at the field near to critical the electron temperature is about 1-2 eV. It means that if the radius of avalanche $R_a$ , estimated as

$$R_a = 2 \cdot \sqrt{D_e \cdot t}, \tag{2}$$

becomes more than Debye radius $R_D$

$$R_D = 2 \cdot \sqrt{\frac{\kappa_B \cdot T_e}{4 \cdot \pi \cdot n_e \cdot e^2}} \tag{3}$$

($k_B$ - Boltzmann constant, e-electron charge) the polarization electric field will stop the free diffusion of electrons at the central part of the avalanche cloud. It will be at time moment, defined by equation

$$\frac{\exp(v \cdot t_D)}{\sqrt{v \cdot t_D}} = \frac{\kappa_B \cdot T_e}{2 \cdot e^2} \cdot \sqrt{\frac{\pi \cdot D_e}{v}} \tag{4}$$

Accordingly the radius of avalanche at this moment equals

$$R_a = 2 \cdot \sqrt{\frac{D_e}{v} \cdot (v \cdot t_D)} \tag{5}$$

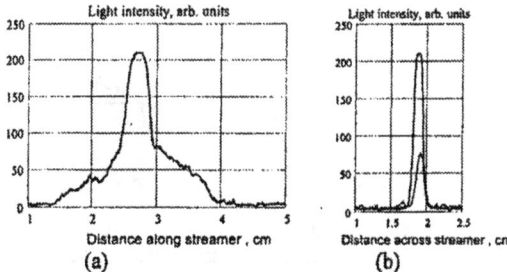

Figure 3. The measured light intensity of the microwave streamer, (a) - distribution along the streamer axis; (b) – distribution across the streamer, solid line – across the corn (z=2.7 cm at (a)), dot – aside the corn (z=3.0 cm at (a)).

# 38

## 2.2. The streamer start

Of course the ambipolar field exchange the free electron diffusion to ambipolar diffusion so one can wait that ionization front velocity must correspondingly decrease. But it was shown in [11] for plane ionization front, that the velocity is defined by free electron diffusion and effective ionization frequency in unperturbed field before the front

$$V_{fr_0} = 2 \cdot \sqrt{D \cdot \nu(E_0 / E_{cr})} \qquad (6)$$

The electron diffusion exponential precursor has very small electron density, so the ambipolar field in the precursor is exponentially small too. Therefore electron diffusion ahead the main front is free. The small electron density in precursor is insufficient for influence on the electric field, thus ionization in precursor corresponds to unperturbed field [11]. It is important to see the process of transformation of avalanche to streamer taking into account the ambipolar field and microwave field changing.

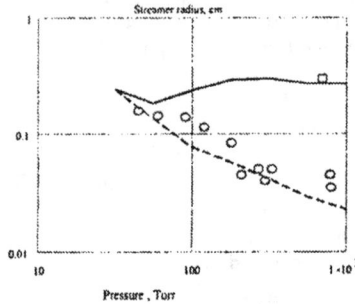

Figure 4. The streamer radius dependence on gas pressure. Circles – the measured values from photo of the streamer by exposed lens, box – usual value of the streamer radius measured by shadow method, lines – the simulation result: solid – the streamer radius after expanding caused by explosion, dash – the radius at moment when inducted current ( and correspondingly light radiation of the streamer) is maximum.

As we have shown earlier the avalanche has spherical symmetry in a spatial homogeneous microwave field. The spherical symmetry is being destroyed first of all by change of the microwave electric field distribution and by the quasi-stationary ambipolar field. Thus for initial stage of the streamer development we can receive the supposition of the spherical symmetry of electron density distribution of developing plasmoid. Because the size of electron avalanche is much time smaller than the wave length of microwave radiation the microwave electric field can be described by the Poisson's equation. It is known that if the object with spherical symmetry distribution of an electrical permeability is located in azimuth symmetry external electric field $E_0$ the solution of the Poisson's equation in spherical coordinates can be fined in form

$$\Phi = f(r) \cdot \cos(\theta)$$

where f(r) is defined by 1-D equation

$$\frac{\partial^2 f(r)}{\partial r^2} + \frac{2}{r} \cdot \frac{\partial f(r)}{\partial r} - \frac{2}{r^2} \cdot f(r) + \frac{1}{\varepsilon(r)} \cdot \frac{\partial \varepsilon(r)}{\partial r} \cdot \frac{\partial f(r)}{\partial r} = 0 \qquad (7)$$

The electric field amplitude can be found by the formula

$$E = -\nabla \Phi = -(\cos(\theta) \frac{\partial f(r)}{\partial r} i_r - \sin(\theta) \frac{f(r)}{r} i_\theta) \qquad (8)$$

The equations (7)-(8) are added by continuity equation for electron density taking into account the ionization, attachment, recombination, diffusion and drift in quasi-stationary ambipolar electric field G and equation

$$\frac{\partial n}{\partial t} = \nu(E) \cdot n + D \Delta n + \frac{e^2}{m \cdot \nu_{tr}} \cdot \nabla(G \cdot n) \qquad (9)$$

$$\frac{\partial G}{\partial t} = -\frac{4 \cdot \pi \cdot e^2}{m \cdot \nu_{tr}} \cdot n \cdot G + 4 \cdot \pi \cdot e \cdot D \cdot \nabla n \tag{10}$$

$$\varepsilon(r) = 1 + i\frac{4 \cdot \pi \cdot n(r) \cdot e^2}{m \cdot \nu_{tr} \cdot \omega} \tag{11}$$

The transfer coefficients correspond to electron temperature 2 eV. The electrical permeability is defined by electrical ohmic conductivity $\varepsilon = 1 + i4\pi\sigma/\omega$ ($\omega$ - cycling frequency of microwave field). The ion mobility is insignificantly small. The equation system was solved with initial and boundary conditions that correspond to distribution (1) and homogeneous external electric field amplitude $E_0 = 1.1E_{cr}$ ($E_{cr}$ - critical value of microwave field).

The results of modeling are shown on Fig.4-10. From the very beginning the electric field amplitude on the pole of the plasmoid increases and at the equator decreases (see Fig. 4).

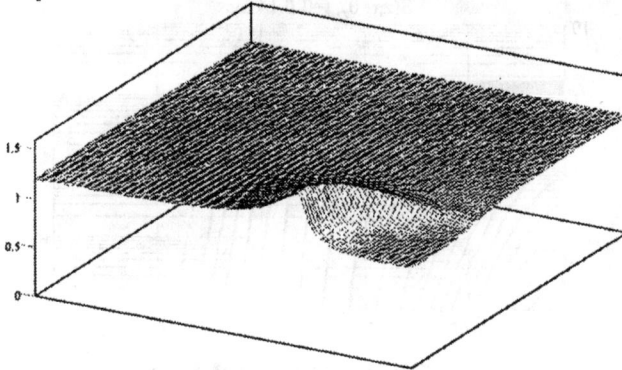

Figure 4. The spatial distribution of electric field amplitude at the streamer on initial stage of development, t=0.071mcs.

It reasons the quick development of plasmoid along the vector of external electric field. The distribution of parameters at front of ionization along the external field at the streamer axis quickly transformed from the distribution (1) to universal profile (see Fig.5). The quasi-stationary ambipolar field G is negligible at precursor region (S<0.1) so it is no wonder that the normalized electric conductivity $S = 4\pi\sigma/\omega$ is being decreased exponentially by law (12)

$$S_{fr} = C \cdot \exp\left(-\frac{z - z_{fr}}{d_{fr}}\right) \tag{12}$$

$$d_{fr} = \sqrt{\frac{D}{\nu\left(E_{fr}/E_{cr}\right)}} \tag{13}$$

where D – free electron diffusion coefficient, where $E_{fr}$ - the local value of the microwave field at the front.

It is important to notice that the streamer development need not the photo-ionization. The free electron diffusion is enough for its propagation in gas without of preliminary ionization. The details of the parameters distribution at front of the streamer are shown on Fig.6-7.

The ambipolar field G exists in a thin layer at the front only and exponentially small at the precursor. The microwave field E and ambipolar field G are maximum at the same point. The potential U of the quasi-stationary field is rising up to several times of the electron temperature $T_e$: $U = (5-6)T_e$. The microwave field amplitude on the pole of plasmoid and near ahead the front at precursor is more than $E_0$. Therefore the streamer front velocity is defined by the ratio (11) [11],

$$V_{fr} = 2 \cdot \sqrt{D \cdot \nu(E_{fr}/E_{cr})} \tag{14}$$

The computed dependence $V_{fr}(t)$, the front velocity estimated by Eq. (14) and Eq. (6) are compared in Fig.7. The computed volume is near to the (14).

The noticed peculiarities allows to declare that the characteristic depth $d_{fr}$ of the front (13) is the same as the radius of the streamer head. It is because the microwave field $E_{fr}$ on the top of the streamer is inversely proportional to the head radius and it will being decreased up to the depth of the front. Using the method presented in [1] based on the supposition that the front velocity is defined by the local value of the electric field one can show that plane front of ionization is unstable for the waved perturbations. The theory shows that the instability increment is proportional to the wave number of perturbation up to value of the front depth. It means that at nonlinear stage the instability will form the long thin tongue with radius that equals to the front depth (13). It is very important result because it gives us the possibility of the correct estimations and designing of the simple models based on the knowledge of the radius of the streamer.

S(z,r=0), t=0-0.14 mcs

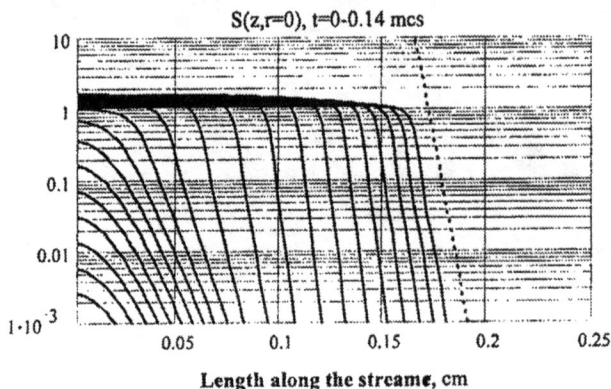

Length along the streame, cm

Figure 5. The temporary evolution of the normalized electrical conductivity distribution along the streamer, time durance 0 – 0.14 µs, The Eq. (12) – dot line

The performed study of the ambipolar field influence and the main factors of the streamer development allow us to design the different models which is able to describe the main stage of the streamer evolution, namely, resonance stage, when the energy of the microwave radiation is being absorbed intensively. One of the designed models (and results of investigations of the streamer development) is described in [2]. The other example of the more advanced simple model is submitted below.

## 3. THE SIMPLE THEORETICAL MODEL OF THE RESONANT SREAMER GISCHARGE

The streamer development is the very complicated nonlinear process. Its adequate theory description can be performed by the computer simulation only. The complete model is too complicated for the brief and wide investigation. The simple model is taking into account the main factors of the process but several suppositions allow to describe it by the system of the 1D differential equations.

Let us suppose that the streamer discharge plasma is the conducting cylinder with the diameter $2a(z)$ and unlimited length, oriented along external microwave electrical field. The electrical conductivity $\sigma(z)$ is homogenous inside the cylinder along radius and has a distribution along the axis of it, depending on the time. It allows to use for calculation of the current inducted in the cylinder the elements of theory developed for a wire antenna, namely, the integral equation.

$$E(z) = E_0 + i \cdot \int_{-L}^{L} E(z') \frac{S(z')}{2} \Psi(S(z')) \cdot W(z,z') \cdot k \cdot dz' \qquad (15)$$

$$W(z,z') = \frac{\exp(kR(z,z'))}{(kR(z,z'))^3} \cdot \left( (1 - ikR(z,z')) \cdot \left( 2 - 3 \cdot \left( \frac{a(z')}{R(z,z')} \right)^2 \right) + (ka(z'))^2 \right)$$

$$R(z,z') = \sqrt{a(z')^2 + (z-z')^2}$$

$$\Psi(S) = \int_0^0 J_0 \left( kr\sqrt{1 + i \cdot S} \right) \cdot k^2 \cdot r \cdot dr$$

$$I(z) = E(z) \cdot \frac{\omega}{4\pi} \cdot \frac{S(z)}{2} \Psi(S(z)) \qquad (16)$$

where E(z) – the microwave electric field effective amplitude, k=ω/c – wave number (ω - cycling frequency of the microwave field, c – light velocity), Ψ(z) takes into account the skin effect, S=4πσ/ω.

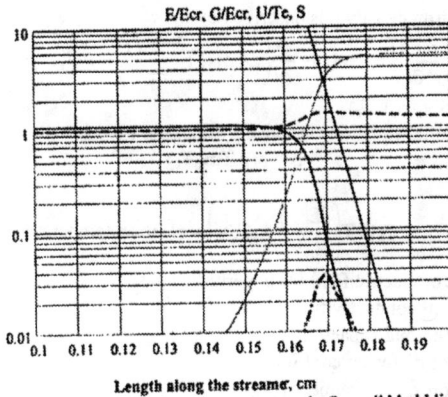

Figure 6. The distribution of the parameters along the streamer axis: S – solid bold line, electric field amplitude E/E_cr – dash line, quasi-stationary polarization field G/E_cr – dot-dash line, its potential U/T_e – solid thin line, asymptotic solution for front (12) –dot line.

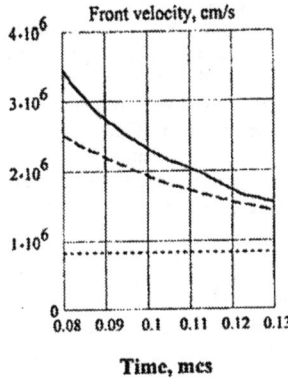

Figure 7. The front velocity temporary dependence. The modeling result – solid, the Eq. (14) – dash, the Eq. (12) – dot line.

The radius of the streamer is defined by the ionization front velocity from one side and gas dynamics from other side as result of the streamer heating. The Sedov's theory of a point explosion in the gas with a finite pressure is used for simulation of the radial dynamics of the streamer.

The temporal behavior of radius of the cylinder a(z) is described by the simple differential equations

$$\frac{da(z)}{dt} = 2 \cdot \frac{D(\rho_0, T_e(z))}{a(z)} + V(z) \tag{17}$$

$$\frac{dV(z)}{dt} = \frac{2}{a(z)} \cdot \left( \frac{(p_0 - p(z))}{\rho_0} - V(z)^2 - \left( \frac{2 \cdot J(z)}{c \cdot a(z)} \right)^2 \cdot \frac{1}{8 \cdot \pi \cdot \rho(z)} \right) \tag{18}$$

$$\frac{dp(z)}{dt} = \frac{1}{\rho(z) \cdot C_v(p(z), T(z), n(z), T_e(z))} \cdot \left( \frac{\omega}{4\pi} S(z) \cdot \frac{c}{k} \cdot |I(z)|^2 - L(p(z), T(z), n(z), T_e(z)) \right) \tag{19}$$

where $p, \rho$, – total pressure, total density of plasma mixture, $T$ - temperature of molecules, atoms and ions, $n, T_e$–number density and temperature of the plasma electrons, $C_v$ – thermal capacity of the gas mixture. The estimations show that at high unperturbed gas pressure the pinch force of the inducted current can be more than the discharge plasma pressure gradient thus the pinch force is included in the movement equation (18).

The evolution of the electron density n is described by the continuity equation

$$\frac{\partial n(z)}{\partial t} = v \cdot n(z) + \frac{\partial}{\partial z} \left( D(\rho, T_e) \frac{\partial n(z, t)}{\partial z} \right) - \beta(T_e, n) \cdot n^2 \tag{20}$$

where $v(\rho, T, T_e) = v_i(\rho, T, T_e) - v_a(\rho, T, T_e)$.

The system of the differential equations is added by the state equation (21),

$$\rho = \rho(p, T, n_e, T_e) \tag{21}$$

and calculated functions

$$T_e = T\left( \frac{|E|}{E_{cr}} \cdot \frac{\rho_0}{\rho}, T, n \right) \text{ - electron temperature,} \tag{22}$$

where $E_{cr}$ – the critical value of electric field for unperturbed gas,

$\sigma = \sigma(\rho, T, n_e, T_e)$ – electrical conductivity, $\tag{23}$
$C_v = C_v(\rho, T, n_e, T_e)$ – gas thermal capacity, $\tag{24}$
$D = D(\rho, T_e)$ – electron free diffusion coefficient $\tag{25}$
$v_i = v_i(\rho, T, T_e)$ – electron ionization frequency $\tag{26}$
$v_a = v_a(\rho, T, T_e)$ – electron attachment frequency, $\tag{27}$
$L = L(p, T, n, Te)$ – the light losses,
$\beta = \beta(T_e, n)$ - effective coefficient of recombination $\tag{28}$
($\beta$ takes into account the several kinds of recombination).

The state equation (21) and the functions (22)- (28) for diatomic molecular gas was calculated by means of quite adequate model in a wide diapason of a temperature for the not equilibrium ionization. So the designed model pretends to be near to reality. It must be noticed that the function (22) is monotonous only if the electron temperature is less than some limit which depends on the gas type (approximately 10-20 eV). The function (22) has the hysteresis loop. If the electron temperature is rising higher than the limit the function (22) must be changed by the differential equation.

The MW generator provides the energy to the open resonator but diffraction and heat losses decrease the stored energy. Both factors is took into account in differential equation for electric field at the resonator focus:

$$\frac{dE_0}{dt} = \frac{E_m - E_0}{\tau} - \frac{4 \cdot \pi^2}{\Omega_{res} \cdot E_0} \cdot \int \sigma E(z)^2 a(z)^2 dz - W(I(z)) \tag{29}$$

where $\Omega_{res}$, $\tau$ - equivalent volume and own time of the resonator accordingly, $E_m$ - maximum field in the resonator, $W(I(z))$ – the diffraction losses. It is supposed that resonator is tuned in resonance with generator frequency.

Thus the evolution of the microwave streamer discharge in the microwave resonator is described by the system of the differential equations (16-20, 29) and functions (21-28).

The task parameters are the wave length (or wave number k) of the microwave radiation, maximum field in the resonator $E_m$, the resonator parameters $\Omega_{res}$ and $\tau$, the gas initial pressure and the moment of appearance of initial free electron.

# 4. THE SIMULATION RESULTS AND ITS COMPARISON WITH EXPERIMENTAL DATA

The task parameters were the same as in the experiment. Maximum field in resonator in stationary regime $E_m = 35$ kV/cm, own time of the resonator $\tau = 5\mu s$. The initial free electron appears in the focus of resonator when electric field in the focus have achieved the critical value $E_{cr}$ depending on gas pressure. The gas pressure was varied from 50 Torr to 760 Torr. The results of simulation for discharge at gas pressure 760 Torr are presented below. The picture of the calculated light losses from the streamer is shown on Fig. 8. One can see the good qualitative accordance with the experimental photo on the Fig. 2b.

The time dependence electric field $E(z,t)$ distribution is shown on the Fig.9. The electric field is maximum on the ends of the streamer. The field on the ends is achieving the maximum if the streamer length is resonant, that is approximately half wavelength. At resonance the inducted current I and heating power are maximum.

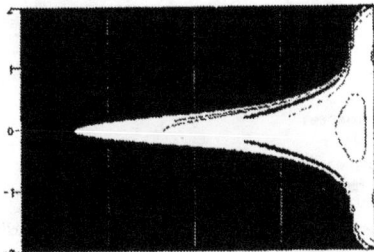

**Figure 8.** Computed spatial-temporal distribution of light radiating from the streamer (time - on a horizontal, length - on a vertical).

The simulation shows that independently on the gas pressure if electric field exceeds the critical value the electron avalanche starts to develop and forms the streamer. The streamer length, electric conductivity and inducted current rise up to the maximum value limited by finite energy stored by the resonator. Electric field in the resonator decreased quickly, it is breakdown. The streamer radius rises initially up to value limited by decreasing electric field inside the streamer and after resonant heating rises quickly because of the explosion. Fig.10a demonstrates this process on the case of the gas pressure 760 Torr. One can see that after heating the electron temperature $T_e$ and gas temperature T are the same and achieve the value 20,000-30,000 K (see Fig.10b) [13].

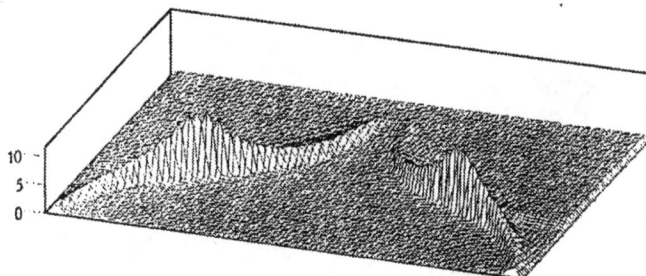

**Figure 9.** Spatial-temporal electric field E distribution along the streamer.

The maximum inducted current and maximum velocity of the streamer ends are near the theoretic limit if the pressure is quite high. The current is limited by the radiation resistance of an ideal resonant vibrator in the critical external field $E_{cr}$. The inducted current is limited by the diffraction losses. The ends velocity is limited by the product of the free electron diffusion and maximum ionization frequency.

The gas temperature is maximum at the middle of the streamer. The calculated intensity of light of the middle of the streamer is strongly higher than at the streamer branches too. The calculated maximum gas temperature exceeds 0.5eV if the pressure is more than 500 Torr. So the observing bright corn at the middle of the streamer can be explained by the high gas temperature at the middle. Thus it is clear why the bright corn in experiment is arising if the pressure is more than approximately 500 Torr.

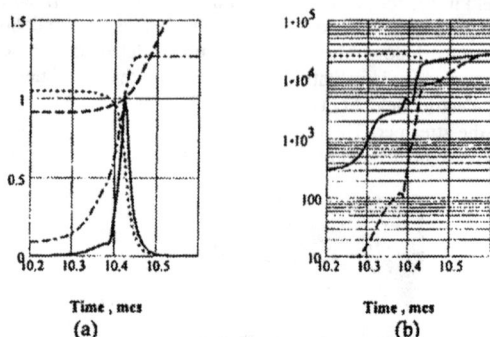

Time , mcs
(a)

Time , mcs
(b)

Figure 10. The example of simulation. Gas pressure is 760 Torr. (a)The solid line - inducted current $J/J(t_1)$, ($J(t_1) = J_{max}$ ),dot– electric field in the resonator $E/E_{cr}$, dash – streamer radius $a(t)/a(t_1)$, dadot – length of the streamer divided on the half wavelength, t – the time after switching on the generator. (b)The solid line – the gas temperature $T(t)$, K, dot – the electron temperature $Te(t)$, K, dash – electric conductivity $4\pi\sigma(t)/\omega$.

## 5. SUMMARY

One can see the quite good accordance between the experimental data and the simulation. It gives us some assurance what designed model has relation to the reality and successfully can be used for the forecast of the discharge parameters even if the gas pressure is more than 1 atm. But the parameters forecast for discharge at the gas pressure about several tens atm requires the much deeper experimental and theoretic study than it is performed here.

### Acknowledgements

The author are grateful to Dr. I.I.Esakov and Dr L.P.Grachev for the providing of the experimental data and for the fruitful discussions.

### References

1. Khodataev K.V. Proc. of Workshop on Weakly Ionized Gases (USAF Academy, Colorado 9-13 June 1997), v.1, p. L-1 .
2. Khodataev K., Ershov A. Proc. supplement of 2-nd Weakly Ionized Gases Workshop. (Waterside Marriott Hotel, Norfolk, Virginia, USA, 24-25 April, 1998), 339.
3. Esakov I.I., Grachev L.P., Khodataev K.V. Proc. supplement of 3-rd Weakly Ionized Gases Workshop. (Waterside Marriott Hotel, Norfolk, Virginia, USA, 1-5 November, 1999), 99, 4821.
4. Grachev L.P., Esakov I.I., Mishin G.I., Khodataev K.V. Preprint of LFTI A.F.Joffe, 1992, №1577.
5. Grachev L.P., Esakov I.I., Mishin G.I., Khodataev K.V.. Tech. Phys.,1994, 39, 130.
6. Grachev L.P., Esakov I.I., Mishin G.I., Khodataev K.V., Tsyplenkov V.V.. Pis'ma v JTF, 1992,18, 34.
7. Grachev L.P., Esakov I.I., Khodataev K.V. JTF, 1998, 68, 33.
8. Grachev L.P., Esakov I.I., Mishin G.I., Khodataev K.V. JTF, 1996, 66, 32.
9. Khodataev K.V. Proc. of the ICPIG XXIII (Toulouse,1997), Contributed papers, IV-24.
10. Khodataev K.V. Proc. of the ICPIG-XX (Piza, 1991), Invited papers, p. 207.
11. Khodataev K.V., Gorelik B.R. Fizika Plasmy, 1997, 23, 236.
12. Khodataev K.V. Fizika Plasmy, 1995, 21, 605.
13. Esakov I.I., Grachev L.P., Khodataev K.V. Proc. of the Int. Workshop "Strong microwaves in plasmas (2-9 August 1999)", Nizhny Novgorod, 2000, v.1, 291.

# THEORY OF HIGH POWER MICROWAVE AND OPTICAL GAS BREAKDOWN.

**M. V. Kuselev and A. A. Rukhadze.**

General Physics Institute RAS, 38 Vavilova str., Moscow, Russia.

**Abstract.** The analytical review of theoretical investigations of high power microwave and optical gas breakdown is presented. The oscillation energy of electrons in intensive radiation is supposed to be much higher than atomic ionization energy. In the microwave frequency range the electric field of radiation is much weaker than atomic field and therefore gas ionization mechanism is determined by electron-atom collisions. On the contrary, in the optical frequency range the electric field of radiation turnes out to be of the order of atomic field and therefore gas ionization mechanism is determined by tunneling of atomic electrons. Nevertheless the distribution functions of electrons in both cases are similar, extremely anisotropic and as a result unstable.

## 1. INTRODUCTION

At the end of 60-s the high power short pulse coherent radiation sources appeared. The power density of such lasers exceeds $10^{14} W/sm^2$ and electric field of radiation is of the order of atomic field $E_a = 5,1 \cdot 10^9 V/sm$. Just by that time the problems of gas ionization and distribution function of created plasma electrons were formulated. The oscillation energy of electrons $I_0$ in the electric field of such radiation is much larger than ionization energy of atoms $I_i$, or

$$I_0 = \frac{e^2 E_0}{2m\omega_0^2} \gg I_i,\qquad(1.1)$$

where $\omega_0$ is the radiation frequency and $E_0$ is a strength of electric field. In such fields of optical radiation the mechanism of tunneling of atomic electons becomes dominant. Just this mechanism of atom ionization was studied in the early works [1-3]. As for distribution function of electrons in the created plasma, its calculation in this period was unsuccessful. The correct calculations of this function were done much later [5-8]. Unfortunately the authors of these papers did not know the theoretical investigations of high power microwave gas breakdown carried out much earlier (see [9,10] and referenses). Contrary to optical breakdown, in the microwave frequency range the electric field of high power sourses usually is much less than atomic field and therefore the ionization mechanism of atoms is determined by electron-atom collisions. Inspite of the difference in the ionization mechanism, the electron distribution functions in both cases are similar; the ionization mechanism influences only on the dynamics of plasma density.

In the microwave frequency range ($\omega_0 \approx 2 \cdot 10^{10} - 2 \cdot 10^{11} rad/s$) the reachieved radiation power density is of the order of $10^8 W/cm^2$ and electric field $E_0 \leq 10^6 V/cm$ is much less than atomic field. Another situation takes plase in the optical frequency range ($\omega_0 \approx 2 \cdot 10^{15} - 2 \cdot 10^{16} rad/s$), where the power density reaches $10^{19} W/cm^2$ and electric field $E_0 \leq 10^{10} V/cm$; the latter turnes out to be of the order of atomic field. In both cases the strong inequility (1.1) satisfies. Moreover electron oscillation energy is comparable to relativistic one.

Below an analytical review of papers [7,9], in which only nonrelativistic cases were considered, is presented. We not only calculate distribution functions but also investigate the stability problem for gas breakdown in high power electromagnetic fields. As for relativistic effects which were firstly considered in [8,10], we will discuss them very shotly in the appendices of our paper only for circular polarization of radiation, which allows analytical solution of the problem.

## 2. ELECTRON DISTRIBUTION FUNCTION AND PLASMA DENSITY DYNAMICS IN GAS BREAKDOWN.

Under the condition (1.1) the thermal (chaotic) motion of electrons in breakdown plasma can be neglected comparing with their oscillatory motion in the electric field of radiation. Besides, we neglect also the perturbations of external field due to the plasma polarization, that is correct if plasma density $n_e$ is less than critical one, or if $\omega_0^2 \gg \omega_{Le}^2 = 4\pi e^2 n_e / m$. Finally, we suppose that plasma density is much less than the desity of neutral particles $n_0$, and therefore $n_0 = const$.

The kinetic equation describing the dynamics of distribution function in general looks as

$$\frac{\partial f_0}{\partial t} + v \frac{\partial f_0}{\partial r} + e \left\{ E_0 + \frac{1}{c} \left[ v B_0 \right] \right\} \frac{\partial f_0}{\partial p} = n_0 W_i \delta(p). \tag{2.1}$$

Here $E_0(\xi)$ and $B_0(\xi)$ are the electric and magnetic fields of external radiation, propagating in the z – direction, or $\xi = \omega_0(t - z/c)$, $W_i$ is the ionization probobity of atoms in external field.

In the nonrelativistic limit the inhomogeneities of $E_0$ and $B_0$ can be neglected. Moreover, magnetic forse in equation (2.1) also turnes out to be negligible. As a result equation (2.1) takes the form

$$\frac{\partial f_0}{\partial t} + e E_0(t) \frac{\partial f_0}{\partial p} = n_0 W_i \big( E_0(t) \big) \delta(p). \tag{2.2}$$

In this limit for the quantity $W_i$ we have very simple expression, which for the microwave breakdown looks as [9]

$$W_i = \int v_i \left( \left| \frac{p}{m} \right| \right) f_0(p) dp, \tag{2.3}$$

where $v_i = |p/m| \sigma_i (|p/m|)$ is the ionization frequency due to the electron-atom collisions and $\sigma_i$ is ionization cros section. For the optical breakdown [11]

$$W_i(E_0(t)) = 4\omega_a \left( \frac{I_i}{I_a} \right)^{5/2} \frac{E_a}{|E_0(t)|} \exp \left[ -\frac{2}{3} \left( \frac{I_i}{I_a} \right)^{3/2} \frac{E_a}{|E_0(t)|} \right], \tag{2.4}$$

where $I_a = \hbar \omega_a$, $2 = 13{,}6 eV$ is the ionization energy of hydrogen atoms and $\omega_a = 5{,}1 \cdot 10^{15} rad/s$ is the atomic frequency. The expressions (2.3) and (2.4) are valid only in nonrelativistic case and in addition they take into account only one ionization level of atoms.

Substituting (2.3) and (2.4) into the equation (2,2) one can notice the differese between the microwave and optical breakdowns. In the first case equation (2.2) turnes out to be homogeneous for $f_0(p,t)$ and its positive eigenvalue $\gamma(E_0)$ determines the ionization rate

$$\frac{\partial n_e}{\partial t} = \gamma(E_0) n_e = n_0 \int \frac{|p|}{m} \sigma_i (|p|) f_0(p,t) dp. \tag{2.5}$$

At the same time for the case of optical breakdown this equation is nonhomogeneous with given external source, which describes temporal dynamics of plasma density

$$n_e(t) = \int_0^t dt' W_i \big( E_0(t') \big) \approx n_0 W_i (E_0) t, \qquad W_i(E_0) = \frac{\omega_0}{2\pi} \int_0^{2\pi/\omega_0} dt W_i \big( E_0(t) \big). \tag{2.6}$$

However under the conditions

$$\omega_0 \gg \gamma(E_0), W(E_0) \tag{2.7}$$

the right part of equation (2.2) / for microwave and optical breakdowns / can be neglected and for $f_0(p)$ one can use Vlasov equation. In this approximation the general solution of equation (2.2) represents the arbitrary function of characteristyics

$$\frac{dv}{dt} = \frac{e E_0}{m} \cos \omega_0 t, \qquad v = v_E (\sin \omega_0 t - \sin \varphi). \tag{2.8}$$

Here $\varphi$ is the initial phase of field /at the moment of electron creation/ and $v_E = eE_0/m\omega_0$. So, we can write

$$f_0(p) = \mathfrak{f}_0(v)n_e(t), \quad \mathfrak{f}_0(v) = \delta(v_\perp)\delta[v_\parallel - v_E(\sin\omega_0 t - \sin\varphi)], \tag{2.9}$$

where $\int dv \mathfrak{f}_0(v) = 1$.

Taking into account (2.7) the expression must be averaged over $\varphi$, which leads to the well-known random-phase distribution [9]

$$\langle \mathfrak{f}_0(v)\rangle = \frac{1}{\pi\sqrt{v_E^2 - (v_\parallel + v_E\cos\omega_0 t)^2}}\delta(v_\perp). \tag{2.10}$$

Now we can determine the plasma density $n_e$. In the case of optical breakdown it is given by the relations (2.6) with $E_0(t) = E_0\cos\omega_0 t$. As for microwave breakdown then according to (2.5) we have[*]

$$n_e = n_{0e}e^{\pi}, \quad \gamma(E_0) = \frac{n_0}{\pi}\int_0^\infty dv \frac{v\sigma_i(v)}{\sqrt{v_E^2 - v^2}}, \tag{2.11}$$

where $\sigma_i(v)$ in the Borns appromaxion is equal

$$\sigma_i(v) = \frac{\beta}{v^2}\ln\left(\frac{v}{\sqrt{2I_i/m}}\right)\eta(v - \sqrt{2I_i/m}). \tag{2.12}$$

The quantity $\beta$ depends on the type of gas; for hydrogen $\beta = 16,3$.

In conclusion very briefly we discuss the validity of the above consideration. According to (2.7) for optical breakdown the consideration is correct only if $E_0 < E_a$ independently of gas pressure, whereas for microwave breakdown our consideration is valid if gas pressure $p_0 < 10 - 100 tor$.

Thus we can conclude, that under the conditions (1.1) and (2.7) electron distribution function in microwave breakdown plasma as well as in optical one with great accuracy coinsides with the random-phase distribution function.

## 3. INSTABILITIES OF A BREAKDOWN PLASMA CREATED IN HIGH POWER ELECTROMAGNETIC FIELDS.

**3.1.** Electron distribution function (2.10) is very anisotropic in velocity space and therefore it must be unstable. First of all, in a such anisotropic plasma the well-known Weibel instability [12] arises. We show this using adiabatic approximation, which is valid when increment of instability is larger than the growth rate of breakdown plasma density. In this approximation the instabilities can be described using well-lknown dispersion equation [13]

$$\left|k^2\delta_{ij} - k_i k_j - \frac{\omega^2}{c^2}\varepsilon_{ij}(\omega,k)\right| = 0, \tag{3.1}$$

where $\omega$ and $k$ are the frequency and wave vector of perturbations and $\varepsilon_{ij}(\omega,k)$ is a dielectric permittivity tensor

$$\varepsilon_{ij}(\omega,k) = \left(1 - \frac{\omega_{Li}^2}{\omega^2}\right)\delta_{ij} + \frac{\omega_{Le}^2}{\omega^2}\int dv\left[v_i\frac{\partial\mathfrak{f}_0(v)}{\partial v_j} + v_i v_j\frac{k}{\omega - kv}\frac{\partial\mathfrak{f}_0(v)}{\partial v}\right]. \tag{3.2}$$

Here we take into account also ion contribution in $\varepsilon_{ij}(\omega,k)$ supposing that ions are colde cold. Besides,

---

[*] In [9] equation (2.2) wos solved numerically for microwave breakdown with experimental dependenses $\sigma_i(v)$ for hydrogen and helium up to $E_0 = 10^6 V/sm$. It was shown that numerical solutions for $\mathfrak{f}_0(v)$ differ from (2.2) no more than 20%. As for $\gamma(E_0)$ then this quantity coinsides with (2.11) exactly.

we neglect the perturbations of external electic field in the kinetic equation for electrons, that is correct if $kv_E \approx \omega_{le} \gg \omega_{Li}$.

Substituting (2.10) into (3.2) we find the following expressions for electrons contribution in $\varepsilon_{ij}(\omega, \vec{k})$

$$\delta\varepsilon_{11} = \delta\varepsilon_{22} = -\frac{\omega_{Le}^2}{\omega^2},$$

$$\delta\varepsilon_{13} = \delta\varepsilon_{31} = \frac{\omega_{Le}^2}{\omega^2} \frac{k_\perp}{k_{II}} \left(1 - \frac{S}{\sqrt{1 - k^2 v_E^2/\omega^2}}\right), \tag{3.3}$$

$$\delta\varepsilon_{33} = -\frac{\omega_{Le}^2}{\omega^2} \frac{k_\perp^2}{k_{II}^2} + \frac{\omega_{Le}^2}{k_{II}^2}\left(2\frac{k_\perp^2}{\omega^2} + \frac{k_\perp^2 + k_{II}^2}{\omega^2 - k_{II}^2 v_E^2}\right)\frac{S}{\sqrt{1 - k^2 v_E^2/\omega^2}},$$

where

$$S = \begin{cases} \text{sgn}\,\text{Re}\,\omega \cdot \text{sgn}\,\text{Im}\,\omega, & \\ 1, & \text{if} \quad \text{Re}\,\omega = 0, \\ 1, & \text{if} \quad \text{Im}\,\omega = 0, k^2 v_E^2 < \omega^2, \\ \text{sgn}\,\omega, & \text{if} \quad \text{Im}\,\omega = 0, k^2 v_E^2 < \omega^2. \end{cases} \tag{3.4}$$

Substituting (3.2)--(3.4) into (3.1) for the purely transverse wave propagation ($k_{II} = 0$) we obtain

$$k^2 c^2 = \omega^2\left[1 - \frac{\omega_{Le}^2}{\omega^2}\left(1 + \frac{k^2 v_E^2}{\omega^2}\right)\right]. \tag{3.5}$$

In the low frequency range, $\omega^2 \ll \omega_{Le}^2$, this equation leads to the growth rate of Weibel instability [9]

$$\omega^2 = -\frac{\omega_{Le}^2 k^2 v_E^2}{2(\omega_{Le}^2 + k^2 c^2)} < -\omega_{Le}^2 \frac{v_E^2}{c^2}. \tag{3.6}$$

It is obvious that this expression is valid only when $\text{Im}\,\omega \gg \gamma(E_0), W_i(E_0)$ for microwave and optical breakdowns accordingly.

3.2. Contrary to the above considered Weibel instability, which takes place for any anisotropic distribution function, below we consider the specific instability for the distribution (2.10) originated by the positive derivative of this distribution over particle energy. This instability takes place under the conditions $\omega_0 \gg kv_E \approx \omega_{Le}$ and is described by the dispersion equation [14]

$$1 + \delta\varepsilon_e'(\omega, \vec{k}) + \delta\varepsilon_i'(\omega, \vec{k}) + \left[1 - J_0^2\left(\frac{k_{II} v_E}{\omega_0}\right)\right]\delta\varepsilon_e'(\omega, \vec{k})\delta\varepsilon_i'(\omega, \vec{k}) = 0. \tag{3.7}$$

Here $\delta\varepsilon_{e,i}'(\omega, \vec{k})$ are partial dielectric permittivities for electrons and ions and they look as [9]

$$\delta\varepsilon_i'(\omega, \vec{k}) = -\frac{\omega_{Li}^2}{\omega^2},$$

$$\delta\varepsilon_e'(\omega, \vec{k}) = -\frac{\omega_{Le}^2}{k^2}\int d\vec{v}\,\frac{\vec{k}\,\frac{\partial f_0(\vec{v})}{\partial \vec{v}}}{\omega - \vec{k}\vec{v}} = -\frac{\omega_{Le}^2 S}{\omega^2\left(1 - k^2 v_E^2/\omega^2\right)^2}. \tag{3.8}$$

Substituting this expressions into (3.7) we find unstable solutions for low frequency ion oscillations

$$\omega = \begin{cases} \omega_{Li}^2\left[1 + i\frac{\omega_{Le}^2 \omega_{Li}}{|kv_E|^3}J_0^2\left(\frac{k_{II} v_E}{\omega_0}\right)\right], & k_{II}^2 v_E^2 \gg \omega_{Le}^2, \\ \frac{\sqrt{3} + i}{2}|k_{II} v_E|\left(\frac{m}{M}\right)^{1/3}, & k_{II}^2 v_E^2 \gg \omega_{Le}^2. \end{cases} \tag{3.9}$$

It is easy to see that these instabilities have kinetic character and are originated by positive derivative of function (2.10) over energy. These instabilities also can be developed only if $\text{Im}\,\omega \gg \gamma(E_0), W_i(E_0)$.

3.3 As a conclusion of this section we consider the instability due to stimulated scattering of external radiation on the plasma density oscillations. For the optical breakdown this phenomenon was investigated in [8]. The theory of stimlated scattering is accounted in many textbooks on plasma physics and therfore we omit all details of calculations and present the final result. The maximal growth rate of instability is reached for back scattering of incident radiation and is equal to

$$\text{Im}\,\omega = \frac{v_E}{2c}\sqrt{\omega_{Le}\omega_0}\,. \tag{3.10}$$

Of course for observing the instability it is necessary that $\text{Im}\,\omega \gg \gamma(E_0), W_i(E_0)$.

Thus we can conclude that the gas breakgown in high power external fields is unstable. Several types of instabilities are possible: anisotropic hydrodynamical instability, which is known as Weibel instability, kinetic instability due to the positive derivative of electron distribution function over energy and stimulated scattering of external radiation on plasma density oscillations.

# 4. CONCLUSIONS

From the above presented analysis one can conclude the following:

1. In the case of microwave gas breakdown as well as optical one under the conditions (1.1) and (2.7) the distribution functions of created plasmas are similar and with great accuracy coinside with random-phase distribution (2.10). The only difference between these two types of gas breakdown is the gas ionization mechanisms, which in microwave breakdown is determined by the electron-atom collisions, whereas in optical breakdown – by the atomic electron tunneling in external field.

2. In the high power electromagnetic gas breakdown the oscillation energy of created plasma electrons is much higher than the ionization energy of atoms. This leads to the additional gas ionization by the plasma electrons. As a result the plasma density may significantly exceed the critical density and moreover plasma can created outside of the region, where external field is localized. Bisides due to the high energy of electrons the process of recombination must be slowed down and the breakdown plasma may occur to be long lived.

3. Gas breakdown plasma in very high power electromagnetic fields becomes unstable. There exist three types of instabilities: hydrodynamical Weibel instability, originated by the plasma anisotropy, kinetical low frequency instabilty, caused by the nonequilibrium positive derivative of distribution function (2.10) over energy and stimulated scattering of external radiation on the plasma density oscillations.

4. Taking into account relativistic effects of electrons motion the average electron velocity parallel to the direction of wave propagation appears. This velocity is originated by the action of cross electric and magnetic fields of electromagnetic wave.

# APPENDIX.

Relativistic effects in the high power gas breakdown.

In order to take into account the relativistic effects in the gas breakdown the equation (2.1) must be solved. As above in the first approximation we neglect the right side and write down its solution as a function of characteristics

$$p_x = mv_E(\sin\psi - \sin\psi_0), \quad p_y = \alpha mv_E(\cos\psi - \cos\psi_0),$$
$$p_z = \frac{1}{2mc}(p_x^2 + p_y^2), \tag{A.1}$$

where

$$E_0(r,t) = \left(E_{0x} = E_0 \cos\psi, E_{0y} = \alpha E_0 \sin\psi, 0\right),$$
$$B_0(r,t) = \left(-E_{0y}, E_{0x}, 0\right),$$

(A.2)

$\alpha$ characterize the wave polarization with phase $\psi = \omega_0(t - z/c)$ and $\psi_0$ is the initial phase (at the moment of electron creation).

Representing $f_0(p,t) = n_e \tilde{f}_0(p)$ , where $n_e(t)$ is given by (2.5) and (2.6), one can write

$$\tilde{f}_0(p) = \delta(p_x - \bar{p}_x)\delta(p_y - \bar{p}_y)\delta(p_z - \bar{p}_z).$$

(A.3)

For the circularly polarized wave $\alpha = 1$ and

$$\delta(p_x - \bar{p}_x)\delta(p_y - \bar{p}_y) = \frac{1}{2mv_E}\delta(p_\perp - mv_E),$$

(A.4)

where

$$p_\perp^2 = p_x^2 + p_y^2 + m^2 v_E^2 + 2mv_E\left(p_x \sin\psi - p_y \cos\psi\right).$$

(A.5)

As a result

$$\tilde{f}_0(p) = \frac{1}{2\pi m v_E}\delta(p_\perp - mv_E)\delta\left(p_z - 2m\frac{v_E^2}{c}\sin^2\frac{\psi - \psi_0}{2}\right).$$

(A.6)

After averaging over $\psi_0$ we obtain

$$\langle\tilde{f}_0(p)\rangle = \frac{1}{2\pi m v_E}\delta(p_\perp - mv_E)\frac{1}{\sqrt{p_z\left(2m\frac{v_E^2}{c} - p_z\right)}}.$$

(A.7)

In order to reach the agreement with /8, 10/ let us average this function also over $\psi$ which leads

$$\tilde{f}_0(p)_{AV} = \frac{1}{2\pi^2\left(p_x^2 + p_y^2\right)}\frac{1}{\sqrt{4m^2 v_E^2 - \left(p_x^2 + p_y^2\right)}}\frac{1}{\sqrt{p_z\left(2m\frac{v_E^2}{c} - p_z\right)}}.$$

(A.8)

From (A.7) and (A.8) follows that the relativistic effects lead to the existens of average electron motion along the direction of wave propagation with momentum of the order of $\approx mv_E^2/c$. It is obvious that such motion leads to new instabilities similar to considered above.

This investigation was initiated by the discussions with the author of [9], for which we express our thanks to him.

**References.**

1. Bunkin. F. V., Prokhorov A. M. JETF 1964, **46**, 1090.
2. Keldysh L. V. JETF 1964, **47**, 1945.
3. Perelomov A. M. et. al. JETF, 1966, **50**, 1393.
4. Afanasiev Yu. A. et. al. JETF, 1972, **63**, 121.
5. Korobkin V. V., Romanovsky M. Yu. JETF Let., 1991, **53**, 493.
6. Chochkov B. N. et. al. Phys. Rev., 1992, **45**, 7475.
7. Bychenkov V. Yu., Tikhonchuk V.T. Laser Phys., 1992, **2**, 525..
8. Soldatov A. V. Plasma Phys. Reports (in press.)
9. Glasov L. G. et. al. In "High Frequency Discharges". Inst.of Appl. Phys.(Gorky) 1988, 63.
10. Glasov L. G.,. Rukhadze A. A. Plasma Phys. Reports 1993, **19**, 1289.
11. Landau L. D., Lifshits E, M. Quantum Mechanics, Moscow: "NAUKA", 1963.
12. Weibel E. Phys. Rev., 1952, **2**, 83.
13. Alexandrov A. F., Bogdankevich L. S., Rukhadze A. A.. "Principles of Plasma Electrodynamics", Springer, 1984.
14. V. P. Silin, "Interaction of High Power Fields with Plasma", Moscow: "NAUKA", 1973.

# DIAGNOSTICS

# SPECTROSCOPIC STUDY OF A SURFACE-WAVE ARGON DISCHARGE AT ATMOSPHERIC PRESSURE

M.C. García, A. Rodero, A. Gamero, A. Sola, M.C. Quintero and C. Lao

Departamento de Física. Universidad de Córdoba. Campus de Rabanales, C-2.
E-14071 Córdoba. Spain.

**Abstract.** The behavior of an argon surface-wave plasma column has been studied by using spectroscopic diagnosis. The axial profiles of electron and gas temperatures and electron density are obtained for plasma columns produced in a quartz capillary tube by coupling up to 110 W of 2.45 GHz electromagnetic power. The analysis of the atomic state distribution function reveals that only the populations of the 5p and higher levels are controlled by the Saha balance, while the 4p levels are underpopulated relative to the Saha equilibrium, mainly close to the end of the column. The spectroscopic study has also included the power interruption technique. The intensity response of the atomic lines when the microwave power is suddenly switched off, in form of abrupt jumps, confirms these results.

## 1. INTRODUCTION

The interest in studying high frequency (HF) discharges has grown in recent years due mainly to the increase of their scientific and technological applications. The surface-wave-sustained discharge (SWD) is a type of HF discharge, where the electromagnetic field of the surface wave sustaining the discharge can only propagate if the electron density exceeds a critical value [1]. These discharges show advantageous properties such as possessing greater dimensions than the HF power coupler, a wide range of operation conditions and high stability and reproducibility. Many theoretical and experimental works have been made on this type of plasma, mainly at low and medium pressures, but at present its behavior at high pressure has not been studied in depth (for example, see [1-3]).

The aim of this work is to contribute to the knowledge of the high pressure surface-wave-sustained discharges. Particularly, argon plasma columns produced at atmospheric pressure are experimentally characterized by using spectroscopic diagnosis methods, and the axial profiles of electron and gas temperatures and electron density are obtained along the plasma column under different experimental conditions.

The atomic state distribution function (ASDF) for a plasma describes how atoms are distributed over their internal states and reflects their microscopic activity. When non-close to local Saha-Boltzmann equilibrium circumstances are detected one must next account for the possible departure from equilibrium in order to interpret the spectroscopic measurements correctly [4]. From the analysis of the ASDF, the stage of departure from equilibrium and the electron temperature are deduced. In this direction, the power interruption technique can be used to complement this analysis, as is shown in section 3.

## 2. LOCAL SAHA EQUILIBRIUM

The thermodynamic equilibrium (TE) requires that each microscopic process be strictly balanced by the inverse one (proper balances). When this applies to all mechanisms, the Maxwell energy distribution for particles, the Boltzmann distribution for the density of excited atoms, the Saha distribution for the density of excited neutral atoms with respect to the density of the ion ground level, and the Planck's law for the radiation equilibrium are verified. The weakest form of TE departure occurs when a part of the emitted radiation is not reabsorbed, escaping from the plasma, and Plank's law is not fulfilled. If the Maxwell, Boltzmann and Saha balances can be maintained by collisions, the ions, atoms and electrons remain locally in equilibrium at the same temperature, which is called local thermodynamic equilibrium (LTE) because of the possible existence of spatial gradients [5].

A next departure from TE occurs when the kinetic energy transfer between electrons and heavy particles is not sufficient to distribute the energy among them. If the interaction among particles of a given type is strong enough, these particles separately obey a Maxwell energy distribution, with a temperature that can depend on the species. A departure from TE occurs, in general, when the electron temperature, $T_e$, is higher than the heavy particle temperature, $T_g$, (neutral and ion temperatures being the same since the effective energy exchange between them). If the degree of ionization in this 2-T plasma is large enough, the free electrons can control by collisions the equilibrium between ionization and ion recombination (Saha balance) and the density of atomic excited states is dominated by transitions induced by electrons (Boltzmann distribution). In this situation, the plasma is said to be in local Saha equilibrium (LSE) and the atomic state distribution function (ASDF) can be described by the Saha-Boltzmann equation. So, the density of an atomic excited state, $n^s(p)$, with a statistical weight $g(p)$, is

$$\frac{n^s(p)}{g(p)} = \frac{n_e \, n^+}{2 \, g^+} \left( \frac{h^2}{2 \pi \, m_e \, k \, T_e} \right)^{3/2} \exp\left( \frac{E_i(p)}{k \, T_e} \right) \tag{1}$$

where $g^+$ is the statistical weight of the ion ground state with density $n^+$, and $E_i(p)$ is the ionization energy of the atomic level.

However, the Saha-Boltzmann distribution applying to all energy levels of a given system (atom, molecule, ion) is rather exceptional because the electrons cannot control the full distribution of the excited state [6]. Energy levels close to the ground state can non-deexcite predominantly by electron-induced transitions, but rather radiatively. Nevertheless, for higher energy levels, the electron-induced rates for ionization and recombination processes are so high that radiative deexcitation and heavy particle-induced transitions can only have a little influence on the electron-controlled Saha balance [7]. As a consequence, only the levels close to the ionization threshold are in partial local Saha equilibrium (pLSE), and are usually named as 'the top' of the atomic system. It can be proved that levels which are populated according to the Saha distribution are also mutually related to each other by an equilibrium Boltzmann balance, which implies partial local Boltzmann equilibrium (pLBE) with an excitation temperature equal to $T_e$.

For plasmas in non-complete LSE, two situations can be observed considering the actual density of the lower levels, $n(p)$, relative to the Saha equilibrium density, $n^s(p)$, i.e. the density that the levels at 'the bottom' of the atomic system would have if they were in equilibrium with the upper levels [5]. If the lower levels are overpopulated, this overpopulation generates an upward net particle flux in the energy diagram of the atom and the term 'ionizing plasma' is used in this case. On the contrary, when the lower levels are underpopulated, there is a downward particle flux in the atom energy diagram and the term 'recombining plasma' is used to designate it. The parameter $b(p)$ is used to characterize the type and the degree of departure from LSE of an atomic system, which is defined as [8]

$$b(p) = n(p) \, / \, n^s(p) \tag{2}$$

Thus, we have $b(p) < 1$ for the lower levels in the recombinig plasma case and $b(p) > 1$ in the ionizing plasma one. It can be proved that $b(p)$ increases monotonically with the excitation energy for the recombining plasma and decreases monotonically for the ionizing plasma. The degree of equilibrium is high or low depending on the lower level $b(p)$ being close to or far from unity. The term 'close-to-LSE' is normally used when $0.1 \le b_1 \le 10$, the parameter $b_1$ corresponding to the atom ground state.

## 3. INTERRUPTION TECHNIQUE

When the microwave power maintaining the discharge is abruptly switched off, the electron temperature suddenly decays towards a value $T_e^*$, presumably close to the gas temperature. Under our experimental conditions, the energy equipartition between electrons and heavy particles takes place in a characteristic time on the order of $10^{-6}$ s [9]. Simultaneously, due to the diffusion and recombination processes, the electron density decreases more slowly, with a characteristic time of about $10^{-5}$ s. During this extinction process the response of the atomic emission lines is dictated by the production/destruction balance governing the population of the corresponding excited atomic state.

If the population of a level $p$ is mainly controlled by a Saha ionization/recombination balance,

$$A_p + e^- \longleftrightarrow A^+ + e^- + e^-,$$

then the population density of this $p$ level will be given by the Saha-Boltzmann equation, shown in Eq.(1). In our case, singly ionized particles, $n_e = n^+$, can be assumed.

When the drop in the electromagnetic field is fast enough, during the first instants the electron temperature decays while the electron density will not undergo appreciable change. So, when the electron temperature reaches the $T_e^*$ value, the population density of the level $p$ is

$$n'(p) = \frac{n_e^2}{2g^+} g(p) \left( \frac{h^2}{2\pi m_e k T_e^*} \right)^{3/2} \exp\left( \frac{E_i(p)}{k T_e^*} \right) \tag{3}$$

Since $T_e > T_e^*$, it is verified that $n'(p) > n(p)$. This change will be reflected in an increase of the intensity emitted by the level, which is proportional to its population density. From the Eqs. (1) and (3), the ratio between the atomic emission line intensities before the electromagnetic power interruption ($I$) and after the electron cooling ($I'$) can be written

$$\ln\left( \frac{I'}{I} \right) = \frac{3}{2}\ln\gamma + E_i(p) \left( \frac{\gamma - 1}{k T_e} \right), \tag{4}$$

where $\gamma = T_e/T_e^*$. In this way, $\ln(I'/I)$ will show a lineal dependence on the ionization energy $E_i(p)$ for the excited levels conforming to the Saha equilibrium. As can be seen in section 5, such a plot allows the evaluation of a possible equilibrium departure of the discharge, as well as the electron temperature change in the cooling process [10].

If on the contrary the population of the level $p$ is controlled by a Boltzmann equilibrium with the atom ground state,

$$A_1 + e^- \longleftrightarrow A_p + e^-$$

then the population density of the level will be given by the Boltzmann equation

$$n(p) = \frac{n_1}{g_1} g(p) \exp\left( -\frac{E(p)}{k T_e} \right), \tag{5}$$

being $E(p)$ the excitation energy of the level. As soon as the electromagnetic power is switched off, the electron temperature decays while the atom ground state density remains unchanged. Now, when the electron temperature equals $T_e^*$ the population density of the level is given by

$$n'(p) = \frac{n_1}{g_1} g(p) \exp\left( -\frac{E(p)}{k T_e^*} \right) \tag{6}$$

Since $T_e > T_e^*$, in this case the intensity emitted by the level will decrease immediately after the electromagnetic power is removed.

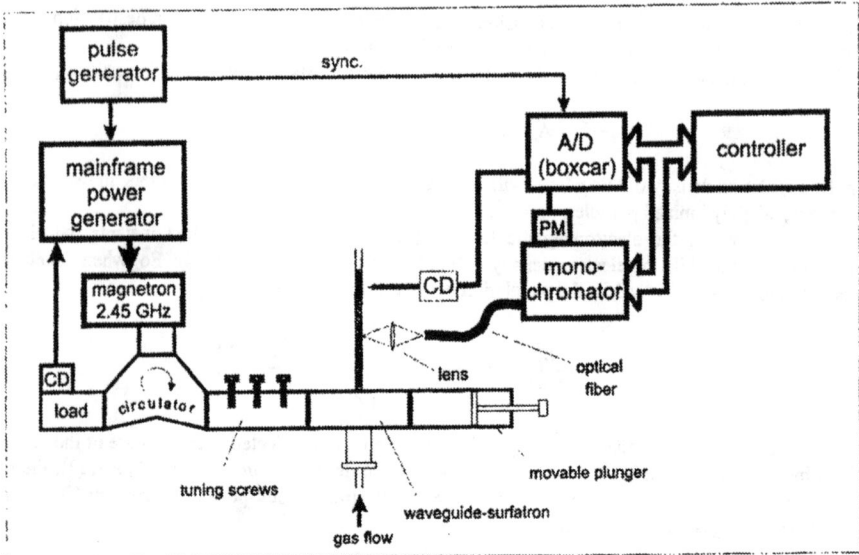

Figure 1. Experimental set-up for the stationary case and the pulse regime.

In general, the population of an excited level is controlled by more than one balance, so the instantaneous response of the intensity emitted due to changes in the electron temperature will be a combination of the previous responses.

## 4. EXPERIMENTAL SETUP

The surface-wave-sustained plasma columns were generated by using the waveguide-surfatron launcher device [11], as shown in figure 1 for the stationary case and the pulse regime. This excitation structure is made up of a standard rectangular waveguide (WR-340) with a lateral coaxial line, which allows the coupling of the 2,45 GHz microwave energy supplied by the power generator (Sairem, GMP 12 KT) to the discharge. The plasma columns were produced within a quartz capillary tube of 1 mm inner diameter with one open end. Under these conditions, the surface wave only propagates in the $m = 0$ mode, of azimuthal symmetry [12]. The columns lengths were in the range 11 - 14 cm, depending on the microwave input power (60 - 110 W) and the argon flow-rate (0,5 - 1,0 l/min.).

The spectroscopic measurements were obtained radially viewing the light emitted by the plasma (figure 1). A 1:1 image of the plasma column, obtained by using a fused silica lens, was focused onto an optical fiber that transmitted the light to the entrance slit of the monochromator (Jobin Yvon, THR-1000S; 1200 grooves/mm). The spatial resolution of the measurements corresponded to the diameters of the fiber core (1 mm), which was similar to the radial dimensions of the discharge and prevented the obtaining of any radial information by Abel inversion. Consequently, all intensity values obtained were radially integrated and corresponded to different axial positions along the column by moving the optical fiber. For convenience, the origin of the axial position z along the stationary discharge was chosen at the end of the column. The spectroscopically measured magnitudes had to be considered as 'apparent' ones, associated to the highest intensity zone or the highest density zone. Thus, the 'apparent' values are representative of the central zone of the discharge. In order to obtain absolute intensities of the emitted lines, the optical system was calibrated by using a tungsten ribbon lamp with a known standard emission spectrum.

The block diagram of the instrumentation used to produce and diagnose the discharge in pulse regime is also depicted in figure 1. In this case, the 2,45 GHz microwave generator provides power modulated by the rectangular pulses coming from a pulse generator. The rise time of the signal was less than 10 μs, the fall time about 1 μs and the repetition period 10 ms. The time between pulses, "off" intervals, were from 36 to 166 μs, which was conditioned for the instrumentation and to minimize the jitter of the ignition lag of the discharge. The average electromagnetic input power was 110 w.

To analyze the pulse regime, the signal provided by the detector was sampled along the time by means of a boxcar averager (Stanford Research Systems, SR-280+250+245+CIM) synchronized with the power pulses. The minimum temporal gate was 50 ns and the average was made from 1000 to 5000 measurements in order to improve the signal-noise ratio.

The detector of the light intensity, photomultiplier and socket combination, has a temporal response in the nanosecond range. In this case, the external component $E_r$ of the surface-wave electric field sustaining the discharge is also obtained by means of an antenna along the plasma column. Again the signal proportional to $E_r^2$, coming from a crystal detector, is time resolved by the boxcar. In the pulse regime, the origin of the axial position z is located at the end of the stationary column.

## 5. DIAGNOSTIC METHODS

The gas temperature of the discharge $T_g$, is determined from the rotational temperature of the OH radical, which resulted from the dissociation of water traces present in the plasma gas. The highly favorable energy exchange between heavy particles in the plasma and the internal vibrational-rotational states of this molecule via collisions allows us to consider the gas temperature equal to the rotational temperature. The rotational temperature values are obtained from the Boltzmann-plot of the rotational spectrum for the $Q_1$ branch of the (0-0) band [13], on the assumption of a Boltzmann distribution for ro-vibrational levels. The relative intensities of the rotational spectrum are sensitive enough to use OH as reliable thermometric species for $T_g$ lower than 1800 K [14]. The calculated experimental dispersion in the $T_g$ values was less than 15%

The electron density is calculated through the well-known Stark broadening technique for the Balmer $H_\beta$ line. The hydrogen is present as an

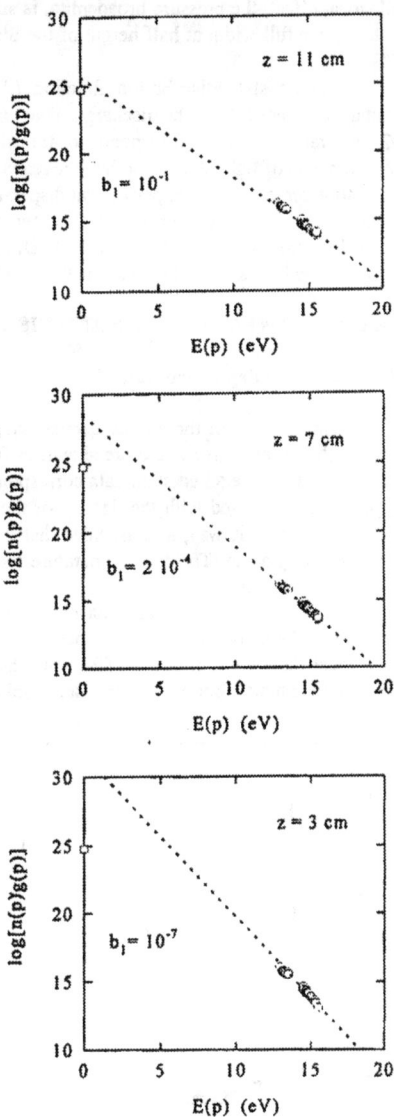

Figure 2. Boltzmann-plots at different axial positions (110 W, 1,0 l/min)

impurity in the discharge. The Stark broadening is determined from the Voigt profile fitted to the detected line, by separating the Lorentzian and Gaussian contributions. In this way the Gaussian contributions (the Doppler broadening and the instrument function) can be eliminated. Under our conditions the Lorentzian part of the profile is mainly produced by the Stark effect due to the other collisional effect, the pressure broadening, is small compared with it. The electron density thus obtained, related to the full width at half height of the Stark broadening [15], shows an experimental dispersion of 20%

The atomic state distribution function (ASDF) is determined from the absolute line intensities spontaneously emitted by the discharge. That is way, 58 atomic emission lines are measured, in the 394 to 912 nm range, which correspond to transitions coming from 4p up to 7d excited levels [16]. Measurements of light absorption have revealed that the plasma is optically thin for all lines in the radial observation direction. The experimental dispersion of the excited state population densities is about 40%.

As will be discussed later, the electron temperature can be estimated from the excitation temperature obtained via the Boltzmann-plot of this ASDF. In this direction, the analysis of the equilibrium departure for the atomic levels is needed to know those levels which conform to a Saha-Boltzmann distribution.

## 6. EXPERIMENTAL RESULTS AND DISCUSSION

### 6.1. Equilibrium departure state

The Boltzmann-plots of the atomic excited state densities obtained at different axial positions along the plasma column for a given case are shown in figure 2. The group of the points located to the left in the assembly of atomic excited state data corresponds to the 4p levels. In order to analyze the ASDF shape, the plots are completed with the density of the ground state, $n_l$, which can be determined by using the ideal gas law. That is way, it is assumed that the atoms in the ground state are by far the most dominant species in the plasma. The gas temperature used is the previously calculated temperature as described above.

If the plasma was in local Saha equilibrium (LSE) all points in the Boltzmann-plot should fit a straight line, the slope of which is associated to the excitation temperature. Nevertheless for a plasma in partial local Saha equilibrium (pLSE), only the levels on the top of the atomic system conform to the Saha-Boltzmann distribution. In our case, in spite of experimental dispersion of the data, from the straight

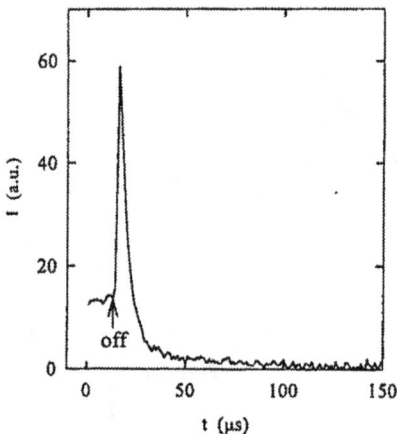

Figure 3. Power interruption response of the 763.51 nm Ar line intensity

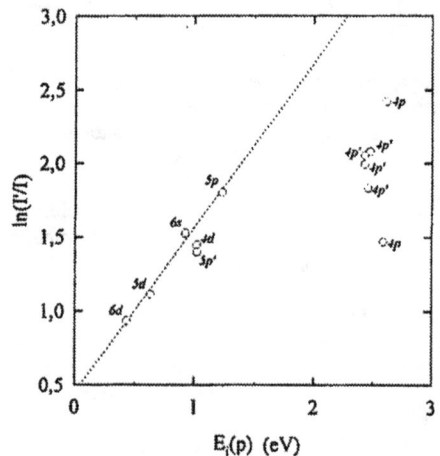

Figure 4. Relative intensity jump versus the ionization energy of the levels

line fitted through these points, it looks as if only the levels corresponding to 5p and above obey to the local Saha equilibrium.

In all cases studied the plasma was found to be in a pLSE state. In each position, the ground state density, $n_1$, permits the estimation of the degree of the equilibrium departure via the $b_1$ parameter. Even taking in to account the experimental dispersion of the data, as can be seen in figure 2, at positions close to the end of the column, the ground state is underpopulated relative to the Saha equilibrium density, $n_1^s$. This recombining character of the discharge gradually decreases with decreasing distance from the coupling device. This behavior is observed under all our experimental conditions.

As was discussed in section 3, the power interruption technique has been used in order to confirm this assumption deduced from the shape of the ASDF, and which levels also correspond to the top and the bottom of the atomic system. Figure 3 shows the temporal behavior of the 763.51 nm ArI line, corresponding to the 4p-4s' transition, when the electromagnetic power maintaining the discharge is switched off. After the surface-wave power is removed, the line intensity undergoes an upward jump in the first instants. This instantaneous response is followed by a temporal evolution with a slower fall. This behavior is also observed for all ArI lines analyzed in the different axial positions along the column.

The jump observed during the first microseconds of the instantaneous response of the lines is a consequence of the electron cooling process and reflects the difference between electron and gas temperatures in the initial stationary state of the plasma. Moreover, the Saha-like behavior of this instantaneous response reveals, as was seen in section 3, that the populations of the corresponding starting levels (from 4p to 6d in this case) are controlled, at least partly, by Saha balances.

The relative intensity jump, $I'/I$, measured for the different lines versus the ionization energy of the corresponding level, $E_i(p)$, is depicted in figure 4. As can be observed, the points corresponding to the levels nearest the ion ground state (5p and higher) are fitted to a lineal dependence. So, it is expected that those levels are initially close to the Saha equilibrium, according to the discussion in section 3. However, the jumps of lower levels (4p) are away from the lineal dependence, which reveals that for them there is a departure from the Saha equilibrium before the cooling. This result is in agreement with the previous analysis, based on the ASDF shape, establishing that in the steady state the levels 5p and above are in partial local Saha equilibrium.

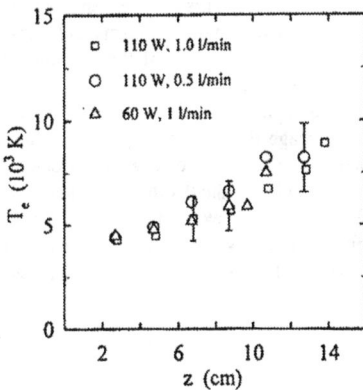

Figure 5. Axial profiles of the electron temperature for different plasma columns.

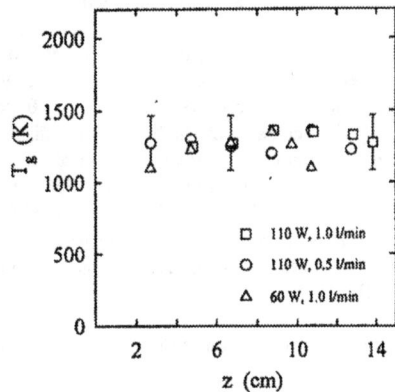

Figure 6. Axial profiles of the gas temperature for different plasma columns.

## 6.2. Axial profiles

The electron temperature is obtained from the Boltzmann-plot method, including only the levels on the top of the atomic system (5p and higher). The axial profiles of the electron temperature obtained in this way, with 20% of experimental dispersion, for different experimental conditions are shown in figure 5. The electron temperature decreases towards the end of the column and ranges from 4500 to 8500 K. For a given axial position, z, from the column end, within the experimental dispersion, the electron temperature is the same not depending on the gas flow-rate and power used. When the power increases a new portion of the discharge is added to the axial profile close to the launcher, at the high z values region.

These $T_e$ values agree with those reported for surface-wave argon plasmas under comparable conditions [17].

Figure 6 shows the gas temperature obtained along the plasma column for different experimental conditions. As it can be observed, these temperatures are much lower than the electron temperatures and the $T_g$ values remain constant about 1300 K along the discharge for all cases, within the experimental dispersion.

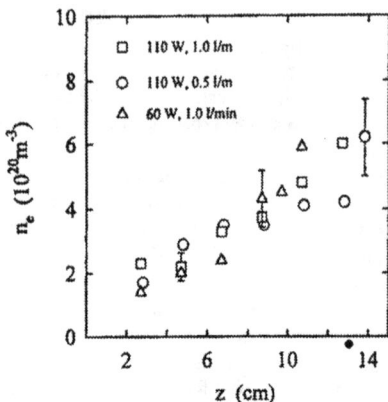

**Figure 7.** Axial profiles of the electron density for different plasma columns.

The electron density axial profiles obtained from the $H_\beta$ Stark broadening under the same experimental conditions are depicted in figure 7. The profiles are decreasing towards the end of the column, as expected for a surface-wave-sustained discharge [2]. The electron density varies from 1 to 7 x $10^{20}$ m$^{-3}$ along the plasma column. From the electron density and gas temperature values, the degree of ionization of the discharge is estimated to be between $10^{-4}$ and $10^{-5}$, which suggests that the excitation kinetics in the plasma is electron-controlled (EEK plasma) [18].

For a given axial position z measured from the end of the column, the same electron density is obtained, within the experimental dispersion, independent of the gas flow-rate and power used to generate the discharge, as observed in the electron and gas temperature profiles too. This independence allows us to assume that, under our experimental conditions, the surface-wave-sustained plasma columns are equivalent, provided that the origin of the axial position z is chosen at the end of the columns. This characteristic has been previously shown for low pressure surface-wave discharge [19].

The power absorbed per electron, the $\theta$ parameter, required to maintain the discharge is usually employed to characterize high-frequency discharges [1]. The average $\theta$ value for a plasma column is determined from the microwave power absorbed and the total number of electrons, which is calculated from the electron density axial profiles. For a flow-rate of 0.5 and 1.0 l/min, the $\theta$ value is 2.8 and 2.6 pW respectively with 110 W, and 2.1 pW in both cases with 60 W. Although the experimental error estimated is about 20%, the value of $\theta$ is higher with higher electron density (higher power). This dependence on $n_e$ and the $\theta$ values themselves are in agreement with that theoretically obtained under similar conditions by Sà [20].

### 6.3. Temporal evolution of the electron temperature

When the electromagnetic power maintaining the discharge is suddenly interrupted, the electrons will cool down quickly towards a final value of the electron temperature, $T_e^*$, coming closer to the gas temperature. The temporal evolution of the electron temperature can be determined from the Boltzmann plots obtained at different times during the first instants of the extinction process. That is way, it is assumed that in this time the type of equilibrium is maintained while the electron temperature changes. In this transitory case, the number of lines with enough intensity which be used is low. The Boltzmann-plot is made with 14 lines, including some lines from 4p levels. The error introduced in this way is not important due to the 4p levels are close to LSE, as is seen above, specially at the beginning of the column.

The results obtained at z = 13 cm, 110 W of power and 1.0 l/min of flow-rate, are shown in table I. The electron temperature suddenly falls in the first 2 - 3 μs, which is the time that the maximum of the intensity jump of the lines appears. After that, the electron temperature tends to be stabilized about 5000 K.

The intensity jump of the lines during the extinction of the discharge can be used to contrast such a behavior. From the equation (4), the initial electron temperature, $T_e$, and the final one, $T_e^*$, could be obtained from the intensity jump for different lines corresponding to levels in pLSE on the top of the atomic system. The values obtained in this way show a large experimental dispersion around 60% due to the sensitive error in determining the point of intersection with the ordinate axis (figure 4).

However, the electron temperature obtained for the steady-state plasma can be used to determine the value to which the electron temperature decays, combined with the slope of the equation (4). The results thus obtained, with less than 25% of experimental error, for 110 W of power and 1.0 l/min of flow-rate, are shown in table II. The $T_e^*$ value for z = 13 cm agrees with the previous one shown in table I. As it can be seen, the $T_e^*$ values are higher than the gas temperature in the steady state plasma. Therefore, it seems that the electrons do not thermalize with the heavy particles, during the first step of the extinction process after the electromagnetic power is removed. So, an additional electron heating mechanism, not related to the electromagnetic power, must exist in the discharge. This phenomenon has been previously observed in argon ICP discharges, and it was ascribed to three-body recombination processes [21].

Table I. Temporal evolution of $T_e$ obtained from the Boltzmann-plot method at different times after the power interruption (110 W, 1.0 l/mim., z= 13 cm).

| t (μs) | $T_e$ (K) |
| --- | --- |
| 0 | 7100 |
| 0.8 | 5700 |
| 0.9 | 5500 |
| 1.0 | 5500 |
| 1.1 | 5700 |
| 1.2 | 5500 |
| 1.3 | 5500 |
| 2.3 | 5100 |
| 2.8 | 5100 |
| 3.8 | 5000 |
| 4.2 | 5000 |
| 4.6 | 5000 |

Table II. Initial ($T_e$) and final ($T_e^*$) electron temperatures in the decay process obtained from the intensity jump method, at three axial position for 110 W of power and 1.0 l/mim. of gas flow-rate.

| z (cm) | $T_e$ (K) | $T_e^*$ (K) |
| --- | --- | --- |
| 13 | 7600 | 5200 |
| 11 | 6700 | 4100 |
| 9 | 5600 | 4700 |

## 7. CONCLUSIONS

This work shows a spectroscopic study of an argon surface-wave-sustained discharge at atmospheric pressure, produced by the waveguide-surfatron launcher. The axial profiles for electron density, electron temperature and gas temperature are obtained under different power and gas flow-rate conditions. Since these three axial profiles coincide, under our experimental conditions the plasma columns are equivalent when the column ends are chosen as the reference, the same as at low pressure.

The electron density decreases along the plasma towards the end of the column, as a result of the surface wave power flow decreasing with increasing distance to the coupler, which is typical of surface-wave-sustained plasmas. The average power per electron, θ, to maintain the discharge is about 2.4 pW, and the experimental results suggest that θ increases with increasing electron density.

The analysis of the atomic state distribution function in the steady state shows that the populations of the top levels in the atomic system, 5p levels and above, are controlled by Saha equilibrium. The 4p levels, at the bottom of the atomic system, are underpopulated relative to the Saha equilibrium. This 'recombining character' is greater at the end of the plasma column and gradually approaches to local Saha equilibrium as the distance from the launcher decreases.

The previous results on the departure equilibrium have been confirmed by using the power interruption technique. The instantaneous response of the atomic argon line intensities, when the microwave power is switched off, shows an upwards jumps during the first 2 - 3 microseconds, as a consequence of the electron cooling process. The relative intensity jumps for different lines suggest again that only the levels 5p and higher conform to Saha equilibrium.

On the other hand, the analysis of the extinction process of the discharge, using both the Boltzmann plots for different times and the instantaneous responses of the Argon lines, reveals that the temperature reached by the electrons after the power is removed is higher than the gas temperature. So, an additional electron heating mechanism, unrelated to the electromagnetic field, must be involved.

## Acknowledgments

This work was carried out with the partial support from the DGES of the Spanish Ministry of Education and Culture, in the framework of the project no. PB96-0508.

## References

1. Moisan M., Hubert J., Margot J., Sauvé G. and Zakrzewski Z., in *Microwave Discharges: Fundamentals and Applications*, Ed. C.M. Ferreira and M. Moisan, Plenum Press, New York & London, 1993, 1.
2. Moisan M. and Zakrzewski Z., in *Microwave Excited Plasmas*, ed. M. Moisan and J. Pelletier, Elsevier, 1992, 123.
3. Ferreira C.M., in *Microwave Discharges: Fundamentals and Applications*, Ed. C.M. Ferreira and M. Moisan, Plenum Press, New York & London, 1993, 313.
4. Gamero A., in *Microwave Discharges: Fundamentals and Applications*, Ed. C. Boisse-Laporte and J. Marec, EDP Sciences, Paris, 1998,339.
5. Van der Mullen J.A.M., Phys. Rep., 1990, **191** (2&3), 109.
6. Van der Mullen J.A.M., Benoy D.A., Fey F.H.A.G., van der Sijde B. and Vlcek J., Phys. Rev. E, 1994, **50**, 3925.
7. Van der Mullen J.A.M., Benoy D.A., Fey F.H.A., in *Microwave Discharges: Fundamentals and Applications*, Ed. C.M. Ferreira and M. Moisan, Plenum Press, New York & London, 1993, 395.
8. Bates D.R., Kingston A.E. and McWhirter R.W.P., Proc. R. Soc. London, 1962, **267**, 297.
9. Garcia M.C., Ph.D. Thesis, University of Córdoba, 1999.
10. Bydder E.L. and Miller G.P., Spectrochim. Acta, 1988, **43B**, 819.
11. Moisan M., Chacker M., Zakrzewski Z. and Paraszczac J., J. Phys. E: Sci. Instrum., 1987, **20**, 1356.
12. Margot-Chacker J., Moisan M., Zakrzewski Z.,Glaude V.M. and Sauvé G., Radio Sci., 1988, **23**, 1120.
13. Mermet J.M., in *Inductively Coupled Plasma Emission Spectroscopy, Part II: Applications and Fundamentals*, Ed. P.W.J.M. Boumans, Wiley-Interscience, New York, 1987, 353.
14. Ricard A., St-Onge L., Malvos H., Gicquel A., Hubert J. et Moisan M., J. Phys. III France, 1995, **5**, 1269.
15. Czernikowski A. and Chapelle J., Acta Phys. Pol. A, 1983, **63**, 67.
16. Wiese W.L., Smith M.W. and Miles M., *Atomic Transition Probabilities* vol.I, NSRDS, Washington DC, 1969.
17. Calzada M.D., Moisan M., Gamero A. and Sola A., J. Appl. Phys., 1996, **80**, 46.
18. Biberman L.M., Vorobev V.S. and Yakubov I.T., Sov. Phys. Usp., 1979, **22**, 411.
19. Sola A., Gamero A., Cotrino J. and Colomer V., J. Phys. D: Appl. Phys., 1988, **21**, 1112.
20. Sá A.B., in *Microwave Discharges: Fundamentals and Applications*, Ed. C.M. Ferreira and M. Moisan, Plenum Press, New York, 1993, 75.
21. de Regt J.M., van Dijk J. van der Mullen J.A.M. and Schram D.C., J. Phys. D: Appl. Phys., 1995, **28**, 40.

# HOT ELECTRONS AND EEDF-ANISOTROPY IN LARGE-AREA SURFACE-WAVE DISCHARGES

**J. Kudela, T. Terebessy* and M. Kando***

Satellite Venture Business Laboratory, Shizuoka University,
Johoku 3-5-1, Hamamatsu 432-8561, Japan
*Graduate School of Electronic Science and Technology, Shizuoka University,
Johoku 3-5-1, Hamamatsu 432-8561, Japan

**Abstract.** In this work, we present an experimental study focused on the hot electrons in low-pressure large-area microwave discharges. At gas pressures below about 50 mTorr, a flux of hot-electrons directed away from the waveguiding plasma-dielectric interface has been observed. Energies of these electrons attain values of some 60 eV and they are believed to be originating from the resonantly enhanced electric field region localized near the dielectric. The hot-electron flux appears to play a significant role in the discharge heating mechanism, which is demonstrated by plasma parameter profiles. In the profiles of slightly overdense discharges, a localized hot-electron region has been observed near the dielectric in the place of critical plasma density. The results are supported by the measurements of the electron energy distribution functions. The existence of both presented phenomena, the hot-electron flux and the localized hot electrons is explained on the basis of the transit-time heating in the resonantly enhanced electric field. The presented experimental results provide an evidence that the plasma resonance region plays an active role in the maintenance of low-pressure microwave discharges.

## 1. INTRODUCTION

Since the pioneering work by Komachi and Kobayashi [1] steadily increasing attention has been paid to the so-called large-area surface-wave (SW) plasma sources [1-5]. These sources work without the use of static magnetic fields and produce plasmas of high densities ($>10^{11}$ cm$^{-3}$) over large areas, which makes them attractive for the industrial applications. The plasma is produced as a result of energy absorption of an electromagnetic wave propagating along the plasma-dielectric interface. At low gas pressures, where the electron-neutral collision frequency is much lower than the applied field frequency, the main mechanism of the energy absorption turns out to be collisionless rather than the conventional Joule heating. For the SW-discharges sustained in thin cylindrical dielectric tubes [6], this phenomenon has been theoretically described by Aliev *et al.*[7-9] as follows: In a cylindrical SW-plasma column, due to the radial plasma inhomogeneity, a region with enhanced radial electric field exists near the plasma column boundary where the local plasma frequency equals the applied frequency. By quasi-linear wave-particle interactions, this enhanced field accelerates the electrons down the plasma density gradient (to the wall), which results in the formation of a hot tail in the electron energy distribution function (EEDF). These hot electrons are reflected back from the negatively charged wall and sustain the discharge.

Obviously, the region of the resonantly enhanced electric field near the plasma-dielectric boundary is expected to exist also in the large-area SW plasmas, which has already been discussed in a few works [10-13]. Taking into account the above-mentioned theory [7-9], in our work, we have focused on the experimental observation of the hot electrons in the low-pressure microwave discharges and their role in the discharge heating.

In this contribution, we present an overview of our recent results [14,15] extended by the measurements of the EEDFs. The paper is organized as follows: Section 2 describes the experimental apparatus. In section 3.1, the observation of a hot-electron flux directed away from the waveguiding plasma-dielectric interface is presented. The presence of the hot-electron flux is supported also by the plasma parameter profiles shown in section 3.2. Section 3.3 presents the detection of hot electrons localized in the plasma resonance region near the dielectric. In section 3.4, the EEDFs are provided in a form of the second derivatives of the probe characteristics confirming a hot-electron tail formation with anisotropic distribution. The presented experimental results are supported by the explanation based on the transit-time heating in section 4. Finally, a summary with the impact of the observed phenomena on the processing plasmas and the outlook to forthcoming work are given in section 5.

## 2. EXPERIMENTAL APPARATUS

The experiments were carried out in argon gas in a large-area microwave planar plasma source [5,16] schematically shown in Fig. 1. The vacuum part consists of a stainless steel chamber, cylindrical in shape, with 312-mm diameter and 350-mm height. The top wall of the chamber is a 15-mm thick quartz plate. The microwaves are coupled to plasma via an azimuthally symmetrical microwave applicator placed on the quartz plate on the axis of the source. The applicator is a tunable cylindrical cavity, designed for the $TM_{011}$-mode, with an annular slot at its bottom wall. The cavity is made of copper cylinder with the inner diameter of 110 mm and the wall thickness of 5 mm. The bottom wall of the cavity is a 5-mm-thick and 90-mm-diameter disk, which together with the cylindrical body of the cavity forms an annular slot with 10-mm width and 110-mm outer diameter. The microwave power is fed into the cavity from a rectangular waveguide (WR-430) by a high power coaxial cable (Andrew, LDF5-50A). The cable enters the cavity from its top center and is terminated by a copper coupler with the diameter equal to the coaxial cable inner conductor diameter (9 mm). The rectangular waveguide line is equipped with a water-cooled isolator, two directional couplers and two power detectors enabling the monitoring of the incident and reflected powers. The whole system is powered by a stabilized 2.45 GHz magnetron-generator (Nissin, MPG-30) with the output power in the range from 200 to 3000 W.

The vacuum chamber is pumped by a turbomolecular pump backed by a two-staged mechanical pump with the pumping speed of 900 $m^3$/hod. The base pressure of the whole vacuum system is less than $10^{-6}$ Torr. The working pressure during the experiment is controlled by the mass flow controller and the capacitance diaphragm manometers.

In the experiments, two Langmuir probes were used. The probe-1 was inserted into the chamber through a side-port in a 100-mm distance under the quartz plate and the probe-2 through a bottom-port along the side-wall as shown in Fig.1. Both probes were movable in the axial and azimuthal directions.

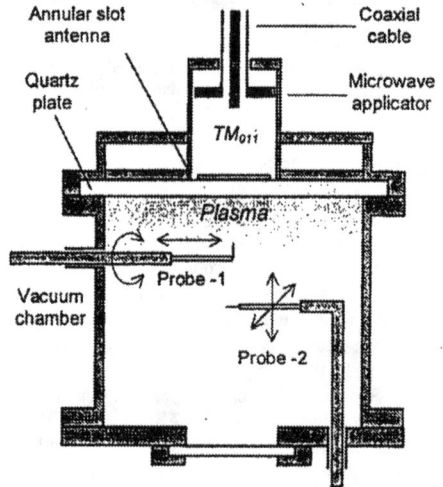

Figure 1. Schematic view of the plasma source.

## 3. MEASUREMENTS AND RESULTS

### 3.1. Hot-electron flux in low-pressure discharges

The existence of a directed electron flux in the discharges was investigated firstly by rotating the probe-1 (Fig. 1) with an L-shape-bent probe tip (Fig. 2) around the probe axis. By this technique, a probe tip with length much larger than diameter, in different orientations, exhibits different surfaces to the predicted electron flux. Consequently, this should lead to the changes in probe characteristics. In the experiment, an 8-mm-long probe tip made of 0.4-mm-diameter tungsten wire was used. Measurements for the tip in two orientations, parallel and perpendicular to the quartz plate, placed on chamber axis 100 mm under the quartz in a 10 mTorr argon discharge are shown in Fig. 3. As it can be seen from the figure, deep in the ion saturation current region ($V_p \leq -50$ V), both characteristics are identical with the fits [17] implying ion density $N_i = 1.9 \times 10^{12}$ cm$^{-3}$. However, a significant difference is observed in the remaining part of

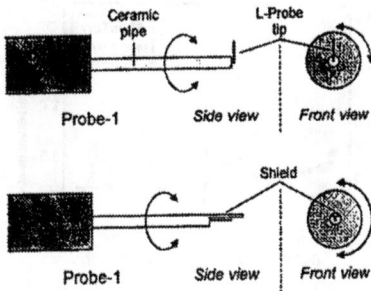

Figure 2. Details of the L-shape-bent and one-side-shielded probe tips of the probe-1 used for the observation of the hot-electron flux from the plasma-dielectric interface.

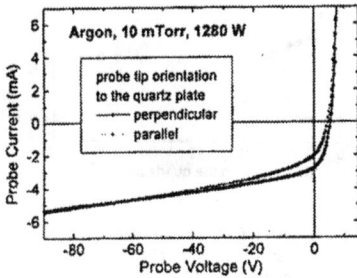

Figure 3. Part of probe characteristics measured at two different orientations of the L-shape-bent probe tip indicating the presence of hot-electron flux from the region near the quartz plate in a 10 mTorr argon discharge sustained by 1280 W. Probe was placed on chamber axis 100 mm under the quartz plate.

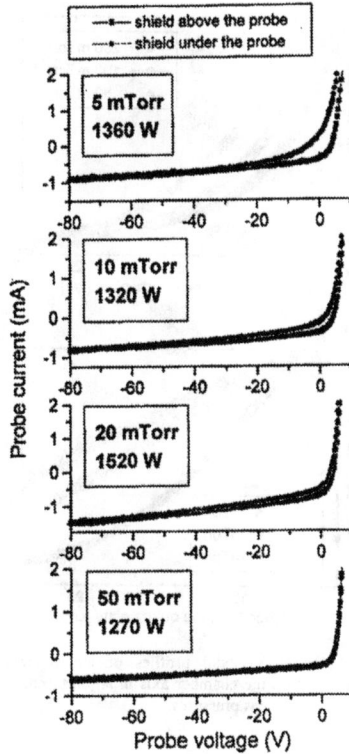

Figure 4. Part of probe characteristics measured at different gas pressures with the shield positioned above and under the probe tip. Probe was placed on chamber axis 100 mm under the quartz plate.

characteristics. The lower values of probe current $-I_p$ measured by the probe with tip parallel to the quartz plate confirm the presence of axial flux of hot electrons to the probe. The estimated flux current density is roughly 30 mA/cm$^2$. The differences in currents at probe biases as high as about $-50$ V and the value of the space potential $V_s = 12.6$ V together suggest hot electron energies attaining some 60 eV. We believe that these hot electrons originate from the predicted resonantly enhanced electric field region localized near the quartz. To confirm the electron flux direction another probe was used. This probe had a straight, 4-mm-long, one-side-shielded probe tip with the radial distance between the shield and probe surface 0.8 mm (Fig. 2). As in the previous measurement, the probe was rotated into two positions, with the shield up and down, respectively. The presence of hot electrons in probe characteristic was observed when the shield was under the probe tip (Fig. 4). The effect of gas pressure, also seen in Fig.4, provides additional support for the phenomenon. With increasing gas pressure the presence of directed flux of hot electrons becomes weaker and above some 30-50 mTorr, in a 100-mm-distance from the quartz, it is no more detected. This occurs apparently due to the collisions with neutrals: the mean free path of hot electrons can be estimated as $\lambda_{mfp} = v_e / v_{en}$ , where $v_e = (2E_e / m_e)^{1/2}$ is the electron velocity and $v_{en} = 3 \times 10^9\, p[\text{Torr}]$ is the electron-neutral collision frequency. Assuming the electrons with energies $E_e = 60$ eV, the estimated mean free paths are 306 mm at 5 mTorr, 153 mm at 10 mTorr, 76 mm at 20 mTorr and 31 mm at 50 mTorr, which is in reasonably good agreement with the experimental observations.

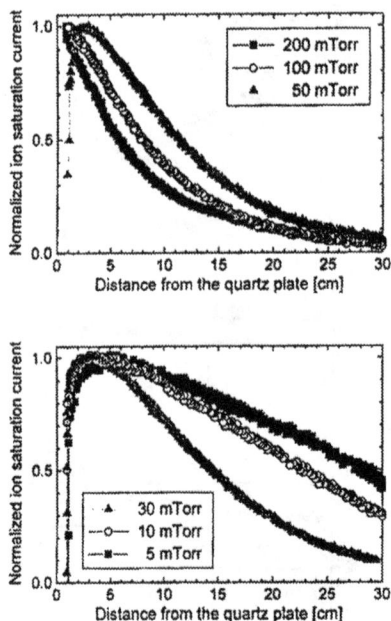

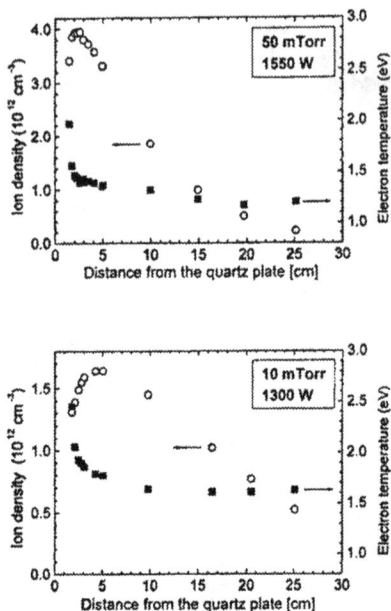

Figure 5. Normalized axial profiles of ion saturation current measured on the chamber axis in argon discharges sustained at different gas pressures.

Figure 6. Axial profiles of ion density and electron temperature measured on the chamber axis in argon discharges sustained at 50 and 10 mTorr.

## 3.2. Axial plasma parameter profiles

It is reasonable to expect that the interaction of the hot electron flux with the neutral gas should affect the shapes of axial plasma parameter profiles. The profiles were measured by the probe-2 (Fig 1) with a conventional cylindrical probe tip. The probe was moved axially along the chamber axis to a minimum distance 10 mm from the quartz plate. First, the ion saturation current profiles ($V_p = -50$ V) measured for pressures ranging from 5 to 200 mTorr are presented in normalized forms in Fig. 5. It should be noted that the ion saturation current profiles, quantitatively, do not exactly correspond with the plasma density distributions due to axial variations of the space potential and the effective electron temperature, which will be shown below. However, for illustrative purposes, they provide sufficient information about the plasma density changes. At 200 mTorr, the profile exhibits maximum at the nearest measured point to the quartz plate and a relatively steep exponential-like decay away from it. At lower pressure, 100 mTorr, this decay becomes less steep and, at 50 mTorr, profile with a peak in the vicinity of the quartz is observed. With further decrease of pressure, the peak becomes wider and moves away from the quartz. Between 50 and 5 mTorr, the maximum values of ion saturation current are measured in the axial positions of about 20 to 50 mm from the quartz. Considering skin depth for the applied electromagnetic field in these discharges being only a few millimeters ($\lambda_{sk} = 5.3$ mm at $N_e = 10^{12}$ cm$^{-3}$), such profiles support the assumption of plasma heating by the hot electron flux. To justify the profiles, complete probe characteristics were measured and analyzed for different probe positions. Figure 6 shows the actual variation of the ion density along with the temperature of Maxwellian electrons (determined from the linear part of ln$I_e$-$V$ curves) for pressures 50 and 10 mTorr. The temperature exhibits decreasing profile with stabilized values in distances larger than mean free path of hot electrons.

In view of the above observations, three distinctive space regions can be recognized in the axial direction. Starting from the quartz plate, there is the region of spatially localized enhanced electric field,

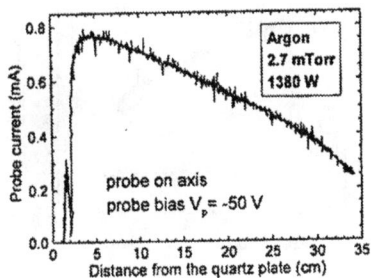

Figure 7. Typical axial probe current profile ($V_p = -50V$) observed on the chamber axis in argon discharges at very low gas pressures (< 3 mTorr).

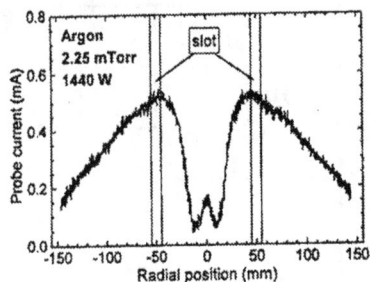

Figure 8. Radial probe current profile ($V_p = -50V$) measured 3 cm under the quartz plate in argon at 2.25 mTorr and 1440 W.

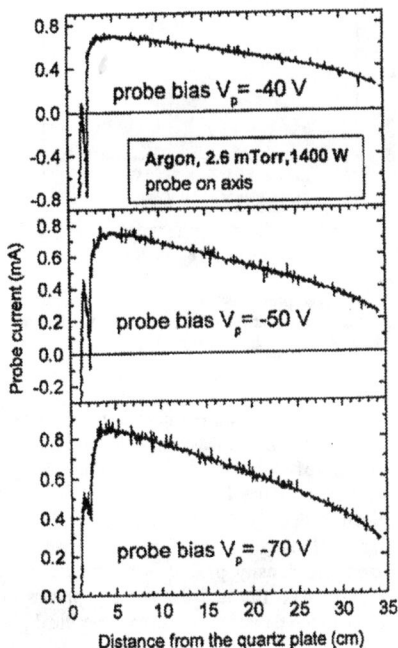

Figure 9. Axial probe current profiles measured on the chamber axis in argon at 2.6 mTorr and 1400 W of absorbed microwave power at different probe biases.

where hot electrons are generated. Then the region follows, in which the hot electrons lose their energy. This region exhibits enhanced plasma density and decaying electron temperature. Finally, in a distance larger than the hot electron mean free path, the region of thermalised diffusive plasma exists.

### 3.3. Localized region of hot electrons

The axial plasma parameter profiles at lower gas pressures, typically under 3 mTorr, exhibit different behavior than the ones mentioned in the preceding section. Figure 7 shows a probe current profile measured at 2.7 mTorr on the chamber axis at the probe bias $V_p = -50$ V. As it can be seen in the figure, a trough of the probe current with about 0.5-cm-width is observed near the quartz plate. The trough is observed always at the same value of the probe current (measured at the same probe bias) and it disappears when the plasma source operates in the underdense-plasma conditions. Therefore, we believe that the trough is related to the plasma resonance region. As expected, the phenomenon can be observed also in the radial probe current profiles (Fig. 8). The influence of the probe bias on the axial profile shown in Fig. 9 suggests the explanation for the local decrease in the probe current. Under the same discharge conditions, the depth of the trough increases with decreasing negative probe bias and the current in the minimum can even change the sign. Such a behavior indicates that the trough is caused by the current of hot electrons to the probe collected in the resonantly enhanced electric field region. Further decrease of the probe current to negative values in the vicinity of the quartz is due to hot electrons generated by the strong electric field at the interface. The assumption of the trough formation is justified by axial probe current profile measured under the slot antenna (Fig.10). As mentioned above, the hot electron generation is related to the perpendicular electric field component $E_z$, which is not expected to exist under the slot

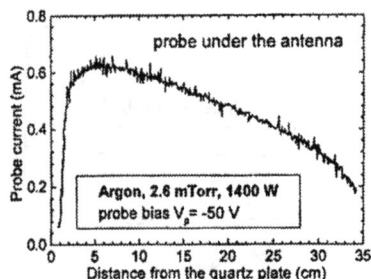

Figure 10. Axial probe current profile measured under the slot antenna in argon at 2.6 mTorr and 1400 W of absorbed microwave power. Probe bias $V_p = -50V$.

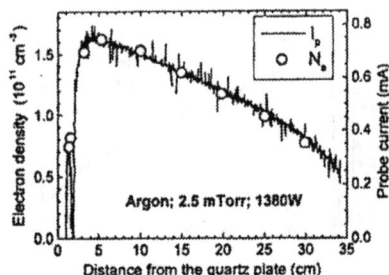

Figure 11. Axial plasma density and probe current ($V_p = -50V$) profiles measured on the chamber axis in argon discharge at 2.5 mTorr and 1380 W.

antenna. Indeed, the trough does not appear in the profile. Substantial experimental confirmation is provided by the measurement and complete evaluation of probe characteristics along the chamber axis. Figure 11 shows the axial variation of the electron density $N_e$ [18] and the probe current measured along the whole plasma chamber. As expected, the density profile is consistent with the probe current profile and the trough appears in the place where the density is close to the critical plasma density ($7.4 \times 10^{10}$ cm$^{-3}$). Figure 12 provides more details on the spatial variations of plasma parameters over the plasma resonance region. The density profile exhibits a slight depletion in the plasma resonance region, which also contributes to the localized decrease of the probe current. However, as mentioned above, the main contribution to the formation of the probe current trough is due to hot electrons localized in the plasma resonance region. This is indicated by profiles of the floating potential $V_f$ and the temperature of Maxwellian electrons $T_e$. The floating potential sharply decreases to less than –50 V in the plasma resonance region, while being

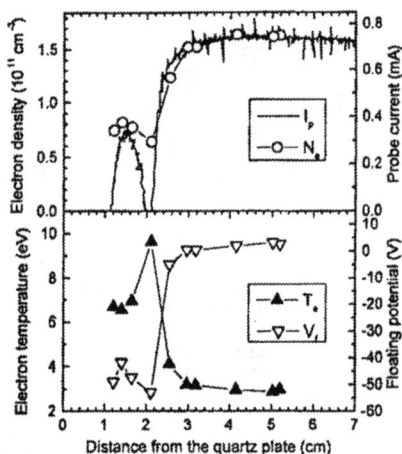

Figure 12. Axial plasma parameter and probe current ($V_p = -50V$) profiles measured on the chamber axis in argon discharge at 2.5 mTorr and 1380 W.

about +5 V in the bulk plasma. This effect, along with the electron temperature profile, confirms the presence of a localized hot-electron region. Considering the value of the plasma potential $V_s = 20$ V and the fact that the trough is well observed up to about $V_p = -50$ V, the energy of the hot electrons is believed to be as high as 70 eV.

The explanation of the origin of localized hot electrons with such energies can be provided on the basis of the heating mechanism mentioned above [7-9] as follows: the electrons entering the plasma resonance region can gain energy from the enhanced high-frequency field and the efficiency of the power transfer depends on their transit time through the region. The energy gain is efficient only if the transit time $\tau$ is short compared to the field period $T$ and it reaches its maximum value at $\tau_{opt} \approx 0.7 \ T/(4\pi)$. Under such a condition, the electrons are resonant (from the viewpoint of wave-particle energy transfer) and are accelerated out of the region transferring energy to the bulk plasma. However, if the transit time is comparable to the field period, the electron energy gain decreases since the electrons become non-resonant and are accelerated and afterwards decelerated by the reversing field when moving through the region of plasma resonance. Thus, the hot electrons "remain" localized in the plasma resonance region.

## 3.4. Electron energy distribution function

The above-presented phenomena can be described the best by the spatial evolution of the EEDFs. The distribution functions here are presented in the form of second derivatives of the probe characteristics [19]. In the measurements, the probe-2 (Fig. 1) with a 5-mm-long probe tip made of 0.2-mm-diam. tungsten wire was used. The probe tip was placed on the chamber axis in both, the horizontal and vertical positions. The probe characteristics were recorded by using a PC-based data acquisition system with a 12-bit resolution and 250-kHz sampling frequency. Each data point of the total 1000 points was averaged 4000 times in order to reduce the noise and to achieve the smooth probe characteristics. In addition, a numerical smoothing method based on the Savitzky-Golay filters [20] by using second order polynomials [21] was used.

The spatial (axial) evolution of the second derivatives measured by the horizontally-oriented probe tip at gas pressures 100, 50 and 5 mTorr are shown in Figs. 13-15. At 100 mTorr (Fig. 13), the linear slopes in semi-logarithmic scale plots suggest a Maxwellian-like electron-energy distribution. Similar behavior is observed at 50 mTorr (Fig. 14), however, a slightly enhanced hot-electron tail appears with decreasing distance from the quartz. As expected, at 5 mTorr, this hot-electron tail becomes well pronounced (Fig.15). To confirm the anisotropy in the EEDF-tail formation at 5 mTorr, the same measurements were carried out also by the vertical probe tip (Fig. 16). Indeed, the pronounced hot-electron tail is not observed in these measurements indicating the anisotropic electron-energy distribution. The hot-electron tail formation in the EEDF is still under the investigation and more data will be provided in near future.

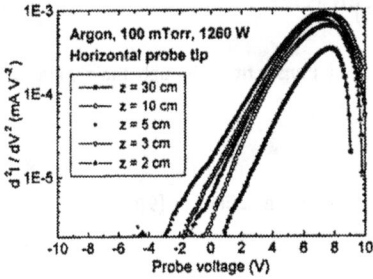

Figure 13. Second derivatives of the probe characteristics measured in a 100-mTorr argon discharge by the horizontally oriented probe tip. The tip was placed on the chamber axis in different axial positions.

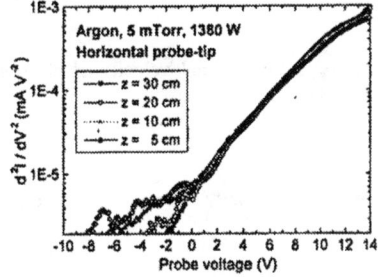

Figure 15. Second derivatives of the probe characteristics measured in a 5-mTorr argon discharge by the horizontally oriented probe tip. The tip was placed on the chamber axis in different axial positions.

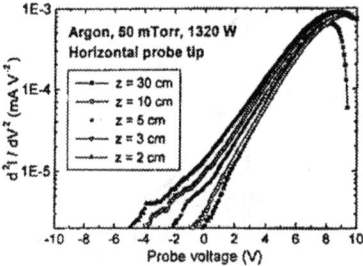

Figure 14. Second derivatives of the probe characteristics measured in a 50-mTorr argon discharge by the horizontally oriented probe tip. The tip was placed on the chamber axis in different axial positions.

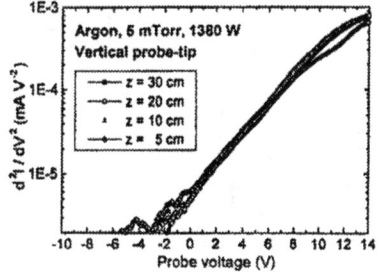

Figure 16. Second derivatives of the probe characteristics measured in a 5-mTorr argon discharge by the vertically oriented probe tip. The tip was placed on the chamber axis in different axial positions.

## 4. DISCUSSION

The considerations of the preceding sections can be strengthened by the estimation of the electron transit time through the plasma resonance region. The distribution of the resonantly enhanced electric field component $E_z$ in the region (Fig. 17) can be expressed by the formula [22]:

$$E_z = \frac{E_0}{-(z - z_{res})/L + i(v_{eff}/\omega)}$$

Here $E_0$ is the electric field amplitude and $L$ is the density scale-length at the resonance point $z_{res}$, $v_{eff}$ is the effective collision frequency and $\omega$ is the applied field frequency. The density scale-length is given as [9]:

$$L = \left(\frac{d\ln n}{dz}\bigg|_{z=z_r}\right)^{-1}$$

and, assuming a linear density profile near the quartz, it corresponds to the distance of the resonance region from the quartz. In the collisionless case, $v_{eff}$ is determined by the field convection caused by thermal electron motion effect given by the formula [9]:

$$v_{eff} = \omega(v_{Te}/\omega L)^{2/3}$$

where $v_{Te}$ is the thermal velocity of electrons. The electron transit time through the region can be calculated by the formula:

$$\tau = \Delta / v_h$$

where $v_h$ is the velocity of hot electrons and $\Delta$ is the width of the region defined as [9]:

$$\Delta = \frac{v_{eff}}{\omega} L$$

As mentioned in section 3.3, the efficiency of the energy transfer to the electrons in the resonance region strongly depends on the electron transit time through the region. This dependence is plotted in Fig.18 [7].

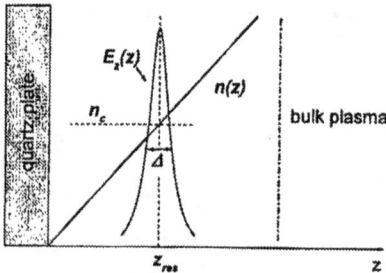

Figure 17. Enhancement of the electric field in an inhomogeneous plasma in the place of critical plasma density.

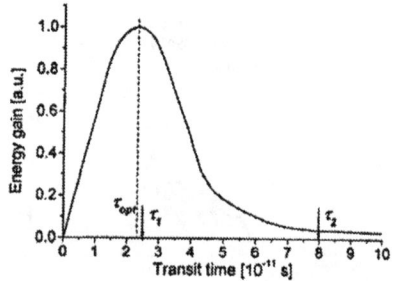

Figure 18. Efficiency of the energy transfer as a function of the electron transit-time through the plasma resonance region.

In our experiments, the typical conditions for the strongest hot-electron flux (the strongest light emission and power absorption) are: $p=20$ mTorr, $L=0.1$ cm, $T_e=2$eV, $v_h=4.6\times10^8$ cm/s (corresponding to 60 eV), which gives $v_{eff}=1.8\times10^9$ s$^{-1}$ (determined by the field convection) and $\varDelta=1.1\times10^{-2}$ cm and, consequently, the electron transit time is estimated to $\tau_l=2.5\times10^{-11}$ s. This value corresponds well to the optimum transit time $\tau_{opt}\approx2.3\times10^{-11}$ s (see Fig.18). When the localized hot-electron region is observed the typical experimental conditions are as follows: $p=2.6$ mTorr, $L=2.1$ cm, $T_e=4$ eV and $v_h=5\times10^8$ cm/s (corresponding to 70 eV). Under these conditions, the estimated parameters are: $v_{eff}=2.9\times10^8$ s$^{-1}$ and $\varDelta=4\times10^{-2}$ cm, which gives the electron transit time $\tau_2=8\times10^{-11}$ s. This value is comparable to the field period $T=4\times10^{-10}$ s and it is about 3.5 times larger than the optimum time $\tau_{opt}$, thus the condition for the effective power transfer to the electrons is not fulfilled. Indeed, under these experimental conditions, we do not observe the hot-electron flux.

## 5. SUMMARY

An experimental study on hot electrons in low-pressure large-area microwave discharges has been presented. At gas pressures below about 50 mTorr, a flux of hot electrons directed away from the waveguiding plasma-dielectric interface was observed. These electrons are believed to be originating from the region of resonantly enhanced electric field near the dielectric. They contribute significantly to discharge heating, which was demonstrated by the axial plasma parameter profiles. At pressures below 3 mTorr, a trough in the probe current profiles was observed near the quartz plate. The trough appears in the place of critical plasma density and we believe that it is caused by the localized hot electrons generated in the resonantly enhanced electric field. The formation of both, the hot-electron flux and the localized hot-electron region is explained on the basis of the transit-time heating. The presence of hot electrons in the discharges was investigated also by the EEDF measurements, which are inevitable to describe the phenomena. These measurements are in progress, however, the preliminary results indicate a hot-electron tail formation with an anisotropic distribution. The presented experimental results can be considered as an evidence that the plasma resonance region plays an active role for the heating mechanism in the low-pressure microwave discharges.

We believe that the phenomena may be an important factor in processing plasmas affecting the discharge chemistry in the volume plasma and on the processing surface as well. Particularly, the hot electron flux depositing the energy and charge on the processing surface may have a strong impact. This can be either positive or negative depending on the required process. Moreover, the occurrence of the hot-electron flux can also affect the discharge stability, which appears to be an issue for the large-area microwave discharges.

Since the main stream of research is the production and control of microwave discharges on large areas, we pay a particular attention to the hot electrons. We believe that the detailed measurements of the spatial profiles of the EEDF, which are the subjects of our present and the near-future work, will help to clarify the heating mechanism at low gas pressures.

## Acknowledgement

The authors are indebted to Prof. Yu. M. Aliev of Lebedev Physical Institute of Russian Academy of Sciences for his essential contribution to the presented work. The discussions with Dr. D. Korzec of Microstructure Research Center of University of Wuppertal (Germany), Prof. J. Marec of University of Paris-Sud (France) and Dr. E. Stamate of Nagoya Institute of Technology (Japan) are also acknowledged. This work was supported by a Grant-in-Aid for scientific Research (B)(2) from the Ministry of Education, Science, Sports and Culture of Government of Japan.

## References

1. Komachi K. and Kobayashi S.: J. Microwave Power Electromagn. Energy, 1989, 24, 140.

72

2. Kimura T., Yoshida Y., and Mizuguchi S.I. Jpn. J. Appl. Phys., Part 2, 1995, **34**, 1076.
3. Werner F., Korzec D., and Engemman J. Plasma Sources Sci. Technol., 1994, **3**, 473.
4. Nagatsu M., Xu G., Yamage M., Kanoh M., and Sugai H. Jpn. J. Appl. Phys., Part 2, 1996, **35**, 341.
5. Odrobina I., Kudela J., and Kando M. Plasma Sources Sci. Technol., 1989, **7**, 238.
6. Moisan M., Margot J. and Zakrzewski Z. *High Density Plasma Sources*, ed. O.A. Popov (Noyes Publication, Park Ridge, 1996), 191.
7. Aliev Yu.M., Bychenkov V.Yu., Maximov A.V., and Schlüter H. Plasma Sources Sci. Technol., 1992, **1**, 126.
8. Aliev Yu.M., Maximov A.V., Kortshagen U., Schlüter H., and Shivarova A. Phys. Rev. E, 1995, **51**, 6091.
9. Aliev Yu. M., Schlüter H., and Shivarova A. *Guided-Wave-Produced Plasmas* (Springer-Verlag, Berlin, 2000), 206.
10. Rauchle E. J. Phys. IV France, 1998, **8**, 99.
11. Sugai H., Ghanashev I. and Nagatsu M. Plasma Sources Sci. Technol., 1998, **7**, 192.
12. Ghanashev I., Sugai H., Morita S., and Toyoda N. Plasma Sources Sci. Technol., 1999, **8**, 363.
13. Yasaka Y., Nozaki D., Koga K., Ando M., Yamamoto T., Goto N., Ishii N., and Morimoto T. Jpn. J. Appl. Phys., Part 1, 1999, **38**, 4309.
14. Kudela J., Terebessy T., and Kando M. Appl. Phys. Lett., 2000, **76**, 1249.
15. Terebessy T., Kando M., and Kudela J. Appl. Phys. Lett., 2000, **77**, 2825 (in print).
16. Kudela J. Doctor thesis, Graduate School of Electronic Science and Technology, Shizuoka University, July1999.
17. Peterson E.W., Talbot L. AIAA J., 1970, **8**, 2215.
18. Chen F.F. in: Plasma Diagnostic Techniques, eds. R.H. Huddlestone and S.L. Leonard (Academic Press, New York, 1965) p.127
19. Druyvesteyn M.J. Z. Phys., 1930, **64**, 781.
20. Press W.H. and Teukolsky S.A. Comput. Phys., 1989, **3**, 63.
21. Fujita F. and Yamazaki H. Jpn. J. Appl. Phys., Part 1, 1990, **29**, 2139.
22. Ginzburg V. L.: *The Propagation of Electromagnetic Waves in Plasmas* (Pergamon Press, Oxford, 1964) p. 260.

# ADDITIONAL MICROWAVE DIAGNOSTICS OF AFTERGLOW WITH ESR SPECTROMETER

**V. Kudrle, A. Tálský, J. Janča**

Department of Physical Electronics, Faculty of Science, Masaryk University
Kotlářská 2, CZ 61137 Brno, Czech Republic

**Abstract.** When a small amount of admixture ($O_2$, $H_2$, Ar, Ne) is added to a nitrogen gas, one may observe increased concentration of atomic nitrogen in the discharge afterglow. It is accompanied by increase in the electron concentration, too. In this work we present the dependence of concentrations of both isotopes $^{14}N$ and $^{15}N$ on admixture type and amount. For measurement of atomic nitrogen concentration and electron concentration we employed electron spin resonance spectrometer.

## 1. INTRODUCTION

In recent years there may be observed an increase in interest in non-equilibrium kinetics of low pressure plasmas in nitrogen, oxygen and their mixtures. The applications range from industrial (oxidation, nitridation of surfaces), scientific (basic research) to ecology applications (better understanding of the processes in the upper atmosphere, cleaning of pollutants, etc.). In many studies various authors observed that even a small admixture to a background gas changes radically the production of atoms (for list, see e.g.[1]). Some authors explained this phenomenon by complicated reactions between radicals but today it is generally supposed that surface reactions, influencing recombination rate, play a dominant role in this phenomenon [2]. This effect is often accompanied by the changes of electron concentration in afterglow.

## 2. EXPERIMENTAL

A microwave discharge is produced in a quartz discharge tube with inner diameter of 13 mm. Plasma launcher of surfatron type is powered by 100 W magnetron working at 2.45 GHz. Our experiments being carried out in a flowing regime, the afterglow may be observed in 1 m long quartz tube with inner diameter of 8 mm. To minimize the influence of discharge radiation on the reactions taking place in the afterglow, the discharge and the afterglow tubes are connected by a knee with a horn (see Fig.1).

The gases are led from the standard gas bottles through the reduction valves and mass flow controllers to a cold trap which removes traces of impurities (hydrocarbons, water, etc.). The manufacturer of gas (Linde) states a purity of 99.998% for nitrogen, 99.995% for oxygen, 99.999% for hydrogen, 99.9995% for argon and 99.999% for neon. Flow rate of admixture was varied between 0 and 10 sccm, nitrogen flow rate being kept at 25-100 sccm. The pressure in the discharge was 300 Pa for 25 sccm $N_2$ flow, 450 Pa for 50 sccm $N_2$ flow and 600 Pa for 100 sccm nitrogen flow. Viton seals being used, baking out is not performed but the whole vacuum system was carefully heated up to 100 °C during several hours. After this procedure the total desorption and leak rate was better than $10^{-3}$ sccm. Back diffusion of hydrocarbon impurities from a rotary oil pump is prevented by the second cold trap.

Concentration of various species in the afterglow is determined by means of the electron paramagnetic/spin resonance (EPR/ESR) spectrometer JEOL JES-3B. The detailed description of EPR use for observing the gas phase reactions may be found elsewhere [3]. To obtain the absolute values of concentrations, additional calibration by molecular oxygen is necessary [4]. Distance between the discharge and a measuring resonator (X band) of EPR spectrometer may be varied and thus reaction rates determined

EPR spectrometer works by measuring resonant absorption of microwave energy by magnetic dipole transitions between Zeeman split levels. The specimen is placed in slowly varying static magnetic field and changes in quality of measuring resonator (due to absorption) are recorded.

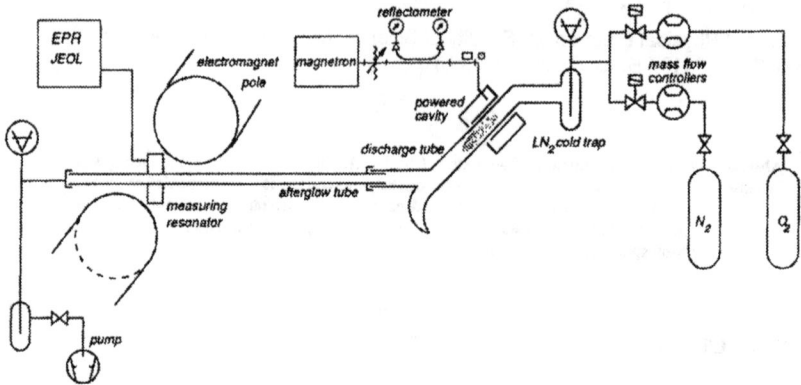

**Figure 1.** Drawing of experimental setup.

In comparison with other techniques for measurement of concentrations of various species, like mass spectrometry or actinometry, the EPR spectroscopy has the major advantage in its non-invasivity. In comparison with optical spectroscopy, we measure directly the concentration of ground state. If sufficiently sensitive spectrometer is available, it is even possible to directly observe also the excited states with non zero total magnetic momentum and thus describe the distribution of excited states.

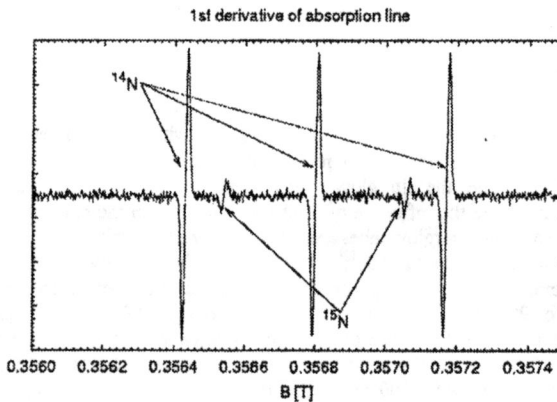

**Figure 2.** Typical EPR spectrum of atomic nitrogen. Due to a different nuclear spin of isotopes of nitrogen, one can clearly distinguish the triplet of 14N and the doublet of $^{15}$N.

Due to the employed mode of measuring cavity ($TE_{011}$) we can find a resonance (see Fig. 3) between a periodic motion of electrons in magnetic field and the microwave frequency. So obtained [5] electron cyclotron resonance curve is after the calibration by microwave resonator method used to determine the electron concentration at given position in the afterglow.

1st derivative of absorption line

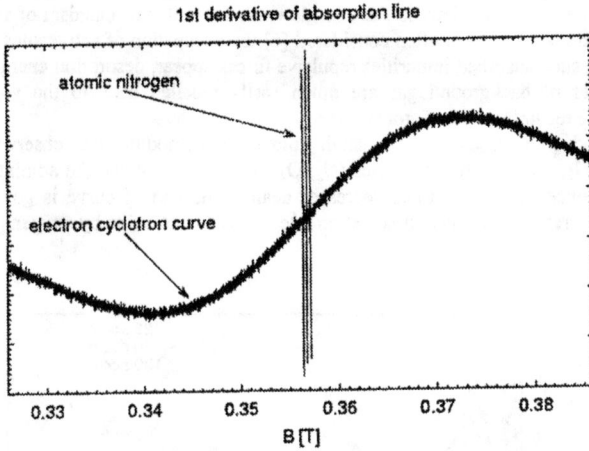

Figure 3. Shows electron cyclotron resonance curve measured by EPR spectrometer. In the center it overlaps with the EPR spectrum of atomic nitrogen (see previous figure).

## 3. RESULTS AND DISCUSSION

Addition of small amount of admixture into the nitrogen flow prior to its passage through the discharge gives rise to more than ten-fold increase in N atom concentration in afterglow (see Fig.4). On the other side, when the same admixture is added after the discharge (into an afterglow), one does not observe this phenomenon.

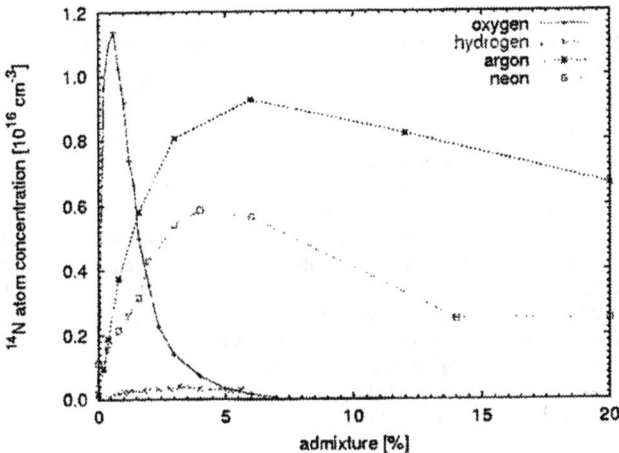

Figure 4. Rise of production of atoms in 50 sccm nitrogen, measured 37 cm from the discharge.

Thus some process in the discharge is responsible for this behavior. It may be explained as a reduction of probability of recombination of atomic nitrogen on the walls of the discharge tube. Due to an intensive bombardment of walls by electrons, ions and fast neutrals, the influence of discharge tube walls is much higher than the influence of walls later in the afterglow. As a possible mechanism of such change of the surface recombination coefficient is often considered [6] the occupation of active sites of the surface by the impurities. Near such adsorbed impurities repulsive forces appear, desorption energy is lowered and physisorbed particles of background gas are much easily repelled back to the volume, effectively inhibiting the surface recombination of atoms.

The rise of production of atoms with small amount of admixture was observed also for other background gases ($N_2$, $O_2$, $H_2$) and impurities ($N_2$, $O_2$, $H_2$, Ar, Ne). When the admixture abundance is further increased, concentration of atoms decreases again. This part of curve is governed by volume processes, such as decrease of nitrogen dissociation rate due to de-excitation by oxygen [7].

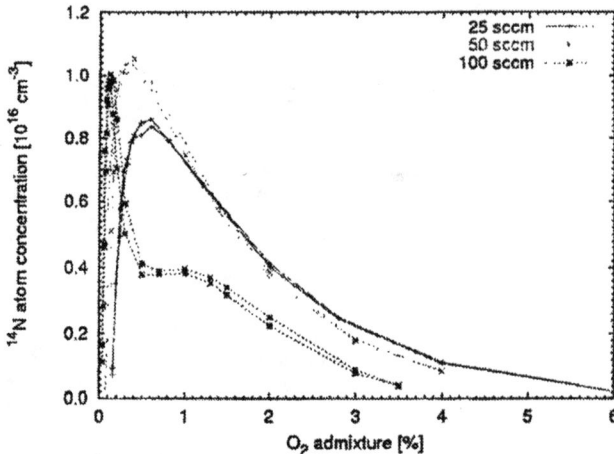

**Figure 5.** Influence of oxygen admixture on concentration of atomic nitrogen for three different flows of nitrogen gas.

In Fig.5 we may observe, that the maximum production of atoms occurs not for the same percentage of admixture, but for the same absolute amount of admixture. This supports a hypothesis that surface reactions are responsible for observed rise in atom concentration.

The capability of nitrogen to store the energy is well known and characteristic Lewis - Rayleigh afterglow may be easily observed even for very long afterglow times. Several investigators [8] have demonstrated the presence of free electrons in activated nitrogen. It is generally accepted [9] that Lewis-Rayleigh afterglow is caused by recombination of atomic nitrogen. Our experiments shown that when the apparatus was thoroughly cleaned and pure nitrogen was used, concentration of nitrogen atoms in the afterglow was very low and no L.-R. afterglow might be observed.

Among the studied admixtures, which all increased atomic nitrogen production, only the oxygen admixture radically increased the concentration of electrons in the afterglow. The results are presented in the Fig.6.

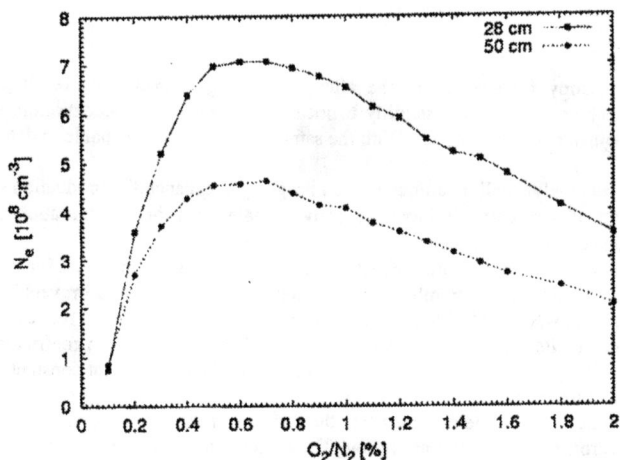

**Figure 6.** Dependence of electron concentration on the amount of oxygen admixture. Two curves, measured at different distances from discharge, are presented.

We observed an interesting fact, that isotope composition of nitrogen atoms depends on admixture content, too. The ratio $^{14}N/^{15}N$ is shown in Fig.7 where one can observe that it is almost always under the natural level (0.38% of $^{15}N$ in nitrogen). e.g. the addition of admixture enhances the $^{15}N$ fraction. It seems, that surface reactions of heavier isotope are substantially different from those of $^{14}N$. But very often the isotope is used to track chemical reactions, and thus it is supposed that rate constants are the same. This discrepancy is not yet clearly explained.

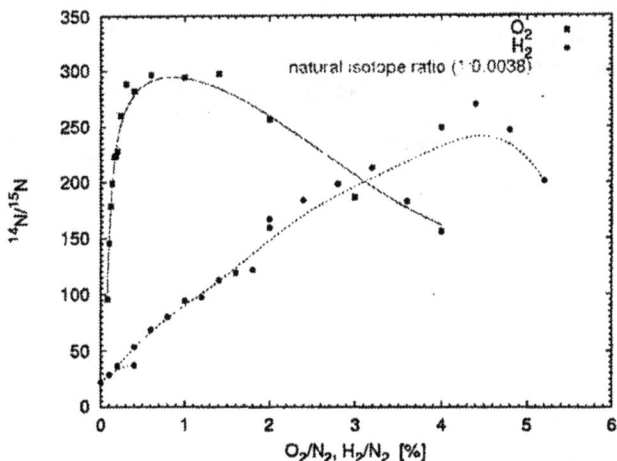

**Figure 7.** Variations of isotope composition of nitrogen atoms with changing the amount of oxygen or hydrogen admixture.

## 4. CONCLUSION

The use of EPR spectroscopy for analysis of the afterglow may give interesting results, not easily achievable with other techniques. It has possibility to measure not only the concentrations of various atoms, but also to distinguish between isotopes. With the same apparatus one is capable to determine also the electron density.

Impurities added in relatively small quantities to the nitrogen gas enhance the production of nitrogen atoms. When the amount of admixtures is increased above certain level the concentration of nitrogen atoms decreases again due to volume processes.

Surface processes on the walls of discharge tube are suggested as the most important for this increase. We suppose that surface contains certain number of adsorption sites (reversible and irreversible) and that repulsive interaction exists between unlike atoms. Thus admixture atom sorbed at wall reduces the time of residence of nitrogen atoms in its vicinity, effectively reducing surface recombination coefficient.

Both studied isotopes of nitrogen ($^{14}N$ and $^{15}N$) behave similarly but their ratio is not constant. It depends on type of the impurity and its amount.

When oxygen admixture is used, one observes that the changes of atomic nitrogen concentration are closely followed by electron density in the afterglow. These electrons do not originate directly in the discharge, but are produced by reaction of atomic nitrogen with oxygen in the afterglow. ·

### Acknowledgments

This work was partially supported by Grant Agency of Czech Republic under contract No.106/96./K245 and by Czech Ministry of Education, contracts No. VS96084 and CEZ:J07/98:143100003.

### References

1. Zvoníček V. Thesis, Masaryk University Brno 1997.
2. Gordiets B.et al. J.Phys.D: Appl. Phys. 1996, **29**, 1021.
3. Westenberg A. A. Prog. React. Kinet. 1973, **7**, 23.
4. Krongelb S., Strandberg H. W. P. J. Chem. Phys. 1959, **31**, 1196.
5. Ultee C. J. J. Phys. Chem. 1960, **64**, 1873.
6. Zvoníček V. Proceedings of ICPIG'97 Toulouse, 1997, vol.4, 160.
7. Kudrle V., Tálský A., Kudláč A., Křápek V., Janča J. Czech. J. Phys. 2000, **50**, 305.
8. Rayleigh: Proc. Roy. Soc. (London) 1951, **A180**, 140.
9. de Monchy A. R. Thesis, Universiteit van Amsterdam 1970.

# INFRARED ABSORPTION DIAGNOSTICS AND CHEMICAL MODELLING OF MICROWAVE PLASMAS CONTAINING HYDROCARBONS

L. Mechold, J. Roepcke, D. Loffhagen and P.B. Davies[*]

Institut für Niedertemperatur-Plasmaphysik e.V., 17489 Greifswald, F.-L.-Jahn-Str. 19, Germany

[*]Department of Chemistry, University of Cambridge, Lensfield Road, Cambridge CB2 1EW, U.K.

**Abstract.** The molecular concentrations of the methyl radical and ten related stable molecules, $CH_4$, $CH_3OH$, $C_2H_2$, $C_2H_4$, $C_2H_6$, $CH_2O$, $HCOOH$, $CO$, $CO_2$ and $H_2O$, have been determined in $H_2$-$Ar$-$O_2$ microwave discharge plasmas ($f \approx 2.45$ GHz) containing small amounts of methane or methanol by means of infrared tunable diode laser absorption spectroscopy. The fragmentation rates of methane and methanol and the respective conversion rates to several measured molecular products have been estimated for different ratios of molecular hydrogen to oxygen concentrations. The rate of fragmentation of methanol - $R_F$ ($CH_3OH$)= 1.0-2.2 x $10^{16}$ molecules/J - was found to be larger than that of methane - $R_F$ ($CH_4$)= 0.6-1.1 x $10^{16}$ molecules/J - for all ratios of hydrogen to oxygen concentrations. The conversion rates to hydrocarbons were less than $10^{14}$ molecules/J and decreased exponentially with growing oxygen admixture. The largest rates of conversion were determined for water, carbon monoxide and carbon dioxide. The measured species concentrations are compared with results of model calculations for discharge plasmas containing methane. Good agreement between measured and calculated species concentrations was obtained. Main chemical reaction pathways are discussed and an appropriate reaction scheme is established.

## 1. INTRODUCTION

Molecular microwave plasmas have been used extensively for plasma processing and technology like diamond deposition and surface modifications for biotechnological purposes. Another challenging subject of plasma technology is the effective conversion of natural gas, which contains mainly methane, to higher hydrocarbons. Low pressure microwave plasmas with molecular source gases are characterized by high degrees of dissociation of the precursor species and high chemical reactivity due to the large concentrations of transient and stable chemically active neutral species present. Thus, the in-situ monitoring of precursor molecules as well as of stable and transient reaction products, in particular, the measurement of their ground state concentrations, is the key to discover detailed properties of plasma chemistry and kinetics. Its knowledge is essential for the control of molecular non-equilibrium plasmas for a specific purpose.

In the last decade, a variety of papers has been published studying the conversion of hydrocarbons by different experimental and theoretical approaches. Only some recent examples are given here. A more comprehensive discussion can be found in the literature [1-3]. Bugaev et al. [4] investigated the oxidative conversion of a mixture of natural gas and oxygen in a barrier-discharge by monitoring stable reaction products using gas chromatography. Radio-frequency plasmas were used by Hsieh et al. [5] to convert methane into acetylene, ethylene and ethane. These stable species were detected by Fourier transformed infrared spectrometry.

For the investigation of highly chemically active plasmas non-invasive diagnostic methods have also been developed, particularly, emission and absorption spectroscopy as well as specific laser spectroscopy techniques. Tunable infrared diode laser absorption spectroscopy (TDLAS) was shown to be a very useful technique because it provides ground-state species concentrations of radicals and related transient and stable plasma species. This diagnostic technique has been applied for different groups of substances like hydrocarbons in microwave plasmas [2]. In plasma conditions close to thin carbon film depositions tunable diode lasers were used to measure the absolute concentrations of the methyl radical and related molecules. In fact, only the determination of species concentrations has permitted a more detailed understanding of the neutral plasma chemistry.

There have been several reports in the literature concerning the modelling of the chemistry of hydrocarbon plasmas under different conditions [6-14]. These studies have mainly been focused on rf plasmas, hot filament reactors and plasma jets. Recently, extensive modelling has also been performed for moderate pressure microwave plasmas applied for diamond depositions [15,16]. Mainly based on calculated species concentrations some of the authors were able to develop a simplified reaction scheme for the volume processes, also in combination with the growth mechanism at surfaces. The calculated molecular concentrations were only partly compared with experimental results of e.g. mass spectrometric measurements. In the majority of examples methane is the precursor, but also other hydrocarbons like $C_2H_4$ or $C_2H_6$ were considered.

For hydrocarbon-containing microwave plasmas at low pressure no comparable theoretical studies have been performed so far. To overcome this situation extensive experimental and theoretical investigations were done. In the present article studies of $H_2$-Ar-$O_2$ plasmas containing small amounts of methane or methanol in planar microwave discharges are reported. TDLAS has been used to measure the methyl radical ($CH_3$) and the stable molecules methane ($CH_4$), methanol ($CH_3OH$), acetylene ($C_2H_2$), ethylene ($C_2H_4$), ethane ($C_2H_6$), formaldehyde ($CH_2O$), formic acid ($HCOOH$), carbon monoxide ($CO$), carbon dioxide ($CO_2$) and water ($H_2O$). Molecular concentrations were measured dependent on various hydrogen to oxygen ratios. Based on the experimental data fragmentation rates of the precursor hydrocarbons and conversion rates to the plasma product molecules have been determined. In addition to experiments for the relatively high gas flow regime, measurements under static conditions have been performed.

The experimental studies are accompanied by model calculations, in particular, for methane-containing plasmas under static conditions. In the framework of the model an extensive set of rate equations for the various species has been solved to determine theoretically concentrations of the measured species. The experimental and theoretical results are compared and a detailed analysis of the main reaction pathways is given.

## 2. EXPERIMENT

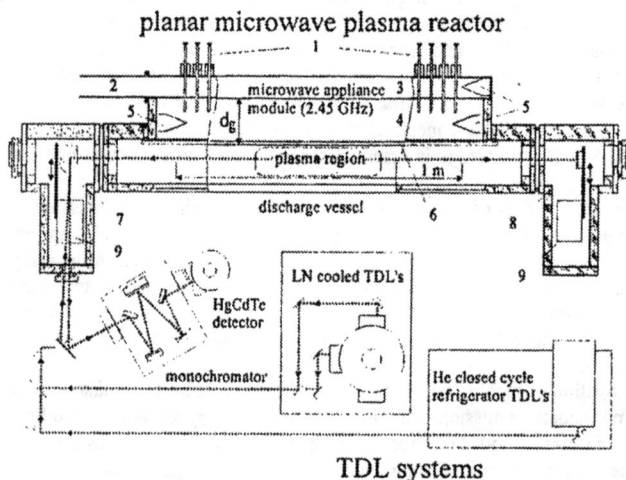

Figure 1. Experimental arrangement of planar microwave plasma reactor (side view) and tunable diode laser (TDL) infrared sources. The path of the diode laser beam is indicated by dotted lines. 1- tunable coupling elements, 2- to generator, 3- upper waveguide, 4- interface waveguide, 5- absorber, 6- microwave window, 7- periscope optics, 8- retroreflector, 9- stepper motor.

The design of the planar microwave plasma reactor and the tunable diode laser (TDL) assemblies is shown in Fig.1. The numbers in brackets in the following remarks refer to this scheme. The main features of the concept of planar microwave discharges were described in references [17,18]. Only the specific properties of the planar microwave plasma used are given. The applicator contains two T-shaped rectangular waveguides (3) and (4) to guide the microwave power into the discharge chamber. The upper waveguide (3) represents a travelling-wave slotted waveguide for distributing the microwave power over the length of the microwave window (6). Coupling holes in the contact surface between the slotted and the interface waveguide (4) with a distance of λ/4 are used together with tunable coupling elements (1) inside the holes for improving the tuning range and the efficiency of the microwave power coupling behaviour. The metal walls alongside the gap $d_g$, the slotted surface and the microwave window (6) with the conducting plasma below form the interface waveguide (4). Microwaves can propagate in both directions to the ends of the waveguides. Therefore, specific absorbers (5) are installed to damp reflections. The power transfer surface to the plasma chamber has been a rectangular area of about $100 \, cm^2$. Measurements were performed at a microwave frequency of 2.45 GHz and a power of 1.5 kW. The effective applied power yielded 1.2 kW because of losses due to power reflections and wall heating processes. This corresponds to a power flux in the TDL observation plane of about $12 \, W \, cm^{-2}$. The planar discharge configuration has the advantage of being well-suited for end-on spectroscopic observations because considerable homogeneity can be achieved over relatively long plasma path lengths.

Two additional vacuum side ports are mounted at both front ends of the discharge vessel. These ports contain the periscope optics (7) and the retroreflector (8) moved by stepper motors (9) for vertical changes of the optical axis. Two different diode laser assemblies were applied using the same optical path through the plasma source. The infrared diode laser beam from one of the two TDL source assemblies entered the plasma chamber via a KBr window. Its distance from the microwave entry window was adjusted using the periscope mirror system in combination with a retroreflector. Measurements reported here were made at a fixed distance of 20 mm from the microwave entry window with the laser beam making two passes through the plasma. After leaving the plasma reactor the diode laser beam first passed through a mode selection monochromator before detection with a HgCdTe detector. Stable molecules are assumed to be distributed homogeneously in the optical path and transient molecules like the methyl radical are assumed to exist only in the plasma region.

Gas flows were arranged by mass flow controllers and the gases were mixed before entering the reactor. They were pumped out via a port in the reactor wall. The gas mixture supplying the reactor consisted of $x$ sccm $H_2$ + 60 sccm Ar + $y$ sccm $O_2$ + $z$ sccm $CH_4$ or $CH_3OH$. The total flow rate and pressure were kept constant at 555 sccm and 1.5 mbar, respectively. When oxygen was added, the proportion of hydrogen was reduced correspondingly to maintain a constant flow rate. The pumping speed was adjusted with a butterfly valve to maintain a constant pressure in the plasma. For both methane- and methanol-containing mixtures, four different admixtures, namely 5, 15, 25 and 40 sccm, were investigated. Parameters used for the experiments were selected to represent typical conditions in practical plasma-enhanced chemical vapour deposition reactors.

A representative TDL absorption spectrum of a $H_2$-Ar-$O_2$-$CH_3OH$ microwave plasma at a pressure of 1.5 mbar and an incident microwave power of 1.5 kW is shown in Fig. 2. The spectrum includes lines due to methanol, methyl and carbon dioxide in the plasma. Additional lines of $N_2O$ are caused by a reference gas cell placed in the beam path.

The identification of the lines and the measurement of their absolute positions were carried out using well-documented reference gas spectra and an etalon of known free spectral range for interpolation [19-21]. The absolute species concentrations are calculated according to the Lambert-Beer absorption law [22]. It determines the reduction of the radiation intensity of a parallel beam by a sample. The law can be expressed as

$$I = I_0 \exp(- N \sigma l) \tag{1}$$

where $I$ is the transmitted intensity and $I_0$ the incident intensity of the infrared radiation. Furthermore, $N$ is the molecule number density, $\sigma$ is the absorption cross section and $l$ is the interaction length.

82

wavelength [μm]

16.499  16.498  16.497  16.496  16.495  16.494

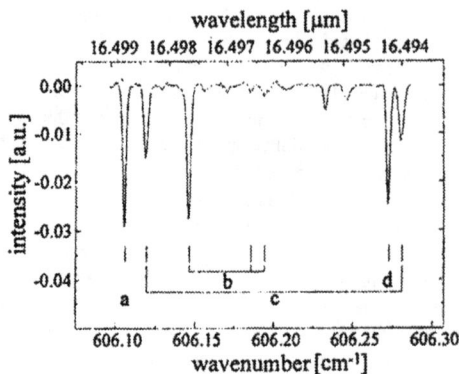

Figure 2. TDL absorption spectrum of some molecules in a $H_2$-Ar-$O_2$ plasma containing methanol under flowing conditions. a: $CH_3OH$, b: $N_2O$, c: $CH_3$, d: $CO_2$. [2]

## 3. PLASMA CHEMISTRY MODELLING

Experimental investigations were accompanied by model calculations. Theoretical studies have been focused to $H_2$-Ar-$O_2$ plasmas containing methane. The aim of the present calculations is to improve the physical understanding of the trends of methyl and associated species concentrations. The model is based on a set of rate equations for 24 species and 125 of their reactions. In order to reduce the number of unknown or poorly known parameters, a static cell model was assumed. Theoretical results are compared with the experimental ones at static conditions (sc). This means, there is no gas flow in the plasma and temperature and volume are set constant. In the framework of the model electron impact dissociation of molecules and neutral-neutral collisions were taken into account. The rate coefficients for various chemical reactions are averaged over space and time and have mainly been taken from the literature [1,12-14,23-25]. The rate coefficients for the electron collision processes have been determined by solving the time-dependent Boltzmann equation for given values of the reduced electric field, microwave frequency and mixture composition up to the establishment of the periodic state. This electron kinetic equation has been solved by means of the multiterm method described in [26]. Respective cross sections for electron impact collisions for hydrogen, argon, oxygen and methane were taken from the established literature [27-30]. The model does not include ionic processes and a rate equation for electrons. In accordance with experiments in the afterglow, an electron density of $1 \times 10^{12}$ cm$^{-3}$ has been assumed. The rate equations form a system of stiff ordinary differential equations which has been solved using the programming package KINEL [31]. Starting from different mixture compositions the concentrations of the species under static conditions have been determined by means of a time-dependent relaxation procedure. The static condition is reached, if all species concentrations become time-independent.

## 4. RESULTS

The plasma chemistry of molecular microwave discharges has been studied by infrared TDLAS. The present investigation comprises results for the methyl radical and ten stable molecules, i.e., $CH_4$, $CH_3OH$, $C_2H_2$, $C_2H_4$, $C_2H_6$, $CH_2O$, $HCOOH$, $CO$, $CO_2$ and $H_2O$. These species were detected in mixed $H_2$-Ar-$O_2$ plasmas containing small admixtures of methane or methanol. The discharge pressure and the power were kept constant at 1.5 mbar and 1.5 kW, respectively.

In previous studies [2] the molecular concentrations of the above mentioned plasma species and the degree of dissociation of the added precursor were reported as the hydrogen to oxygen ratio was varied while maintaining constant discharge pressure and gas flow rate. Carbon monoxide and carbon dioxide as well as water were found as major plasma products. C-2 hydrocarbons showed an exponential decrease with increasing oxygen amounts. The carbon mass balance in the plasma was governed by carbon monoxide and carbon dioxide. In addition, it was found that at higher oxygen admixture about 50 % of available methane and 80 % of available methanol was converted to CO and $CO_2$.

The present contribution is focused on the measured rates of fragmentation and conversion in $H_2$-Ar-$O_2$ plasmas containing methane or methanol and on model calculations of $H_2$-Ar-$O_2$-$CH_4$ plasmas.

## 4.1. Fragmentation and conversion rates

One of the main subjects of this contribution is to get better insight into the plasma chemical conversion of hydrocarbons and related species. Experimental data were used to calculate absolute fragmentation rates of the precursor molecules and conversion rates of the measured plasma products. These rates were normalized to the discharge power. The fragmentation rate $R_F$ of the precursor molecules $CH_4$ and $CH_3OH$ is introduced in analogy to [3] as

$$R_F = \frac{\Phi_p}{60} \frac{D}{100} \frac{N_0}{P}$$

(2)

where $R_F$ has the unit molecules $J^{-1}$, $\Phi_p$ is the precursor flow given in sccm, $D$ is the degree of dissociation of the precursor molecules given in %, $N_0 = 2.69 \times 10^{19}$ is the number of molecules per $cm^{-3}$ at normal conditions and $P$ is the discharge power in W.

The conversion rate to plasma product molecules is expressed analogously as

$$R_C = n_{molecule} \frac{\Phi_t}{60} \frac{10^3}{p} \frac{1}{P}$$

(3)

where $R_C$ has the unit molecules $J^{-1}$, $n_{molecule}$ is the measured molecular concentration given in molecules $cm^{-3}$, $\Phi_t$ is the total gas flow given in sccm and $p$ is the pressure in mbar.

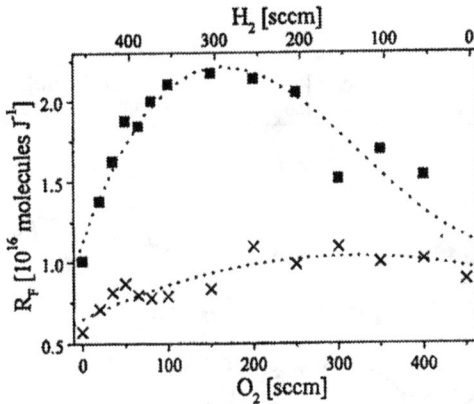

**Figure 3.** The rate of fragmentation of $CH_4$ (X) and $CH_3OH$ (■) as a function of oxygen flow. $\Phi(CH_4) = \Phi(CH_3OH) = 40$ sccm, i.e., (7.2 %), P= 1.5 kW, p= 1.5 mbar, $\Phi_{total}$= 555 sccm.

Figure 3 shows the respective rate of fragmentation of methane and methanol in $H_2$-Ar-$O_2$ plasmas containing 7.2 % of the hydrocarbon precursor. For $H_2$-Ar-$CH_3OH$ plasmas the fragmentation rate is about twice as large as that for the oxygen-free $H_2$-Ar-$CH_4$ plasma. When adding oxygen to the plasma while keeping the total gas flow and the discharge pressure constant, i.e., varying the ratio of hydrogen to oxygen concentration, both rates of fragmentation increase up to a maximum value. It is about a factor of two larger than the fragmentation rate in the oxygen-free plasma. The maximum fragmentation rate is found at $H_2/O_2$ mixtures of 300/150 for the methanol-containing plasma and of about 150/300 for the plasma containing methane. When increasing the oxygen content further both the fragmentation rates decrease. In Ar-$O_2$ plasmas containing methane and methanol, respectively, the rates of fragmentation

84

have nearly the same value. Similar results have been obtained for various admixtures of hydrocarbon precursors, where smaller admixtures of hydrocarbons resulted in lower rates of fragmentation.

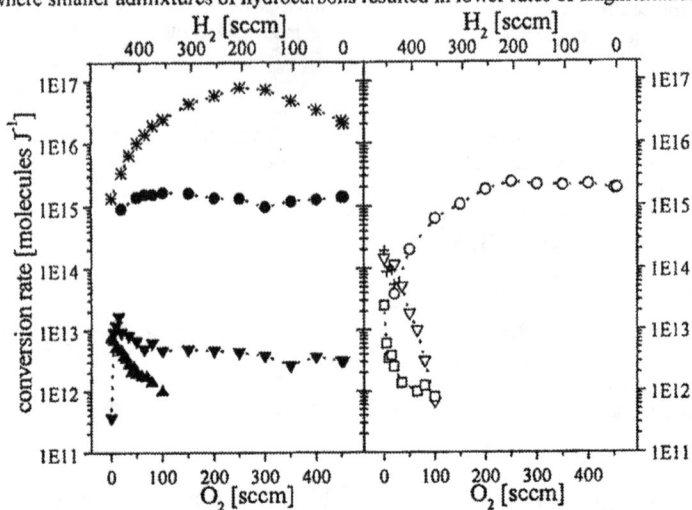

**Figure 4.** Rates of conversion as a function of oxygen flow rate in $H_2$-$O_2$-Ar-$CH_4$ microwave plasmas. Data points have been joined by dotted lines as a guide to the eye. *- $H_2O$; ● - CO; ○ - $CO_2$; ▲ - $CH_3$; ▼ - $CH_2O$; ▽ - $C_2H_2$; □ - $C_2H_4$; + - $C_2H_6$.

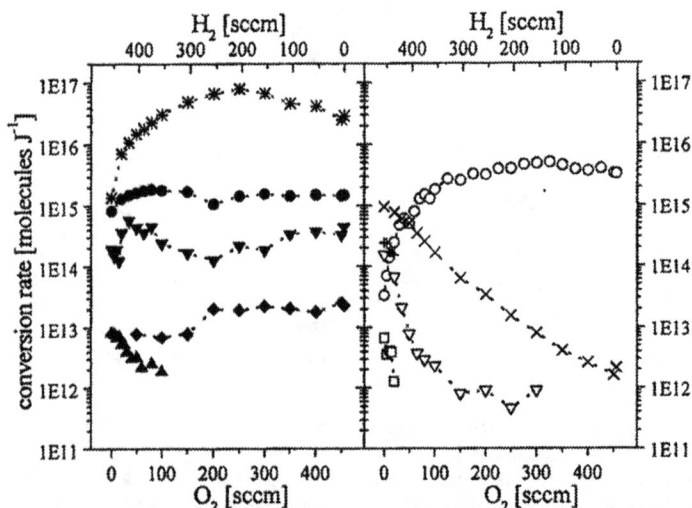

**Figure 5.** Rates of conversion as a function of oxygen flow rate in $H_2$-$O_2$-Ar-$CH_3OH$ microwave plasma. Data points have been joined by dotted lines as a guide to the eye. *- $H_2O$; ● - CO; ○ - $CO_2$; ✕ - $CH_4$; ▲ - $CH_3$; ▼ - $CH_2O$; ◆ - HCOOH; ▽ - $C_2H_2$; □ - $C_2H_4$; + - $C_2H_6$.

Figure 4 shows the rates of conversion of the methyl radical and the related species $C_2H_2$, $C_2H_4$, $C_2H_6$, $CH_2O$, CO, $CO_2$ and $H_2O$ in $H_2$-Ar-$O_2$ microwave plasmas containing methane. The corresponding conversion rates for $H_2$-Ar-$O_2$ plasmas containing methanol are presented in Fig. 5. The rates of conversion of reaction products are based on the measurement of absolute concentrations. They range over six orders of magnitude and are more uniformly distributed over the calculated range for methanol than for methane. The largest value was always found for water. It rises steeply to a maximum at the ratio of hydrogen to oxygen concentration of 1:1. The carbon conversion showed the highest values for carbon monoxide and carbon dioxide. This oxidation behaviour of carbon is similar to that of the combustion chemistry. The rate of conversion of formaldehyde is much higher in methanol-containing plasmas than in methane-containing plasmas. The same product molecules were not always found for both precursor hydrocarbons. Thus, methane was detected in methanol plasmas, but methanol was not detected in methane plasmas, because the methanol concentration was most likely below the detection limit of about $10^{11}$ molecules $cm^{-3}$. On the other hand, formic acid was produced in methanol-containing plasmas only. The most noticeable oxygen-dependent changes occurred for $C_2H_2$, $C_2H_4$, $C_2H_6$ and $CH_3$ at low oxygen concentrations. Their rates of conversion decreased exponentially with increasing oxygen amounts.

## 4.2. Results of model calculations

Model calculations have been performed for the $H_2$-Ar-$CH_4$ and $O_2$-Ar-$CH_4$ microwave plasmas with a composition of 82/10.8/7.2 %. In low pressure plasmas electron impact collisions are acknowledged to be major processes contributing to the dissociation of molecular species. The electron impact dissociation of methane is most likely the process which produces the primary free radicals $CH_3$, $CH_2$ and $CH$. Moreover, the electron impact impact dissociation of molecular hydrogen and oxygen leads to the production of atomic hydrogen and oxygen. These atomic species also play an important role in the complex molecular microwave plasmas considered.

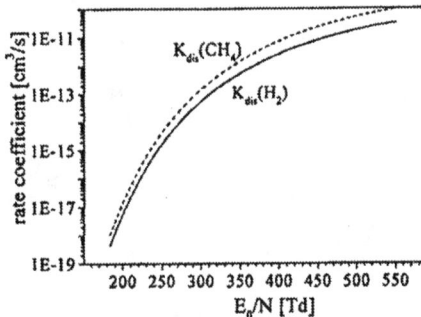

Figure 6. Calculated rate coefficients for electron impact dissociation as a function of the reduced field strength in $H_2$-Ar-$CH_4$ plasmas

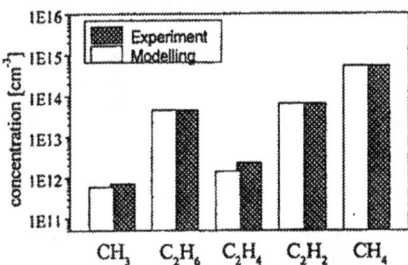

Figure 7. Measured and calculated species concentrations in $H_2$-Ar-$CH_4$ plasmas

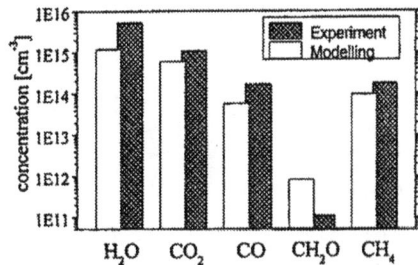

Figure 8. Measured and calculated species concentrations in $O_2$-Ar-$CH_4$ plasmas

Figure 6 shows the rate coefficients for electron impact dissociation of molecular hydrogen $K_{dis}(H_2)$ and methane $K_{dis}(CH_4)$ as a function of the amplitude of the reduced microwave field strength $E_0/N$. The rate coefficients have been determined from the solution of the electron Boltzmann equation as detailed above. For the model calculations of the methane-containing plasma the rate coefficients for $E_0/N= 280$ Td have been used.

The calculated concentrations of the methyl radical, ethane, ethylene, acetylene and methane in $H_2$-Ar-$CH_4$ plasmas are shown in Fig. 7. In addition the respective measured concentrations are given in this figure. The calculated and measured species concentrations agree well for $CH_4$, $C_2H_6$ and $C_2H_2$. Slight differences are found for $CH_3$ and $C_2H_4$.

In Fig. 8 the comparison between the experimental and theoretical results for water, carbon monoxide, carbon dioxide, formaldehyde and methane is presented for the $O_2$-Ar-$CH_4$ plasma. Satisfactory agreement between the measured and calculated species concentrations is found.

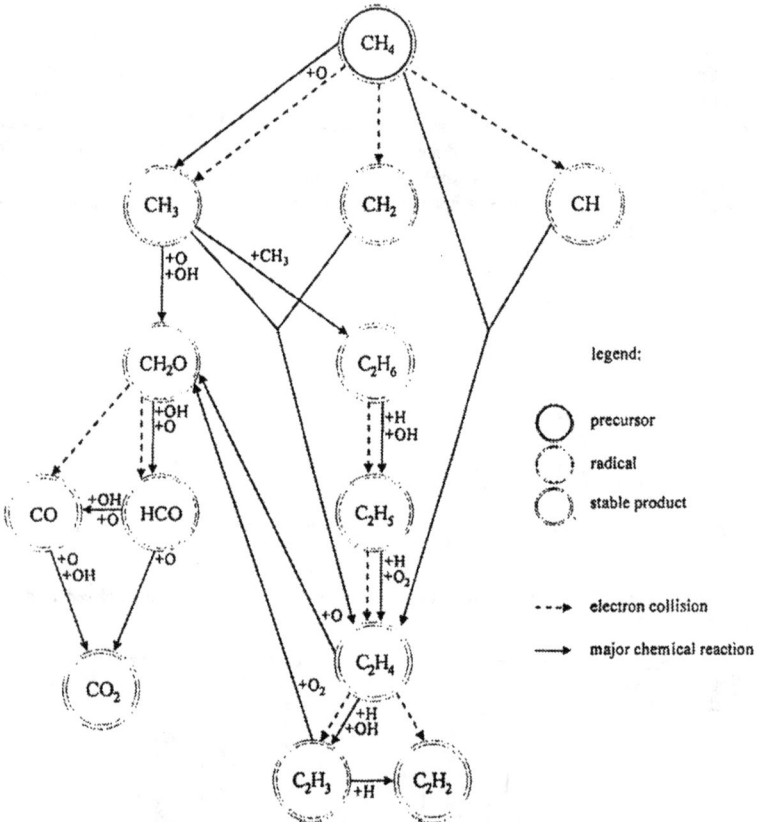

Figure 8. Major reaction paths linking the plasma chemistry of $H_2$-$O_2$-Ar-$CH_4$ microwave plasmas

In order to get a deeper understanding of the $H_2$-Ar-$O_2$-$CH_4$ microwave plasmas, a detailed analysis of the various production and consumption processes has been performed for each species considered in the model. The resulting scheme of the main reaction pathways is presented in Fig. 8.

It is widely accepted that the methyl radical is the most abundant radical in methane-containing plasma. Indeed $CH_3$ was measured by TDLAS in microwave plasmas at various oxygen flows. Because the concentration of $CH_3$ decreases exponentially with increasing admixture of oxygen, the strong reaction activity of the methyl radical is obvious. The central role of the methyl radical, as deduced from measurements, was verified by model calculations.

In $H_2$-Ar-$CH_4$ microwave plasmas ethane, ethylene and acetylene are the stable hydrocarbon products. In investigations of rf plasmas using mass spectrometry higher hydrocarbons produced from stable and transient C-2 hydrocarbons were found, too [7,13]. A related tendency with small concentrations of higher hydrocarbons can also be assumed for microwave plasmas. Ethylene is mainly produced by chemical reactions of methane with the primary radicals $CH_3$, $CH_2$ and CH. In contrast to ethylene the concentration of $C_2H_6$ is mainly balanced by the recombination of the methyl radical and reactions of $C_2H_5$ with hydrogen only. The electron collisions of $C_2H_4$ contribute to the formation of $C_2H_2$, but its concentration is dominated by reactions of the radical counterparts $C_2H_3$ and $C_2H$.

It is important to point out the role of hydrogen in the plasma chemistry. Reactions of hydrogen atoms with hydrocarbons, including $CH_3$ and $CH_2$, contribute to the formation of ethylene and acetylene. Generally, reactions with atomic hydrogen lead to sequential reductive formation of $C_2H_2$ starting from $C_2H_6$. Reactions with molecular hydrogen also produce $C_2H_6$ sequentially starting from $C_2H_2$ although having fairly low rate coefficients. These counteractive processes stabilize the entire chemistry in the plasma.

When oxygen is present in the plasma, further reaction channels are opened. In particular, reactions of $CH_3$ with oxygen or hydroxyl radicals (OH) strongly influence the direction of the plasma chemistry and lead to the formation of e.g. formaldehyde, carbon monoxide and carbon dioxide. Although the degree of methane dissociation is increased now, the amount of hydrocarbons clearly decreases in favour of oxygen-containing molecules. When increasing the oxygen content more carbon monoxide and carbon dioxide occures representing final products of oxidation processes. The main loss channels of C-2 hydrocarbons to formaldehyde and furtheron to carbon monoxide appear from $C_2H_4$ or $C_2H_3$. Although $C_2H_4$ is stable, it represents a very important intermediate molecule. Reactions of all hydrocarbons with atomic oxygen or OH radicals support the reduction of $C_2H_6$ in favour of $C_2H_4$ and $C_2H_2$.

## 5. Conclusion and outlook

Low pressure molecular microwave plasmas are characterized by intensive chemical conversion of precursors to many molecular products. TDLAS was shown to be a very useful diagnostic technique to measure ground state concentrations of various stable and transient molecular species. Progress has been made in probing the species present within a wide range of ratios of molecular hydrogen to oxygen concentrations. Experimental data were used to determine the rate of fragmentation of hydrocarbon precursors and the rate of conversion of various related molecules. The most significant results are that (i) higher rates of fragmentation of methanol lead to higher rates of conversion of carbon monoxide and carbon dioxide, (ii) highest rates of conversion were found for water and (iii) rates of conversion of $CH_3$, $C_2H_6$, $C_2H_4$, $C_2H_2$ decrease exponentially when increasing the oxygen admixture.

The measurements have been accompanied by model calculations. Kinetic modelling has been performed for $H_2$-Ar-$O_2$-$CH_4$ plasmas. An elaborate set of chemical reactions for the neutral species of the complex plasma was used. The comparison of the measured and calculated species concentrations showed good agreement for $H_2$-Ar-$CH_4$ plasmas and satisfactory agreement for plasmas containing Ar-$O_2$-$CH_4$. The analysis of the theoretical results allowed to derive a scheme for the main reaction pathways in $H_2$-Ar-$O_2$-$CH_4$ plasmas. The appropriate reaction scheme of these complex mixture plasmas is presented in this paper for the first time.

Although a wide range of species concentrations has been monitored in $H_2$-Ar-$O_2$ plasmas containing small amounts of methane or methanol, it is obvious that extension of the present studies to other intermediates is desirable. This includes the measurement of other free radicals the most important of which are OH, HCO, $HO_2$ and $C_2H$. The extension of TDLAS measurements to these species is feasible and in progress. To improve the modelling potential much more kinetic data on elementary reactions are required as well as quantitative measurements of the electron density and electron energy distribution function. In addition, ionic species and spatially dependencies should be included in the model.

**References**

1. Fan W Y, Knewstubb P F, Käning M, Mechold L, Röpcke J and Davies P B J. Phys. Chem. A, 1998, **103**, 4118.
2. Röpcke J, Mechold L, Käning M, Fan W Y and Davies P B Plasma Chem. Plasma Process., 1999, **19**, 395.
3. Mechold L, Röpcke J, Duten X and Rousseau A Plasma Sources Sci. Technol., 2001, **10**, 52.
4. Bugaev S P, Kozyrev A V, Kushinov V A, Sochugov N S and Khryapov P A Plasma Chem. Plasma Process., 1998, **18**, 247.
5. Hsieh L-T, Lee W-J, Chen C-Y, Chang M-B and Chang H-C Plasma Chem. Plasma Process., 1998, **18**, 215.
6. Bohr S, Haubner R and Lux B Appl. Phys. Lett., 1996, **68**, 1075.
7. Dagel D J, Mallouris C M and Doyle J R J. Appl. Phys., 1996, **79**, 8735.
8. Gardner W L J. Vac. Sci. Technol. A, 1996, **14**, 1938.
9. Harris S and Weiner A M J. Appl. Phys., 1990, **67**, 6520.
10. Pauser H, Schwärzler C G, Laimer J and Störi H Plasma Chem. Plasma Process., 1997, **17**, 107.
11. Schwärzler C G, Schnabl O, Laimer J and Störi H Plasma Chem. Plasma Process., 1996, **16**, 173.
12. Tachibana K, Nishida M, Harima H and Urano Y J. Phys. D: Appl. Phys., 1984, **17**, 1727.
13. Kline L E, Partlow W D and Bies W E J. Appl. Phys., 1989, **65**, 70.
14. Frenklach M and Wang H Phys. Rev. B, 1991, **43**, 1520.
15. Hassouni K, Leroy O, Farhat S and Gicquel A Plasma Chem. Plasma Process., 1998, **18**, 325.
16. Hassouni K, Duten X, Rousseau A and Gicquel A Plasma Sources Sci. Technol., 2001, **10**, 61.
17. Ohl A in Microwave Discharges: Fundamentals and Applications, Ed. C.M. Ferreira & M. Moisan, New York: Plenum Press, 1993, 205.
18. Ohl A J. Phys. IV France, 1998, **8**, Pr7-82.
19. Rothman L S et al. J. Quant. Spectr. Radiat. Transfer, 1992, **48**, 469.
20. Guelachvili G and Rao K N Handbook of Infrared Standards, 1986, Academic Press, Orlando.
21. Husson N et al. J. Quant. Spectr. Radiat. Transfer, 1994, **52**, 425.
22. Struve W S Fundamentals of Molecular Spectroscopy, 1989, John Whiley & Sons, New York.
23. Zelson L S, Davidson D F, Hanson R K J. Quant. Spectrosc. Radiat. Transfer, 1994, **52**, 31.
24. Baulch D L et al. J. Phys. Chem. Ref. Data, 1992, **21**, 411.
25. Hsu W L J. Appl. Phys., 1992, **72**, 3102.
26. Loffhagen D and Winkler R J. Phys. D: Appl. Phys., 1996, **29**, 618.
27. Buckman S J and Phelps A V JILA Information Center Report, University of Colarado, Boulder, 1985, 27.
28. Hayashi M Plasma Material Science Handbook, Ed. Japan Society for the Promotion of Science, Ohmsha: Ltd Tokio, 1992, 748.
29. Lawton S A and Phelps A V Phys. Rev., 1978, **69**, 1055.
30. Davies D K, Kline L E and Bies W E J. Appl. Phys., 1989, **65**, 3311.
31. Levchenko A and Alexeev G KINEL Manual, Moscow, 1994.

# MICROWAVE PLASMA SOURCES

# MICROWAVE DISCHARGE IN SUPERSONIC AIRFLOW AND RELATED PHENOMENA

V.Brovkin, D.Khmara, A.Klimov, Yu. Kolesnichenko,
A. Krylov*, V. Lashkov*, S. Leonov, I. Mashek*, M. Ryvkin**

IHT RAS - MRTI, Moscow, 127 412, Izhorskaya 13/19
*St.-Petersburg State University, 198 904, Bibliotechnaya pl. 2
**ARSRI Radio Apparatus, St.-Petersburg, 199 106, Shkiperskiy protok, 19

**Abstract.** The free localized microwave (MW) discharge in a linear and circular polarized beam is generated for the first time in a supersonic flow. The discharge parameters at the end of MW pulse are obtained: gas temperature $240\pm10K$, vibrational temperature $1500\pm200K$, electron temperature $2.1$-$2.3$ eV, specific energy input in the discharge 0.4 J/(cm$^3$·atm), average concentration of electrons in the discharge $3\cdot10^{12}$cm$^{-3}$. The possibility of using MW plasma technology in a plasma-airflow interaction area, in particular, to change an aerodynamic drag of a body is demonstrated. The existence of phase of bow shock wave instability under its interaction with MW plasma in the supersonic flow of air is confirmed. Numeric modeling of the kinetic processes during MW discharge and in afterglow is done.

## 1. INTRODUCTION

One of the most perspective methods of remote plasma creation in front of the moving object is microwave (MW) discharge. The beginning of realization of this method was getting started in 1997 by our work [1], when free localized MW discharge in supersonic airflow was obtained for the first time and the possibilities of MW plasma technologies were demonstrated, in particular for changing of drag reduction.

Studies of interaction of nonequilibrium low-temperature plasma with high speed bodies and airflows were conducting intensively during last 15 years [1,2]. There were revealed effects, such as the dispersion of acoustic waves; shock wave dissipation and splitting; the formation of plasma precursors in front of moving bodies and shock waves; the reduction of drag of bodies and the aerodynamic load on them; the reduction of heat flux to the surface of body moving at hypersonic speed, which can be used in aviation, bringing about increasing range of flight and spare fuel.

In investigations of plasma - shock wave interaction obtaining of the valid full set of plasma characteristic values and formulation of a reliable raw data for full-scale modeling seems to be the most important trend. On this basis our last experimental studies were directed on the determination of MW discharge plasma parameters. Also the kinetic and chemical processes in MW discharge and in afterglow, when the discharge region is drifted downstream to the bow shock wave of AD body are to be analyzed numerically to quantify the evolution of MW discharge region parameters. Not less important is to determine plasma parameters and gas composition in zone of discharge region – BSW interaction and to investigate the detailed vibrational kinetic of nitrogen and oxygen, as well as vibrational-translational relaxation of air molecules on oxygen atom at discharge and afterglow stages. These are the main objectives of the present investigation.

## 2. EXPERIMENTAL

Experimental installation was created for a given method realization. The installation consists of three main blocks: MW, gas-dynamic and diagnostic. MW generator forms pulses with peak power about 200 kW, pulse duration 1.3 to 2.25 μs, repetition frequency up to 1.5 kHz in X-range. The mean power of generator does not exceed 400W. The radio pulses pass through the wave-guide to the gas-dynamic chamber. By means of a horn radiator and parabolic reflector which is placed in the annular nozzle exit section and installed at the angle ~ $10^0$ to the flow axis, MW radiation of linear or circular polarization is focused on the axis of supersonic flow in the area between the working nozzle and the AD model, the wave vector of radiation being parallel to the flow axis. In the region of caustic MW field distribution, measured by means of the half-wave dipole along the central axis, has maximums. In cross-section which is perpendicular to the central axis the intensity distribution decreases monotony with increasing the distance from the axis. The width of the power distribution in each maximum in the direction along electric vector of the wave at the level of minus 3dB is about†15 to 18mm.

The gas-dynamic installation provides a stand operating from air reservoirs of high pressure; its working chamber being like the Eifel's one. The flow with a desired parameters is formed owing to an additional supply of gas into the area of the jet with reduced pressure by means of the pipeline 30mm diameter with the Mach number 1.5, passing through the main nozzle in its axis. The degree of the gas extension for the new central nozzle was established as being negligible, the cooling of gas should then be small, the static pressure be equal to pressure of the main annular jet executing a role of an ejector. The area of the central nozzle outlet section is about 15% of that of the annular geometrical nozzle. Thus, the gas□dynamic installation creates a uniform (within 3-4%) working flow of 2‰ mm diameter with Mach number M=1.45, static pressure 60 Torr and static temperature 200K. The principle scheme of installation is shown at Fig.1.

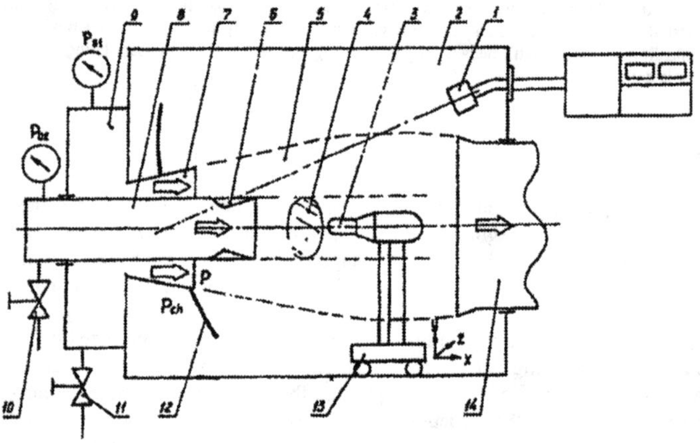

**Figure 1.**
Nomenclature. 1- Radiator; 2- Working chamber; 3- Studied model; 4- Plasmoid; 5- Jet; 6- Working nozzle; 7- Nozzle; 8- Fore-chamber of working flow; 9- Fore-chamber; 10- Locking valve; 11- Valve; 12- Parabolic mirror; 13- Coordinate mechanism; 14- Diffuser; 15- Wave carrier; 16- Microwave generator.

Diagnostic part of the installation includes: Schlieren-system, allowing to visualize the flow under investigation with separation of the needed phase of a process to the account of time gating; spectral system, allowing to obtain emission spectrums of plasma with the temporary resolution near 1 µs; pressure sensors, with the possibility to select a necessary phase of a process.

## 3. RESULTS

### 3.1. MW discharge in supersonic flow

As it was already noted above, exactly on this installation a free localized MW discharge in linear and circular polarized radiation beam in the supersonic air stream was obtained for the first time. The most of experiments were carried out under the peak MW power 180-190kW, pulse duration 1.6-2.2µs and repetition frequency 900Hz.

The discharge existed stable both in the free flow, and at the presence of aerodynamic models. In accordance with maximums in MW field distribution in the region of focusing there were observed two or three denominated areas of discharge light emission (Fig.2), their volume being several cubic centimeters. The nearest to the AD model plasmoid was located just before the front of the bow shock wave. As in the case of MW discharge in the steady-state gas [3], there were observed two main structures of discharge: in the manner of half-wave dipoles, encircled by the shining plasma, and in the manner of curved and branched channels with the base in the form of quarter-wavelength sinusoid. In the majority of experiments the first structure was realized, demonstrating a good spatial reproducibility that is characteristic of discharges with the small threshold excess. Coming from the values of a discharge time formation (measured by means of photo-electric multiplier and MW sensor) equal

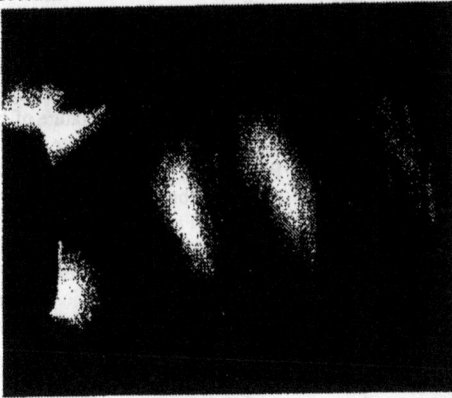

Figure 2.

to 0.2µs for peak MW power 180kW, the initial value of reduced field E/p turned out to be equal to 40V/(cm·Torr), where E is electric field strength of the wave, p is a normalized pressure. Relaxation time of the integral light emission of discharge after the completion of MW pulse comes to 5µs. Breakdown occurred under minimum peak power 120-130kW that corresponds to the value of reduced field in maximums of MW field distribution 32 - 34V/(cm·Torr). Measurements also have shown that the discharge absorbs not more than 30% of the led MW power. Thereby, the specific energy input in the discharge did not exceed several tenth J/(cm$^3$·atm).

## 3.2. Measurement of gas temperature by the optical method

Overview spectrum of discharge in the flow by the end of MW pulse was prescribed in the interval 3700 - 4100 A through 2Å (Fig.3). Each point (on the wavelength) was averaged on 50 realization. When getting the spectrums of MW discharge in the flow registration of light emission took place from the first (upper on the flow) plasmoid. Overview spectrums of air in the bulb under 15 Torr and 50 Torr were received in the same interval of wavelengths. In all overview spectrums sequences $\Delta v = -3$ (up to 4-7 in the flow and 3-6 in the bulb) and $\Delta v = -2$ (up to 2-4) are seen. In the flow transition 4-8 is also seen. In spectrums of air flow and of air in the bulb under 15Torr 0-0 transition of the first negative system of nitrogen ion is observed.

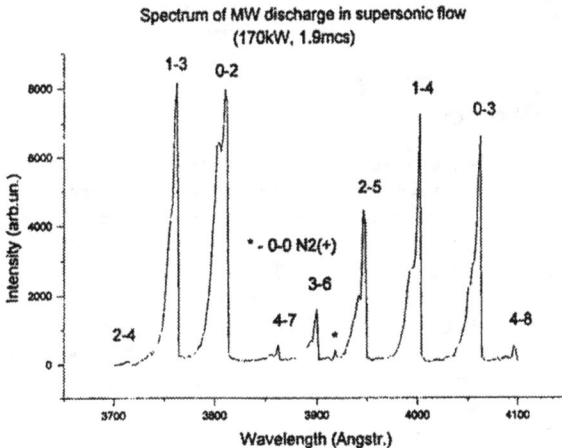

Figure 3. Spectrum of MW discharge in supersonic flow (170 kW, 1.9 mcs)

When getting the spectrums of band edgings, 200 points were fallen on the interval 50-70 A, herewith each point was also shown by the result of averaging on 50 realization (Fig.4). The rotational temperature of gas in the discharge in the flow by the end of MW pulse was defined using vibrational-rotational transitions 0-3, 1-4, 0-2, 0-0 of the second positive nitrogen system. For processing the spectrums the program was developed, allowing to define rotational temperature of gas on unresolved rotational structure of a spectrum. Processing the spectrums has confirmed that distribution over rotational levels is Boltzmann (Fig.5). Determination of rotational temperature was produced in the interval from 10 up to 20 rotational levels of R-branch. In this interval (with provision for low temperature of a gas) the contribution of P- and Q-branches is already insignificant and the inaccuracy in the determination of intensity is not yet too great.

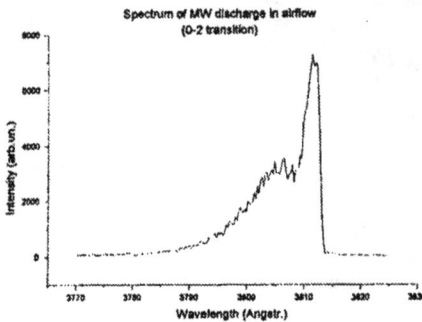

Figure 4. Spectrum of MW discharge in airflow (0-2 transitions)

At MW power radiation 180kW, duration of MW pulse 1.9µs the temperature of the gas turned up equal to 240±10K. Thereby, the maximum temperature raise did not exceed 50K for the pulse. When reducing the peak MW power to 145kW the temperature raise was halved: T =220±10K. We assumed the rotational and translational temperatures to be equal.

Temperature of the gas in the bulb under the same MW power, duration of MW pulse and repetition frequency of the pulses 900Hz and 20 Torr air pressure turned up near 540K. It should be noted that discharge was stratified and looked like a number of «corn-flakes». The mentioned above temperature refers to the one disposed near the bulb wall.

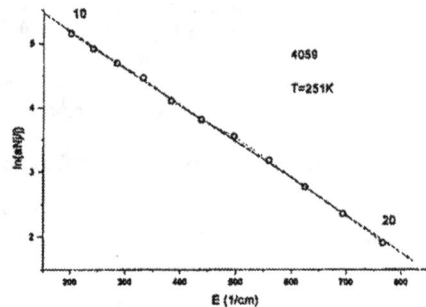

Figure 5. Distribution of rotational levels

## 3.3. Parameters of MW plasma in the discharge

Determination of vibrational temperature of the ground state of nitrogen molecule on relative intensities of edgings of a given sequence ($\Delta v$ =const) is referred to the class of inverse problems (generally speaking, incorrect). For the correct determination of vibrational temperature $T_V$ the necessary condition is an occupying an upper electronic state ($C^3\Pi_u$) by the direct electron impact from the main state ($X^1\Sigma_g^+$). Alternative to the direct electron impact channel of an occupying a $C^3\Pi_u$ state is the process

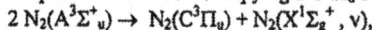

$$2 N_2(A^3\Sigma_u^+) \rightarrow N_2(C^3\Pi_u) + N_2(X^1\Sigma_g^+, v),$$

with the velocity constant $k^* \approx 7.7 \cdot 10^{-11} cm^3/s$ [4]. Comparison of the population fluxes of a $C^3\Pi_u$ level via the both channels under average concentration of electrons in the discharge $3 \cdot 10^{12} cm^{-3}$ and reduced electric field in the plasma near 110Td has shown that there is a basis to consider the population of $C^3\Pi_u$ level by direct electron impact to be the main mechanism.

Processing on the sequence $\Delta v = -3$ (transition 0-3, 1-4 and 2-5) enables to define the relative population of the level v=1 of the nitrogen ground state $X^1\Sigma_g^+$ (due to relatively low vibrational temperature of the ground state only the transitions from $X^1\Sigma_g^+(v = 0, 1)$ to $C^3\Pi_u(v = 0, 1, 2)$ were taken into account) and the corresponding vibrational «temperature», and electron temperature of vibrational levels excitation. Hereinafter on these values the electric field strength in plasma and the specific energy input in the discharge are (independently) calculated, with using these values the average concentration of electrons in the discharge is defined.

The following results were obtained: vibrational temperature $1500\pm200$K, electron temperature $2.1$-$2.3$ eV, specific energy input in the discharge $0.4$ J/(cm$^3$·atm), average concentration of electrons in the discharge $3\cdot10^{12}$cm$^{-3}$.

### 3.4. Study of MW plasma influence on gas-dynamic

By means of Schlieren-system with time gating aerodynamic spectrums of airflow with time step $2$ - $10\mu$s in the interval of delays comparatively MW pulse from 0 to $200\mu$s were obtained. Hemispherical model of 16mm diameter was used, which was installed in the region of the 20mm diameter even kernel of a flow. MW discharge was burned up in front of the model, the nearest to the model plasmoid being located at the distance 4 - 8mm. Experiment has newly demonstrated particularities in the picture of plasma flow around the model. As from $10$-$16\mu$s and up to $80$-$100\mu$s an unstable character of BSW behavior in front of the model was observed. At a moment of MW discharge plasma coming, a division of a BSW into layers occurs. In this process are consecutively observed the phases: twist of the front, its crushing, fluctuations and, finally, division into layers in the area of interaction. A fragment of BSW front is «displaced» and is disposed closer to the surface of the model, simultaneously another shock upstream the flow is formed. The pressed and the moved away shocks are fluctuating in the space. Under the greater delays, i.e. after MW discharge plasma drifting below the main BSW front, a recovering of the main shock and a picture of a flow occurs.

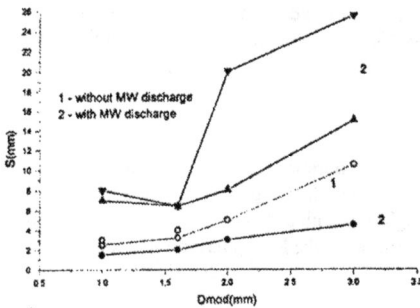

The described above patterns of plasma flow-body interaction are summarized in Fig.6, where the BSW stand-off in a flow without discharge and BSW fragments - the pressed and the moved away shocks in plasma flow are shown. The models are spheres of the appropriate diameter. It is well seen that the amplitude of fluctuation of the moved away shock position increases greatly when the model diameter becomes equal or exceeds the diameter of the working flow.

For the study of plasma influence on the drag of a body, sensors of stagnation and static pressure, working at the base of piezoelectric effect were used. Stagnation pressure sensor had time constant about $1\mu$s and sensitivity 3mV/Torr. System of time gating was used,

Figure 6. Bow shock wave stand-off

time step being $3.2\mu$s, besides every point was a result of equalization over a number of interaction realizations. Zero point refers to MW pulse termination. Investigations of stagnation pressure behavior during interaction with plasma at the leading edge of AD body under different MW conditions were conducted. Reduction of full pressure on the value $\Delta p \cong 30$Torr at time delay near $70\mu$s is fixed. The characteristic time of interaction of plasma flow with the body - near $150\mu$s - correlates well with Schlieren-system data. However, the experimental data obtained by means of the pressure sensor should be considered as preliminary, characterizing sooner qualitative picture of a flow near the model and the value of its drag changing.

The drag force measurements were made with the aid of the aerodynamic balance. AD-models in the form of spheres, blunt cones, hemisphere-cylinders and the special model (20mm hemisphere-cylinder covered by Teflon) were tested. The stable reduction of the drag force of these models due to the MW-plasma action was measured. Drag force reduction for 20mm sphere was 1.6%÷8.6%, and for the special model attained 5.2%±1.4%. A more stable result for the special model can be explained by the better conditions for plasma creation in front of the model.

## 4. NUMERIC MODELING

As it was shown in the previous section already the first experiments with MW discharge in supersonic flow have revealed some peculiarities in discharge region interaction with the AD body: the obtained Schlieren pictures of the AD body streamlining showed every time the unstable behavior of the bow shock wave while the discharge region was passing through the shock; the mean registered changes of the drag were up to 5% , i.e. the instant (during interaction of a discharge region with the shock) drag changes were at least 50%. But processing of the emission spectra gave the temperature rise in a plasma region to be not more than 50K – the value, which could not explain all the data set. The attempt to explain the evolutions of the BSW position by kinetic processes was made in [5] and was based on the important role of O-atom in the processes of VT-relaxation of air molecules. Some assumptions made in that work demanded more careful investigation of nitrogen-oxygen vibrational kinetics. The necessity of lightening of this question as well as to verify the emission spectrum processing were the main motivation of this part of investigation.

### 4.1. Kinetic scheme

The kinetic scheme for the dry air includes 106 charged and neutral components and about 20000 reactions. Among 106 components 25 are: $N_2$, $O_2$, $e$, $O^-$, $O_2^-$, $N_2^+$, $O_2^+$, $N_4^+$, $O_4^+$, $NO^+$, $N_2(A^3\Sigma_u^+)$, $N_2(B^3\Pi_g)$, $N_2(a'^1\Sigma_u^-)$, $N_2(C^3\Pi_u)$, $O_2(a^1\Delta_g)$, $O_2(b^1\Sigma_g^+)$, $N$, $O$, $O(^1D)$, $O(^1S)$, $O_3$, $NO$, $NO_2$, $NO_3$, $N_2O$. The rest are nitrogen and oxygen molecules which can be excited in all vibrational levels – 48 for nitrogen and 33 for oxygen. Electron-neutral particle processes rate constants are calculated on the base of energy distribution function of electrons which is determined by solving the Boltzmann equation. The detailed vibrational kinetics, including VV, VV' (vibrational quantum exchange between nitrogen and oxygen molecules) and VT processes are under consideration. Chemical reactions - both conventional and with participation of vibrationally excited molecules are in the list.

### 4.2. Results of simulation

Calculations were made for $p=100Torr$, ranges of MW pulse duration $\tau =1.6 - 1.9\mu s$ and reduced electric field $E/N=107-110Td$, and relaxation tracing $100\mu s$ and more. Initial gas temperature was taken 300K. According to the experiment [1], the MW discharge region was placed at a distance of about 10mm from the BSW, that is approximately 30μs of a downstream drift after the MW pulse termination.

In a few microseconds after the MW pulse all the excited electronic states of nitrogen are quenched by molecular oxygen with producing of the O-atoms and vibrationally excited $N_2$ molecules in a ground state. We do not consider the so-called fast heating here, as it is a special question of realistic kinetic chain for it and such a scheme is absent now-time. So, only the vibrational kinetics, as well as the kinetics of the main chemical components was of importance at the mentioned above time scale.

<u>4.2.1. Vibrational kinetics of nitrogen.</u> The results of calculation of the distribution function over vibrational levels of nitrogen at time points corresponding to the MW

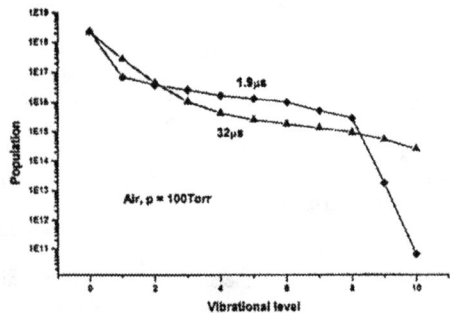

**Figure 7.** Evolution of distribution function over vibrational levels

pulse termination and the beginning of plasma region interaction with the BSW of AD body are shown at Fig.7. It is seen that at the end of MW pulse the VDF is highly non-equilibrium due to excitation of the first 8 vibrational levels of $N_2$ by direct electron impact. Then, VV-relaxation process leads to establishing of the Boltzmann distribution at the several first levels (including zero level). This evolution is demonstrated by Fig.8, where the relative populations (1-0 and 2-1) are presented. VDF at the lower vibrational levels establishes at the time point about $30\mu s$ at a level of $1700$-$1800K$. The total energy input in a gas is about 0.4J/cm$^3$·atm. The effect of VDF non-equilibrium at the end of a short MW pulses and moderate energy inputs should be taken into account while discharge spectra processing.

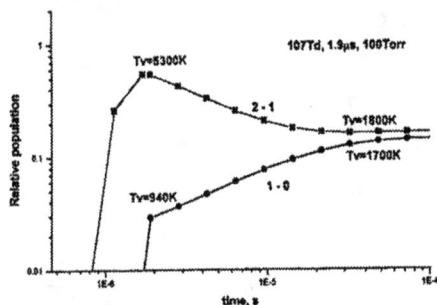

**Figure 8.** Vibrational temperature evaluated over different levels

At a time scale of order 10-100$\mu s$ the VT-relaxation of vibrationally excited $N_2$ on the O-atoms is not important – the gas heating due to this process is of order several degrees Kelvin, but at a longer times or at higher energy input this process may be of principal importance. Thus, consideration of a large AD bodies interaction with discharge plasma regions may demand this process to be taken into account.

### 4.2.2. Vibrational kinetics of oxygen.

Vibrational kinetics of oxygen partially resembles that of nitrogen, but is determined by sufficiently less effective excitation of vibrational levels by direct electron impact (of the first 4 levels) and is strongly influenced by effective VT-relaxation on the O-atom. VV' exchange with molecular nitrogen vibrational reservoir is too slow at such gas temperature and energy storage in $N_2$ vibrations to affect $O_2$ vibrational kinetics. As a result, the energy storage in $O_2$ vibrations is rather small, the VDF establishes practically over all the levels up to $32\mu s$ at a temperature only $560K$. The evolution of the VDF of oxygen molecule is shown at Fig.9.

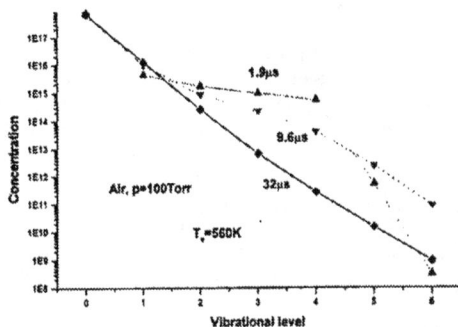

Figure 9. Evolution of distribution function over vibrational

### 4.2.3. Kinetics of the main components.

The kinetics of the main components in a discharge and afterglow is presented at Fig.10. Such a components are: oxygen and nitrogen atoms, ozone, nitric oxide and electrons. Oxygen atom concentration as well as nitrogen one are practically constant in a time range from several microseconds to several hundred microseconds. It is seen that oxygen atom concentration attains 2% of molecular oxygen concentration. It is sufficient for $O_2$ – O VT-relaxation, but not enough for $N_2$ – O VT-relaxation at the time scale under consideration. The process of NO formation takes place via interaction of vibrationally excited nitrogen molecule (with vibrational numbers higher than 15) with O-atom. That is why, the oxidation process goes in spite of low gas temperature, taking place hand by hand with VDF of nitrogen establishing. The rapid conversion of O-atom in the ozone molecule is observed at a time than $100\mu s$.

Electron kinetics is often proposed as very important for different aspects of the problem, but in our case to the moment of interaction with the BSW electron concentration in the discharge region falls to the value about $3 \cdot 10^{10} cm^{-3}$ and seems to be too low for any effect on the BSW.

### 4.3. Results

The detailed analysis of kinetics leads to the following discharge parameters in the region of interaction with the bow shock wave: electron concentration about $3 \cdot 10^{10} cm^{-3}$, concentration of oxygen atom of order $10^{16} cm^{-3}$, vibrational temperatures of nitrogen about 1800K and of oxygen near 600K, gas heating due to VT-relaxation is of order several degrees K

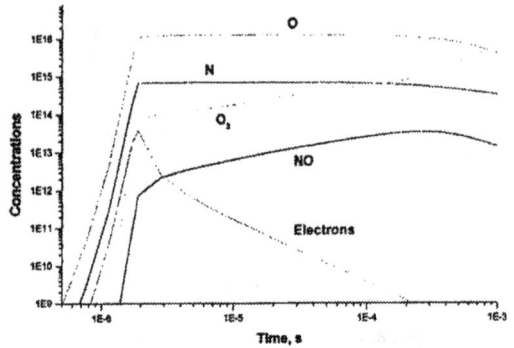

**Figure 10.** Time dependence of main components concentrations

Thus, another reasons for BSW instability and its modification should be proposed, as well as more accurate and detailed experiment is advisable.

## 5. CONCLUSIONS

As it was already noted above, the main aim a given investigation was determination of parameters of plasma of discharge. Self-consistent values of electron and vibrational «temperatures», translational temperature of the gas, average concentration of electrons in the discharge are obtained.

However we consider that the findings are insufficient. To our opinion the following information should be obtained: the distribution of gas and vibrational temperatures over the space occupying by the discharge; relaxation characteristics of electron concentration and vibrational reservoir at time scales tens-hundreds microseconds; also is important to measure oxygen atom concentration. Only after these investigations being carried out it will be possible to formulate a reliable raw data for full-scale gas dynamic modeling and making clear that aspects of interaction of shock wave with the plasma, which did not find hitherto rational explanation.

### Acknowledgments

Authors are grateful to European Office of Aerospace Research and Development (EOARD).

### References

1. Beaulieu W., Brovkin V., Goldberg I., Klimov A., Kolesnichenko Yu., Krylòv A., Lashkov V., Leonov S., Mashek I., Ryvkin M., Serov Yu. Proceedings of the 2nd Weakly Ionized Gases Workshop, AIAA, Norfolk, VA, April 24-25, 1998, pp.193-198.
2. Proceedings of Weakly Ionized Gases Workshop, USAF Academy, Colorado, 9-12 June, 1997.
3. Brovkin V.G., Kolesnichenko Yu.F. J. Moscow Phys. Soc., 1995, 5, 23.
4. Piper L.G. J.Chem.Phys., 1988, 88, 6911.

# ELECTRICAL AND SPECTROSCOPIC CHARACTERISATIONS OF A HELICAL RESONATOR FOR PLASMA : DEVELOPMENT OF A NEW PLASMA SOURCE.

C. Dupret, C. Foissac, P. Supiot, O. Dessaux and P. Goudmand

Laboratoire de Génie des Procédés d'Interactions Fluides Réactifs-Matériaux (GéPIFRéM)
UPRES-EA n°2698 – bâtiment C5 - Université des Sciences et Technologies de Lille – 59655
VILLENEUVE D'ASCQ cédex (France)

**Abstract.** This paper proposes a simple theoretical description of a helical cavity for plasma production. The fields analysis within the coupling device outside the helix is carried out by both electrical and magnetic probes. The plasma is studied by spatial resolved emission spectroscopy in order to evaluate the ion production and the gas temperature. A helical broad band coupling device running from 13.56 MHz to 2450 MHz is also presented.

## 1. INTRODUCTION

Usual microwave cavities [1,2] used as flowing plasma source under pressures ranging from 100 to 1000 Pa do not allow to use discharge tubes with very large volumes ($0.05 - 0.1 \ m^3$). This restriction limits the production of plasma with important volumes and high flow-rates. A key of this problem is the use of slow wave radio-frequencies (13.56 and 27 MHz) structures. These latters used for electronic tubes [3] (travelling wave tube), for telecommunication [4] (helical antenna) or for instrumentation [5,6,7] (electron spin resonance and measurements) have been widely described in literature [8,9]. One of them, the helix [3], simple to build and allowing decelerating the phase velocity, is the most interesting system for the conception of radio-frequencies cavities with important dimension. The helix is used to compress the wavelength. Coupled with high power generators (at 13.56MHz), it allows one to obtain important densities of energy with a coupling device whose plasma tube has an important diameter. Thanks to these techniques, we have realised one plasma source [10] working with a tube of 250 mm diameter at powers at least equal to 10 kW with a Voltage Standing Wave Ratio (VSWR) close to the unity, under a large pressure range.

In this paper, we describe a helical coupling device designed to allow one to characterise the accessible electromagnetic fields and to study the plasma within the helix by spectroscopy. The study is achieved in a flowing nitrogen plasma at 13.56 MHz. The plasma monitoring is performed by emission spectroscopy all along the coupling device length, and also in the afterglow region. The main emissions of $N_2$ and $N_2^+$ are used to study the concentration distributions and the relevant temperatures by mean of previously described methods [11,12]

## 2. THEORETICAL BACKGROUND

The helix is a slow-wave structure, slowing down the electromagnetic field velocity $c$, in free space, which implies a decrease of the wavelength $\lambda = c/f$ ($f$ : electromagnetic wave frequency). This wavelength compression allows one to realise resonators in RF and then to increase the energy density within the cavity. The fundamental helix parameters are actually defined thanks to the fictitious model of the sheath helix [3].
This description is then extrapolated to the real case of a tape helix to determine the electromagnetic fields. The tape helix shown in cylindrical co-ordinates ($r$, $\theta$, $z$) by Fig.1 was studied in details by S. Sensiper [13]. The corresponding parameters are related by the following equation :

$$cot \ \psi \approx \frac{2\pi a}{p}$$

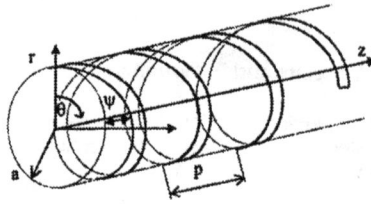

Figure 1. Tape helix geometry : (a) radius, (p) pitch distance and (ψ) pitch angle.

In a structure like the helix, when the wave is propagating along the $z$ axis, the periodicity theorem [14] implies the existence of a spectrum described by the propagation constants :

$$\beta_m = \beta + m \frac{2\pi}{p}$$

where $\beta$ is the propagation constant of the ground mode and $m \in 9$. As a results the fields on the helix can be expressed by linear superposition of all the harmonics. Namely, for the axial electric fields we have the following with $j = \sqrt{-1}$ :

$$E_z = \sum_{m=-\infty}^{\infty} E_m(r, \theta)\, e^{j\omega t - j\beta_m z}$$

$$H_z = \sum_{m=-\infty}^{\infty} H_m(r, \theta)\, e^{j\omega t - j\beta_m z}$$

For use of the helix as a RF plasma source, we wrap it with an external coaxial tube with internal radius, $b$, then realising the helical resonator. For this kind of coupling device, some non-slowed waves with phase velocities ($v_p > c$) can propagate [15,16], but the presence of the external tube allows one to increase the energy density and avoids radiation. In the helical cavity, one can then distinguish two main regions in the transverse plan. First, lets consider the inner part of the helix ($r < a$), where the plasma takes place. If the fused silica discharge tube is neglected, the permittivity in the region is given by :

$$\varepsilon_p = \varepsilon_0 \left(1 - \frac{\omega_p^2}{\omega(\omega - jv)}\right)$$

where $\omega_p$ is the plasma frequency, $v$ the electron-neutral collision frequency and $\varepsilon_0$ the vacuum permittivity. Second, we consider the region separating the helix and the external tube ($a < r \le b$). In that case, the permittivity is equal to $\varepsilon_0$ if we neglect the helix holders. The resolution of the Maxwell equations and the symmetry properties of the helix show that the dependence versus $r$ of the axial fields components is solution of a modified Bessel equation.

For the ground mode $m = 0$ [17], and for $a < r \le b$, we have :

$$E_z(r) = A_1 I_0(h_v r) + A_1' K_0(h_v r) \quad H_z(r) = B_1 I_0(h_v r) + B_1' K_0(h_v r)$$

with $h_v = \left[\beta^2 - \frac{\omega^2}{c^2}\right]^{1/2}$ the transverse wave number in the vacuum, $I_0$ and $K_0$ the zero-order modified Bessel functions of first and second species. For $r < a$, we have the following solution :

$$E_z(r) = A\, I_0(h_p r) \quad H_z(r) = B\, I_0(h_p r)$$

with $h_p = \left[ \beta^2 - \dfrac{\varepsilon_p}{\varepsilon_0} \dfrac{\omega^2}{c^2} \right]^{\frac{1}{2}}$ the transverse wave number in the plasma. Here, the integration constants of $K_0$ are zero, because of the limit condition $r = 0$, $K_0 \to \infty$. According to the Maxwell equations, the transverse fields derived from $E_z$ and $H_z$, are the following :

$$E_r = \frac{j\beta}{h^2} \frac{\partial E_z}{\partial r} \qquad\qquad E_\theta = -\frac{j\omega\mu_0}{h^2} \frac{\partial H_z}{\partial r} \qquad\qquad (1a)$$

$$H_r = \frac{j\beta}{h^2} \frac{\partial H_z}{\partial r} \qquad\qquad H_\theta = -\frac{j\omega\varepsilon_0}{h^2} \frac{\partial E_z}{\partial r} \qquad\qquad (1b)$$

With $h = \{h_v, h_p\}$ and $\mu_0$, the vacuum permeability. In the ground mode, without plasma, we have $\varepsilon_p = \varepsilon_0$ and then for the phase velocity :

$$v_p \approx c\, tg\, \psi \qquad\qquad (2)$$

The study of the helix is similar to that of a classical waveguide for determination of its propagation modes. As for a line, one can define one equivalent lineic inductance $L$ and a equivalent lineic capacitance $C$ [18]. The realisation of such a helical cavity consists, like for the coaxial cavity case, in using a line section. The experiment shows that for $\lambda/4$ coupling device [5,19], the matching conditions are obtained for a helix length :

$$l_H \approx (\frac{2p+1}{4}) \frac{v_p}{c} . \lambda = (\frac{2p+1}{4}) \lambda_H$$

the linear length of the helix being given by :

$$L_H \approx (\frac{2p+1}{4}) \lambda \qquad\qquad (p \text{ natural integral})$$

In a previous paper [20], it is shown that the equivalent circuit of the cavity is modified when a plasma in the helix : a self $L_p$ and a resistance $r_p$, simulating the plasma, are added. To obtain a good transfer of the radio frequency power it is necessary to adapt the impedance of set cavity-plasma ($Z_L = R_L + jX_L$) to the output impedance of the generator $Z_g$ which can be considered as a resistance ($R_g = 50\ \Omega$ ; $X_g \approx 0$).

Various adaptator circuits can be used [21]. The most classical are the gamma (a) and the reversed-gamma (b) circuits shown Figure 2, which require two variable capacitors ($C_1$, $C_2$) and a fixed inductance ($L_0$). Taking into account the equivalent cavity-plasma circuit [20] and the junction cable between the box and the coupling device, a calculation of the plasma and collisions frequencies in the discharge, respectively estimated to $\omega_p \approx 10^9$ s$^{-1}$ and $v \approx 10^8$ s$^{-1}$, leads to a $R_L$ value lower than 50 $\Omega$ . In these conditions, the gamma quadripole is required for a convenient adaptation of the coupling device impedance to the output resistance of the generator [21].

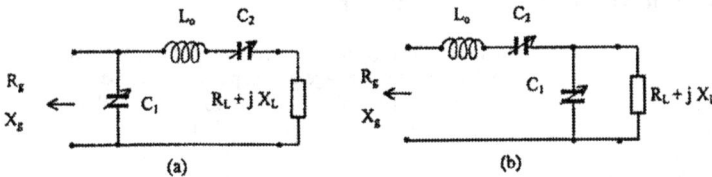

Figure 2. Matching circuits drawing :(a) Gamma quadripole (b) Reversed-gamma quadripole

On the basis of the above development, we have realised a 13.56 MHz plasma source, $\lambda_H/4$, used in a flowing gas to study the electric and magnetic field distribution in the region $a \leq r \leq b$ and also to determine the power distribution within the plasma by a spectroscopic analysis.

## 3. EXPERIMENTAL TECHNIQUES

### 3.1. Helical resonator

The $\lambda_H/4$ helical resonator realised for the study (Figure 3) is 400 mm long (internal dimensions) and is designed for use of a pyrex plasma tube (external diameter ① : 60 mm). The wire diameter used to realise the helix is equal to 5 mm, the turns number is 22 for a helix length of 220mm. The helical coil pitch angle is 2.8 degrees, the helix linear length is 4.5m (less than $\lambda/4$) for compensation of the cavity excitation mode. The inner diameter of the coaxial tube ② is equal to 90 mm. Both extremities of this tube are closed by short circuits which thickness is 10 mm ③. The connection with the radio-frequency generator is ensured by a type N tap ④. The cable used between the coupling device and the matching box is a dielectrical teflon KX 24 one with a 50 $\Omega$ characteristic impedance. The gamma quadripole circuit is settled in a 19 inches IRF box. The capacitors $C_1$ and $C_2$ are variable vacuum capacitors JENNINGS, which capacities are 1300pF and 500pF, respectively. The inductance $L_0$, equal to 2.5 nH, is realised with a 5 mm diameter copper wire. With these values, the matching box components allow to adapt loads with a resistance such as $1.5\Omega \leq R_L \leq 50$ $\Omega$ and a reactance $X_L \leq 170$ $\Omega$ at 13.56 MHz.

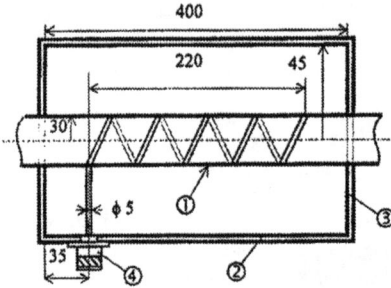

Figure 3. Helical resonator schematic        Figure 4. Schematic of plasma set-up

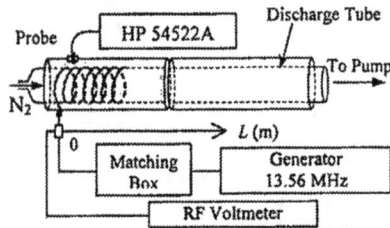

The characterisation of the parameters of the resonator previously described is carried out without plasma, with a network analyser HP 4195A (10 Hz, 500 MHz). The frequency resonance, the Q-factor, the equivalent inductance and capacity of the unloaded coupling device are as follows, respectively: $f_r = 16.74$ MHz, $Q \approx 500$, $L \approx 2.1$ $\mu H$, $C \approx 42$ pF. Measurements were also performed on the cavity loaded with a $N_2$ plasma. For all pressure and incident power conditions, the VSWR is close to unity thanks to the matching box.

### 3.2. Electromagnetic measurements

In order to study the fields and the plasma emissions within the coupling device, we have made a slit along the z axis on the external tube. This slit allows the passage of a capacitive antenna, a inductive loop and an optical fibre for measurements. A set of metal pieces allows to close up the slit where no measurement probe is found. This caution avoids electromagnetic radiation's from the slit and the resulting field perturbation within the cavity. The capacitive antenna takes a signal which is proportional to the local electric field whose lines are parallel to the probe, i.e. $E_r$. This antenna is loaded by a 50 $\Omega$ resistance. The detected signal amplitude measure is made with a numeric oscilloscope HP 54522A. The input signal measurement with a RF voltmeter allows the calibration of the antenna at the input position of the helix inside the cavity, since the dimensions are weak with respect to the fields wave length at 13.56 MHz.

The tension along the helix can also be measured. The inductive loop, designed to allow rotation at 180° around its axis, takes a signal which is proportional to the magnetic field amplitude and is function of its orientation with respect to this field. The resulting signal gets maximum when the loop plan is perpendicular to the magnetic field lines. The signal measurement, on a 50 Ω load, was also made with a HP 54522A oscilloscope.

### 3.3. Plasma conditions and spectroscopic device

The discharge is ignited in nitrogen (Air Liquide, "U" grade). The set-up is sketched by Figure 4. The nitrogen is injected in the pyrex discharge tube upstream the cavity and gas is evacuated by a rotary pump. The total gas pressure is measured by a Pirani PRL-10 Edwards gauge and fixed at 133 and 440 Pa. The flow rates are controlled by mass flow-meters RDM 280 Alphagaz. The spectroscopic tools used for recording of the spontaneous emission coming from the plasma is a HR 460-JY spectrometer equipped with a CCD detector cooled with liquid nitrogen. The emissions are collected with a collimated optical fibre which gives spatial resolution close to 0.005 m. The spectrometer is equipped with two gratings (2400/600 lines/mm) allowing to work from the UV (300nm) to the visible ranges (800nm), respectively.

## 4. EXPERIMENTAL RESULTS

### 4.1. Unloaded coupling device

The results of the measurement without plasma are given by Figure 5 for the radial component, $E_r$, of the electric field and the axial one, $H_z$, of the magnetic field. These measurements are done without load with an effective excitation tension of the coupling device equal to $V_g$=50 V and a current $I_g \approx 0$. The co-ordinates origin along the z axis is given by the N connector position ④ (Figure 3 and Figure 4). The rotation of the inductive loop shows a very weak $H_\theta$ component value. This latter is nevertheless not null because, according to Eq.(1a,1b),we have :

$$|H_\theta| = \frac{\varepsilon_0 \omega}{\beta} |E_r| \qquad\qquad E_r \neq 0.$$

The fact that $H_z$ is the dominant component is verified by the important slit radiation. The superficial currents on the inside part of the external tube are perpendicular to $H_z$. Then they cross the slit, which consequently radiates. Figure 5 shows that the helical cavity behaves as a non short-circuited open line section. The magnetic field cannot be maximum at $L = 0$ (Figure 4) because the current has a finite value at the entrance of the helix in the cavity. From the measurements, one deduces $L = 15$ $\mu H/m$, $C = 300$ $pF/m$ and a phase velocity of $v_p \approx 1.49 \times 10^7$ m/s. This results is in very good agreement with the relation (2) which gives $v_p \approx 1.47 \times 10^7$ m/s, confirming the description of the device as a slow-wave plasma source.

### 4.2. Discharge case

The outer electric field is now studied after plasma ignition for pressures of 133 Pa and 440 Pa with powers of 50 and 100W. The axial field distribution measured in the same way as in section 4.1. is shown by Figure 6. Here, the values are normalised according to their maximum ones. One remarks a global field distribution from $L$=0 to 0.15m similar to that of the unloaded cavity case. This indicates the dielectric behaviour of the gas in that region, where low electron densities are therefore expected. The maximum measured voltage positions ($L \sim 0.17$m) coincide for both the unloaded and loaded cases. The maximum amplitude increases with power what is logical, but also with pressure. This latter phenomenon is somewhat surprising, because the conductive behaviour should increase with pressure (higher electron densities at higher pressures). It could indicate a quite different electric field component radial distribution in this region at higher pressures. Beyond the maximum, we clearly see the effect of the plasma on the field in the cavity, allowing the propagation of the field downstream from the helix to the cavity's end. This effect is more pronounced at low pressure.

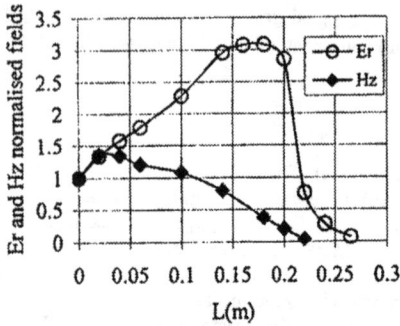

**Figure 5.** Axial distribution of the electric $E_r$ and $H_z$ field components measured by probes in the unloaded cavity. The signal is normalised according to its value at $L = 0$

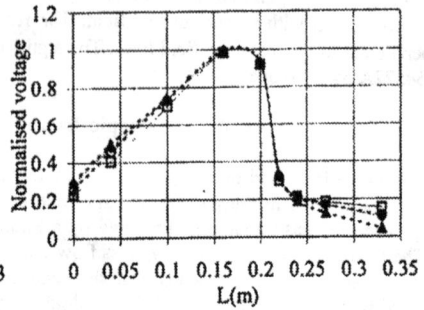

**Figure 6.** Axial distribution of the measured voltage related to $E_r$ component (133Pa : ( ♦ 50 W, $V_{max} = 355$ V) ;( ☐ 100 W,$V_{max} = 410$ V) ; 440 Pa : ( ▲ 50 W, $V_{max} = 440$ V) ; ( ○ 100 W, $V_{max} = 515$ V)

## 4.3. Spectroscopic results

The spontaneous plasma emissions are monitored at different wavelengths in the range 300-780 nm. The emissions spectra of the $N_2(B^3\Pi_g \rightarrow A^3\Sigma_u^+)$ ($1^+$), $N_2(C^3\Pi_u \rightarrow B^3\Pi_g)$ ($2^+$) and $N_2^+(B^2\Sigma_g^+ \rightarrow X^2\Sigma_u^+)$ ($1^-$) transitions are used to study the spatial homogeneity of the plasma. We have focused our attention on the axial profiles of the ion emission intensity and of the temperature.

### 4.3.1. Ion emissions intensity

The ion emissions intensity is monitored on the $N_2^+(B^2\Sigma_g^+,v'=0 \rightarrow X^2\Sigma_u^+,v''=0)$ transition and denoted $I(1^-)$, all along the flow from the position $L=0$ to 0.75 m (see Figure 4). The results are presented by the following figures : Figure 7 and 8 (133Pa, 50W and 100W, respec.) and Figure 9 and 10 (440Pa, 50W and 100W, respec.). The intensities are normalised according to their respective maximum values. It can be seen that the ion intensity profile strongly differs from the low pressure case to that of the higher pressure. Nevertheless, it is worth noting that the intensity maximum always occurs around the same position as that of the measured electric field component, $E_r$ ($L\sim0.2$m) (Figure 6) despite some shift of about 0.01-0.02m in some case. Additionally, the similarity ends to this point if we consider the strong discrepancy of the electric field distribution and the emission intensity profile in the region $0 \leq L(m) \leq 0.15$. It means that the plasma strongly influences the field configuration within the helix. The structure of the discharge is much more complicated probably due to a larger mean free path of the electrons, if we consider that the ion excited state is produced by electron impact.

One notices well the influence of the downstream end of the coupling device ($L=0.35$m) which leads to a stronger intensity decrease (not visible for 133Pa, 100W). The high pressure cases essentially show the same trend but with a more narrow distribution around the maximum. Also, the increase of the power makes the intensity distribution more homogeneous from the helix end to that of the cavity.

### 4.3.2. Rotational and vibrational temperatures

The gas temperature is one of the basic index of the power delivery in the discharge. The use of the rotational temperature , $T_r$, is often possible to evaluate the gas temperature in discharges and post-discharges in molecular gases [11,22,23,24]. This is achieved by implementation of the $N_2(C^3\Pi_u,v'=0 \rightarrow B^3\Pi_g,v''=2)$ and the $N_2(B^3\Pi_g,v'=2 \rightarrow A^3\Sigma_u^+,v''=0)$ transitions, which was shown suitable for gas temperature monitoring. The method of $T_r$ determination was already described in [11,12,24]. The Figures 7 and 8 show the resulting temperature profiles for the 133 Pa cases (50, 100W) while the 440 Pa cases (50, 100W) are illustrated by Figures 9 and 10, like for the ion emission. Common to all curves, the temperatures of both transitions are consistent and confirm again the used spectral method. They also follow the trend of both the ion emissions intensity, excepted for the high pressure case in the region $0 \leq L(m) \leq 0.15$. These results allow to conclude that the regions where both the ($1^-$) intensity and gas temperature profiles converge provide a real information on the power released and the discharge bulk. These conclusions are strengthened by the results obtained from the $N_2(C^3\Pi_u)$ vibrational temperature, denoted $T_v$ [12] (Figures 11 and 12).

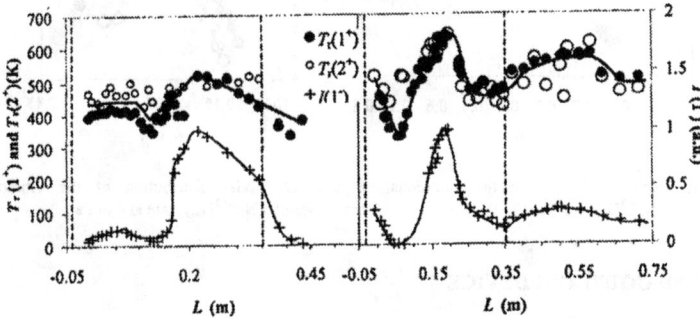

Figure 7. Axial distributions of I($1^-$) and of the rotational temperatures of $N_2(C^3\Pi_u)$ and $N_2(B^3\Pi_g)$ states (133 Pa, 50 W).

Figure 8. Axial distributions of I($1^-$) and of the rotational temperatures of $N_2(C^3\Pi_u)$ and $N_2(B^3\Pi_g)$ states (133 Pa, 100 W)

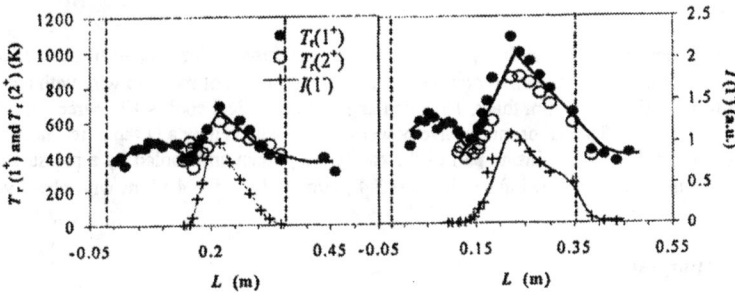

Figure 9. Axial distributions of I($1^-$) and of the rotational temperatures of $N_2(C^3\Pi_u)$ and $N_2(B^3\Pi_g)$ states (440 Pa, 50 W).

Figure 10. Axial distributions of I($1^-$) and of the rotational temperatures of $N_2(C^3\Pi_u)$ and $N_2(B^3\Pi_g)$ states (440 Pa, 100 W)

This also shows a good picture of the local electric field distribution which is completely different outside to inside the helix, especially in the region $0 \leq L(m) \leq 0.15$, where the ion emission intensity, i.e. the roughly plasma density and the power input are low. This result is consistent with a dielectrical behaviour of this region and explains well the small difference between the $E_r$ field distribution for the unloaded (Figure 5) and load coupling device (Figure 6). The absence of the ($2^+$) emissions at high pressure in this region confirms this description and allows to think that the rotational temperature maximum given by $N_2(B^3\Pi_g, v'=2)$ results from reactions between metastable electronically excited states of nitrogen instead of electron impact processes at low electron densities and energies. For the low pressure case, the electric field is allowed to diffuse downstream from the discharge (Figures 7, 8, 11) and leads to a second maximum of the ion density and of the rotational and vibrational temperatures at high power. The kinetics of such a plasma is not presently under consideration because of the difficulty to analyse the data only on the basis of emissions. This will be the subject of further works.

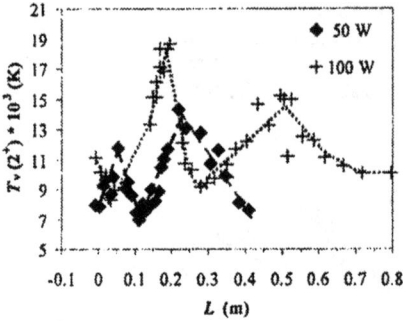

Figure 11. Axial distribution of the vibrational temperature of $N_2(C^3\Pi_u)$ state at 133 Pa

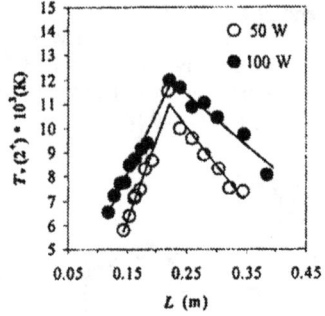

Figure 12. Axial distribution of the vibrational temperature of $N_2(C^3\Pi_u)$ state at 440 Pa

## 5. BROAD BAND COUPLING DEVICE

The study of the excitation frequency influence on the species present in the plasma must be done by varying at least parameters as possible. Matching facilities of the helix [4] [25] on a broad frequency band has motivated us to conceive a device running in a range as large as 13.56-2450 MHz.

### 5.1. Coupling device

This coupling device is studied with a plasma tube mode of fused silica ($\phi_{cx}$ = 30 mm). The internal cavity length is 218 mm and its diameter $\phi$ = 60 mm. The helix, made of metallic wire with a diameter of 3 mm, has 8 turns with step 20 mm thick. The pitch angle of the helical cool is 10 degrees. Its is fixed by mean of Teflon isolators. The RF or microwave power input is achieved by a N tap at its end. For durable characteristics, the helix and the inside part of the cavity are golden and cooled by a pressured air flow. The measurements give the following result : $L$ = 3.4 µH/m and $C$ = 104 pF/m with phase velocity of $v_p \approx 5.3 \ 10^7$ m/s.

### 5.2. Matching purposes

The major problem with this coupling device type is to transfer the electromagnetic energy delivered by various generators (13.56, 433, 2450 MHz) to it with reflected powers, $P_r$, as low as possible. The best matching system must be thought for each frequency band. The preliminary tests made with this cavity are carried out with nitrogen with a pressure of 500 Pa and a 2 L/mn flow rate. The results are presented for each specific transition.

#### 5.2.1. 13.56 MHz case

The use of the matching box detailed in section 3.1 allows a satisfactory matching for incident powers, $P_i$, varying from 50 W to 400 W. Therefore, we obtain a reflected power $P_r$ = 15 W for $P_i$ = 400 W.

### 5.2.2. 433 MHz case

We use the output matching circuit of the generator which is designed in a $\Gamma$ configuration. Its features do not allow a good matching, nevertheless a plasma can be obtained. A suitable $\Gamma$ circuit is under study to solve this problem.

### 5.2.3. 2450 MHz case

The 2450 MHz generators of our laboratory are equipped to carry the microwave energy by mean of rectangular waveguides in the $TE_{10}$ mode. Some guide to coax interface allows the matching of the entire device (transition-helicoidal cavity). The set-up is sketched by Figure 13. Matching is obtained by an assembly of 3 stubs spaced of $\lambda_g/8$, where $\lambda_g$ is the guide wavelength in the $TE_{10}$ rectangular mode. The guide to coax transition is realised with a transverse rod transition ended by a N tap out-put. The results given by this assembly for the standard discharge conditions are very encouraging. We have tested the set-up to incident powers up to 400 W with a $P_r$ value less than 100 W. Adjunction of additional adjusting elements on the transition will allow to optimise the matching to lower $P_r$.

From the above presented results, we prove the feasibility of such a broad band helical coupling device. Once in the final form, the device should allow to study different plasmas obtained just by varying the frequency.

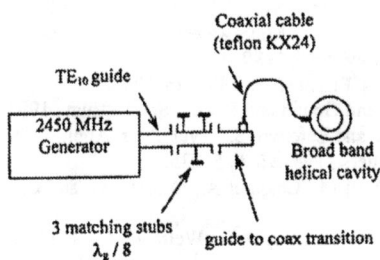

Figure 13. Schematic of the 2450 MHz matching system for the broad band cavity

## 6. CONCLUSION

This paper describes the theoretical and experimental aspects of the study of a helical coupling device. It is the first complete approach of the electric field and power distribution mentioned in the scope of low medium pressures (100-500 Pa) flowing gas discharge with this type of coupling device. The electromagnetic measurements carried out outside the helix show an electric field amplitude maximum around the end of the helix. When filled with a plasma, the cavity essentially shows the same field distribution, with some features downstream the helix due to the local high electron densities. The use of spontaneous emission provides information through $N_2^+(B^2\Sigma_u^+)$ ion emission intensity and rotational and vibrational temperatures measurements about the power locally delivered. A good correlation is obtained by the two types of measurements which shows well under present conditions the non homogeneous feature of the plasma related to the effect of the electric field maximum at the end of the helix.

On the basis of this cavity type, a broad band cavity has been designed and tested from 13.56 to 2450MHz with the same device. One matching system has been studied and applied, allowing satisfactory to very good power transfers.

### References

1. Zander A.T. and Hieftje G.M. Appl. Spectrosc., 1981, **35**, 357.
2. Bluem E. Thesis Paris-Sud University, 1995, 3503.

3. Pierce J.R. Traveling Wave Tube, D. Van Nostrand, Princeton 1950.
4. Kraus J.D. Antennas, Mc Graw-Hill, New York 1950.
5. Diehl R., Wheatley D.M. and Castner T.G. Rev. Sci. Instrum., 1996, 67, 11, 3904.
6. Webb R.H. Rev. Sci. Instrum., 1962, 33, 7, 732.
7. Deri R.J. Rev. Sci. Instrum., 1986, 57, 1, 82.
8. Ramo S. and Whumery J.R. Fields and Waves in Modern Radio, $2^{nd}$ ed. J. Wiley & Sons, New York 1953.
9. Brillouin L, Wave propagation in Periodic Structures, 2n ed. Dover Publications, New York 1953.
10. Dupret C., Mutel B., Dessaux O. and Goudmand P. FP n°95 06301, 1995.
11. Blois D., Supiot P., Barj M., Chapput A., Foissac C., Dessaux O. and Goudmand P. J. Phys. D : Applied Phys., 1998, 31, 2521.
12. Quensierre J D., Dupret C., Supiot P., Dessaux O. and Goudmand P. Plasma Source Sci. Technol., 1998, 7, 491.
13. Sensiper S. Proc. IRE., 1955, 43, 149.
14. Bevensee R.M. Electromagnetic Slow Wave Systems, J. Wiley & Sons, New York 1964
15. Uhm A.S. and Choe J.Y. J. Appl. Phys., 1982, 53, 8483.
16. Ganguli A. and Appala Naidu P. J. Appl. Phys., 1990, 68, 3679.
17. Niaz J.K., Lichtenberg A.J. and Lieberman M.A. IEEE Trans. Plasma Sci., 1995, 23, 833.
18. Tsui J.B.Y. and Krueger C.H. Handbook of Microwave and Optical components, vol.1, Edited by Chang K., J. Wiley Interscience, New York 1989
19. Park J.C. and Kang B. IEEE Trans. Plasma Sci., 1997, 25, 1398.
20. Dupret C., Supiot P., Dessaux O. and Goudmand P. Rev. Sci. Instrum., 1994, 65, 11, 3439.
21. Doughan M., Hayet J.P., David J. and Lefeuvre S. Revue Phys. Appl., 1988, 23, 1243.
22. Plain A. and Ricard A. Physics letters., 1983, 95 A, 5, 235.
23. Blois D., Foissac C., Supiot P., Barj M., Chapput A., Dessaux O. and Goudmand P. C.R. Acad.Sci. Paris, 1998, t 326, Série II b, 441.
24. Foissac C., Campargue A., Kachanov A., Supiot P., Weirauch G., and Sadeghi N., J. Phys. D: Appl. Phys., 2000 at press
25. Dubost G. and Zisler S. Antennes a large bande, Masson, Paris 1976

# SLOT-ANTENNA SURFACE WAVE EXCITATION IN INHOMOGENEOUS PLANAR PLASMAS WITH A LOCAL PLASMA RESONANCE

I. Ghanashev, K. Mizuno, E. Abdel-Fattah, Y. Nakai, H. Sugai

Department of Electrical Engineering, Nagoya University
Furo-cho, Chikusa-ku, Nagoya 464-8603, Japan

Abstract. In the present contribution we present an extension of the multi-mode excitation model for slot-antenna excited planar surface wave plasmas [Ghanashev et al. *Phys. of Plasmas* 7 (2000) 3051] to the case of inhomogeneous plasma with underdense and overdense regions separated by a resonance surface of local electron plasma oscillation. The presence of this resonance surface leads to broadening and overlapping of the individual eigenmode resonance curves, smoother plasma density dependence of the power reflection coefficient and fewer density jumps in the power—density curve. We present also experimental results confirming the occurrence of a local electron plasma oscillation resonance, as already observed earlier in test-wave experiments [Ghanashev et al. *Plasma Sources Sci. Technol.* 8 (1999) 363]. Plasma parameter measurements at the resonance position confirm that the local electron density is equal to the cut-off plasma density.

## 1. INTRODUCTION

Planar surface-wave (SW) plasma sources, as reviewed in [1], have been found to have very good performance for dry processing in the semiconductor industry. Much effort is being made to improve the microwave power coupling to the plasma. This is of crucial importance for the operation of the planar SW plasma sources, because of the jumps of the electron density $n_e$ and the related hysteresis phenomena [2]. The latter were shown [2] to be caused by the resonance minima in the $n_e$ dependence of the microwave power reflection coefficient R. The widely used free-oscillation analysis from the closed-cavity model [3] can predict successfully the resonance densities at which the resonance minima will occur. However, since the coupling geometry is not included in the aforementioned model, it cannot estimate the width or depth of those minima. Therefore, the coupling optimization is still performed experimentally on a cut-and-try basis. Recently [4] we proposed a multi-mode excitation model for slot-antenna excited planar SW plasmas of uniform density, which takes into account the actual coupling geometry. It reproduces correctly the experimentally observed phenomena and gives a theoretical basis for computer-aided optimization. In the present contribution we consider inhomogeneous plasma with underdense and overdense regions separated by a local electron plasma oscillation resonance surface. The latter leads to broadening and overlapping of the individual eigenmode resonance curves, smoother R vs. $n_e$ dependence and fewer density jumps. We present experimental results confirming the occurrence of a local electron plasma oscillation resonance, as already observed earlier in test-wave experiments [5]. Plasma parameter measurements at the resonance position confirm that the local electron density is equal to the cut-off plasma density.

The paper is organized as follows. In the next section we present the experimental results about the local plasma resonance. Then in section 3 we outline the theoretical model of SW plasma excitation by slot antennas. In section 4 we present numerical results on the influence of the axial density profile on the SW excitation by annular

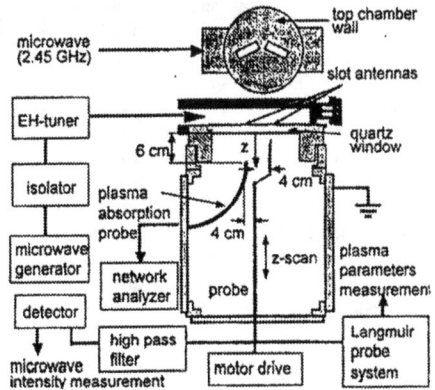

Figure 1. Schematic diagram of the experimental setup for observing the local plasma resonance.

slot antenna in axially symmetric planar SW plasma of finite radius $R$. Finally, section 5 gives a summary and a conclusion.

## 2. EXPERIMENTAL OBSERVATION OF LOCAL OSCILLATIONS AT THE ELECTRON PLASMA FREQUENCY

The local electron plasma frequency resonance ("local resonance" below) is a resonance peak of the electric field of an electromagnetic wave propagating in non-magnetized plasma. It occurs in inhomogeneous plasmas along a surface where the local electron plasma frequency $\omega_p(\mathbf{r}) = e[n_e(\mathbf{r})\varepsilon_0 m_e]^{1/2}$ is equal to the wave frequency $\omega$ (or, in other words, where the local electron density $n_e$ ($\mathbf{r}$) is equal to the cut-off plasma density $n_c = \varepsilon_0 m_e \omega^2/e^2$) [6]. It is frequently held responsible for various phenomena in surface-wave (SW) sustained plasmas, including a turn in the SW phase curve [7] and, most important, non-collisional electron heating at low pressures [8,9]. However, observation of the resonance phenomenon itself is poorly documented in the literature. The local resonance has been identified when a weak, non-ionizing electromagnetic test-wave propagates in an inhomogeneous plasma sustained by some other ionization source [5, 10]. This is a relatively simple experiment, because the test-wave frequency $\omega$ (i.e. the cutoff plasma density $n_c = \varepsilon_0 m_e \omega^2/e^2$) and the density profile $n_e(\mathbf{r})$ can be controlled independently. Thus, one can ensure that the resonance surface $n_e(\mathbf{r}) = n_c$ be located far enough from the vessel walls in order to observe the resonance experimentally. However, experiments in SW plasmas, which are sustained by the electromagnetic wave itself, lack this flexibility. In SW plasmas the bulk plasma densities are usually much higher than $n_c$, so that a resonance peak, if present, would be too close to the wall to be identified within the available space resolution. Therefore there are very few experimental observations of the local resonance in SW plasmas. They were performed either in plasma columns sustained in dielectric tubes [11,12] or around a Goubau line [13], but none of them addresses the case of planar SW plasmas. In the aforementioned measurements, a peak of the microwave intensity at some position in the plasma was identified. However, the electron density was not measured at the resonance position, so that the phenomenon's identification as a resonance caused by local excitation of electron plasma oscillations at $n_e(\mathbf{r}) = n_c$ was based only on indirect considerations.

In this section we present local resonance measurements in a planar SW plasma (the type of SW plasma

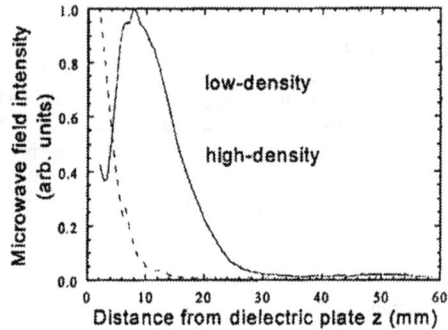

**Figure 2.** Axial profile of the microwave intensity in the plasma as picked-up by the movable probe for the low- and high-density cases (solid and dashed lines, respectively, parameters given in the text).

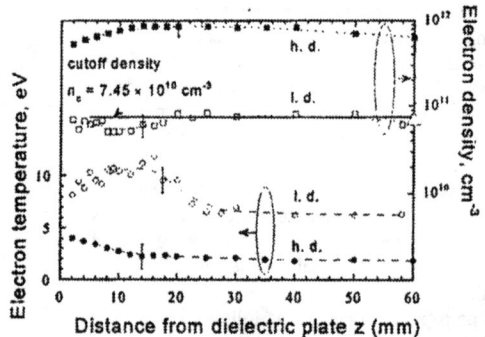

**Figure 3.** Electron density (squares) and temperature (circles) measured by Langmuir probe for the low- and high-density cases (open and full symbols, denoted as l.d. and h.d., respectively, parameters given in the text). The open triangle at $z = 6$ cm indicates the electron density for the low-density case as measured independently by the plasma absorbtion probe [16].

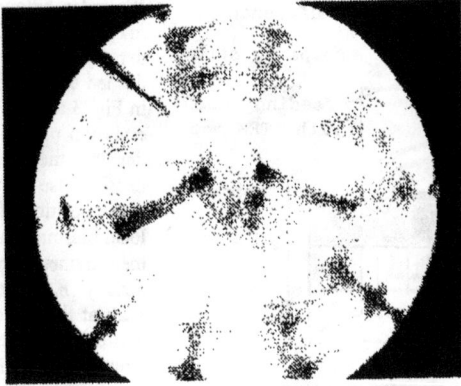

Figure 4. Light emission pattern for the high-density case.

most convenient for dry processing in the silicon industry) accompanied by space-resolved electron density measurements. Thus we were able to confirm that, when a resonance is observed, the electron density at its position is indeed equal to the cutoff density $n_c$.

The planar surface-wave plasma apparatus used in the present experiment is shown in Fig. 1. The upper part of the plasma chamber [14, 15] has a diameter $2R = 22$ cm and is sealed by a quartz plate with a thickness $d = 17$ mm. The microwave ($\omega/2\pi = 2.45$ GHz) is fed through a rectangular wave-guide and enters the chamber through a pair of inclined slot antennas. Space resolved microwave intensity and plasma parameters measurements were performed using an axially movable probe made of a $\varnothing$ 0.6 mm tungsten wire inserted into an insulating ceramic and a shielding stainless steel tubes. The 5 mm long wire tip extending out of those tubes was used as a pick-up antenna to measure the microwave intensity and as a Langmuir probe. It was positioned 4 cm off-axis at an azimuthal angle ensuring maximal microwave signal. Although the probe was directed along $z$, its short length (necessary to ensure reasonable space resolution) results in poor separation of the three electric filed components, because the antenna picks them all up (probably with different sensitivity). Therefore we provide data about microwave signal intensity in the probe (detected after filtering out the DC plasma potential) in relative units, without specifying which field component is meant. In order to be sure that the Langmuir probe measurement is not critically affected by the microwave field, where possible (at lower plasma densities up to $n_e \sim 6 \times 10^{11} \text{cm}^{-3}$), the electron density was independently measured by a non-movable plasma absorption probe (PAP) [16]. Its tip was located 4 cm off-axis at 6 cm from the dielectric plate. All measurements were performed in argon.

Changing the microwave power and/or gas pressure we measured the axial microwave intensity profiles. We were deliberately looking for low bulk plasma density conditions when the microwave intensity had a clear peak [supposedly a local resonance at $n_e(r) = n_c$] located a few centimeters away from the upper dielectric wall (the solid curve in Fig. 2). This happened at gas pressure $p = 12$ mTorr and absorbed microwave power $P_{abs} = 0.3$ kW, and we refer below to this case as the "low-density case". The width of the observed signal peak is determined by the probe space resolution and should not be interpreted as indicating the actual width of the local resonance, which is supposed to be much narrower (less than 1 mm). Then we measured the plasma parameters (the open symbols in Fig. 3) and took a photograph of the light emission pattern. At such conditions we observed more or less uniform plasma emission from the whole plasma cross-section without any special light emission pattern, so that we do not present here the photograph. For comparison we repeated the same measurements increasing the gas pressure and the microwave power until we found conditions when the microwave intensity was decreasing exponentially starting directly (within the available experimental space resolution of a couple

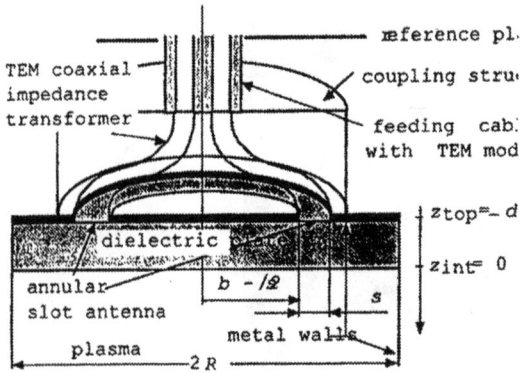

Figure 5. Model of circular planar surface-wave plasma with annular slot antenna excitation.

of millimeters) from the dielectric wall. We refer below to this case ($p = 32$ mTorr, $P_{abs} = 0.4$ kW) as the "high-density case" and it is shown by the dashed curve in Fig. 2, the full symbols in Fig. 3 and the photograph of the light emission pattern in Fig. 4. The plasma density measurements for the low-density case from Fig. 3 confirmed that at the position of the local resonance the local plasma density is equal (within the measurement accuracy) to the cut-off density $n_c$. The electron temperature peak at the same position suggests that non-collisional electron heating by the strong axial microwave electric field may be occurring there.

For the high-density case $n_e(r) \gg n_c$ even very close to the wall, which is consistent with the absence of electric-field-intensity and electron-temperature ($T_e$) peaks in this case. This state has lower $T_e$. The microwave intensity decays exponentially along $z$ with a penetration depth $\gamma_p^{-1} \approx 5.3$mm, so that the transverse wave number $\kappa = j_{mn}/R = (\gamma_p^2 + \varepsilon_p \omega^2/c^2)^{1/2}$ corresponds [3] to that of the $TM_{530}$ pure surface mode: $\kappa R = 15.8 \approx j_{53} = 15.700$ ($j_{mn}$ is the $n$-th root of the $m$-th Bessel function), which is consistent with the light emission pattern from Fig. 4.

The good agreement between measured and expected ($n_c$) local plasma density at the position of the local resonance, the electron temperature peak at the same location and the fact that both peaks disappear when the bulk plasma density is increased permit us to conclude that the plasma density profile observed in relatively low-density SW plasma is attributed to the expected [5, 6] local excitation of electron plasma oscillations. It is reasonable to expect that the same phenomenon should happen (and eventually contribute to the plasma heating) also at the higher plasma densities usually applied for plasma processing, although the resonance peak should not be directly observable due to its location extremely close to the dielectric wall.

## 3. SURFACE WAVE EXCITATION BY SLOT ANTENNAS

The eigenmode theory of finite-area planar SW plasmas in free-oscillation, closed-cavity approximation [3] predicts resonance excitation of standing-wave eigenmodes at some discrete set $n_{eig}^{<n>}$ of plasma densities, called below "eigenmode resonance densities". The density jumps observed when changing the gas pressure [14, 15] or incident microwave power $P_{in}$ [2] and the hysteresis loops in the power-density dependence [2] were successfully explained [2] as eigenmode jumps caused by sharp minima of the reflected power $P_{refl}$ and of the related power reflection coefficient $R \equiv P_{refl} / P_{in}$ near the eigenmode resonance densities $n_{eig}^{<n>}$. This behavior was confirmed by measuring the $R$ — $n_e$ dependence [2,17].

Although explaining many phenomena observed in the planar SW plasma sources, the free-oscillation, closed-cavity approximation [3] has limited applicability. The coupling apertures must be small, the wave attenuation must be low, and the dense part of the eigenmode density spectrum $n_{eig}^{<n>}$ must be avoided. The aforementioned approach cannot be used to study the influence of the antenna form and position on the SW excitation and overall SW plasma source efficiency, because the presence of the coupling elements is entirely neglected.

These shortcomings can be fixed by the model proposed in [4]. It is based on a multi-mode expansion of forced oscillations. By taking into account the actual coupling aperture geometry, it provides a theoretical basis for optimizing the SW plasma coupling. It is briefly outlined in section 3.1 below. Initially [4,18,19] it was applied for the simplest case of homogeneous plasma. Yet the actual plasma is inhomogeneous and a local plasma resonance may occur near the wall. This should influence significantly the eigenmode fields and eventually the overall plasma source performance. Section 3.2 below shows how to include in the model the axial plasma inhomogeneity.

The present approach is not limited to any specific form of the plasma chamber cross-section or slot antenna. The simplest cases arise for structures with high symmetry, e.g. circular chamber with annular slot antenna [4, 18] (experimentally realized in [20]) or two-dimensional plasma slab with an infinitely long linear slot antenna [19, 21]. Below for simplicity we choose the former geometry.

### 3.1. Homogenous plasma

We consider forced electromagnetic oscillations of frequency $\omega/2\pi$ in a plasma chamber of cross-section $S_\perp$, extending from $z = z_{top} \equiv -d$ to $z = +\infty$, with a metal wall and a dielectric plate of permittivity $\varepsilon_d$ and thickness $d$ at the top. The lower part of the chamber contains cold homogeneous plasma of permittivity $\varepsilon_p = 1 - (n_e/n_c)(1 + i\nu/\omega)^{-1}$, where $\nu$ is the electron-neutral collision frequency for momentum transfer (the case of inhomogeneous plasma is discussed in section 3.2 below). In the top chamber wall one or several coupling apertures occupy area $\Delta S$ and provide the energy coupling. The chamber cross-section and antenna shapes may be of any shape. As an example below we consider a cylindrical chamber with an annular slot antenna (Fig. 5) with TM or TEM excitation. This geometry results in relatively simple analytical expressions and has been already experimentally realized in [20]. The transverse (in respect to the chamber axis) microwave electric field in the coupling apertures $\mathbf{E}^{(ex)}_\perp(x,y)$ is externally set and approximated as.

$$\mathbf{E}^{(ex)}_\perp(x,y) = \hat{\rho} E_0 b/\rho. \tag{1}$$

Along the metal part of the top wall the transverse electric field $\mathbf{E}_\perp$ must vanish, and one can write the boundary condition at $z = z_{top}$ as

$$\mathbf{E}_\perp(x,y,z = z_{top}) = \begin{cases} \mathbf{E}^{(ex)}_\perp(x,y) & \text{for } (x,y) \in \Delta S \\ 0 & \text{for } (x,y) \notin \Delta S \end{cases} \tag{2}$$

The microwave field is written as series of eigenmodes

$$\mathbf{E}_\perp(x,y,z) = \sum_n A^{(n)} E^{(n)}_\perp(z) \mathbf{e}^{(n)}_\perp(x,y) \quad Z_0 \mathbf{H}_\perp(x,y,z) = \sum_n A^{(n)} H^{(n)}_\perp(z) \mathbf{h}^{(n)}_\perp(x,y) \tag{3}$$

$$E_z(x,y,z) = \sum_n A^{(n)} E^{(n)}_z(z) e^{(n)}_z(x,y) \quad Z_0 H_z(x,y,z) = \sum_n A^{(n)} H^{(n)}_z(z) h^{(n)}_z(x,y) \tag{4}$$

of known axial ($E_z^{(n)}$, $E_\perp^{(n)}$, $H_z^{(n)}$, $H_\perp^{(n)}$) and transverse ($e_z^{(n)}$, $e_\perp^{(n)}$, $h_z^{(n)}$, $h_\perp^{(n)}$) eigenmode functions with unknown eigenmode amplitudes $A^{(n)}$. The latter can be computed from the boundary condition (2)

$$A^{(n)} = (1/S_\perp) \int_{\Delta S} \left(\mathbf{E}^{(ex)}_\perp(x,y)\right)^* \cdot \mathbf{e}^{(n)}_\perp(x,y) dS_\perp. \tag{5}$$

This expression assumes that the axial and transverse eigenfunctions are normalized by

$$\int_{\Delta S} \left(\mathbf{e}^{(n_1)}_\perp(x,y)\right)^* \cdot \mathbf{e}^{(n_2)}_\perp(x,y) dS_\perp = \delta_{n_1 n_2}, \quad E^{(n)}_\perp(z_{top}) = 1.$$

In (3) and (4) $Z_0$ is the wave impedance of free space.

The explicit expressions for the axial eigenfunctions do not depend on the cross-section and antenna geometry. For TM modes they are

$$\frac{H^{(n)}_\perp(z)}{i k_0 \varepsilon(z)} = \begin{cases} \left(\gamma_d^{(n)} D^{(n)}\right)^{-1} \left[\left(\gamma_p^{(n)}/\varepsilon_p\right) \sinh\left(\gamma_d^{(n)} z\right) + \left(\gamma_d^{(n)}/\varepsilon_d\right) \cosh\left(\gamma_d^{(n)} z\right)\right] & \text{for } z < 0, \\ \left(\varepsilon_p D^{(n)}\right)^{-1} \exp\left(\gamma_p^{(n)} z\right) & \text{for } z > 0, \end{cases} \tag{6}$$

$$E_{\perp}^{(n)} = (1/ik_0\varepsilon)dH_{\perp}^{(n)}/dz, \quad E_z^{(n)} = (\kappa^{(n)}/ik_0\varepsilon)H_{\perp}^{(n)}, \quad H_z^{(n)} = 0, \tag{7}$$

where

$$\gamma_{p,d}^{(n)} = -\left[(\kappa^{(n)})^2 - \varepsilon_{p,d}k_0^2\right]^{1/2}, \quad D^{(n)} = (\gamma_p^{(n)}/\varepsilon_p)\cosh(\gamma_d^{(n)}d) - (\gamma_d^{(n)}/\varepsilon_d)\sinh(\gamma_d^{(n)}d),$$

and $\kappa^{<n>}$ is the transverse wavenumber of the $n$-th eigenmode. Because of the structure and excitation field symmetry only TM$_{0n}$ modes are excited. Their transverse wavenumbers and eigenfunctions are

$$e_{\perp}^{(n)} = -\hat{z} \times h_{\perp}^{(n)} = \hat{\rho}J_1(\kappa^{(n)}\rho)/J_1(\kappa^{(n)}R), \quad e_z^{(n)} = -J_0(\kappa^{(n)}\rho)/J_1(\kappa^{(n)}R), \quad \kappa^{(n)} = j_{0n}/R, \quad n = 1,2,\ldots. \tag{8}$$

Thus all field components, as well as various derived characteristics, including the power reflection coefficient R, can be computed in terms of the externally set excitation field intensity $E_0$.

In order to compute the power—density dependence it was assumed [2,22] that the electron density is proportional to the absorbed power $P_{abs} = P_{in} - P_{refl} = P_{in}(1-R)$

$$\frac{n_e}{n_c} = \frac{P_{in}}{P_c}[1-R(n_e)], \quad \text{or} \quad P_{in} = P_c\frac{n_e}{n_c}[1-R(n_e)]^{-1}, \tag{9}$$

where $P_c$ is the absorbed power necessary to sustain electron density equal to the cut-off plasma density $n_c$. Thus, once the dependence $R(n_e)$ is computed, one can compute the power–density dependence.

### 3.2. Axially inhomogeneous plasma

If the plasma is axially inhomogeneous with permittivity $\varepsilon(z)$ only the expressions for the axial eigenfunctions (6) must be changed. Closed analytical expressions cannot be derived. The axial eigenfunctions are found by solving numerically the differential equation (7a) coupled with

$$H_{\perp}^{(n)} = \left[ik_0\varepsilon/(\gamma_p^{(n)})^2\right]dE_{\perp}^{(n)}/dz. \tag{10}$$

All other equations remain unchanged, but one should consider everywhere $\varepsilon$ and $\gamma$ not as constants, but as functions of $z$.

### 4. NUMERICAL RESULTS OF SW EXCITATION MODELING

We computed the electromagnetic field distributions and several derived quantities for a cylindrical chamber of diameter $2R = 31.2$ cm, dielectric plate with permittivity $\varepsilon_d = 4$ (quartz) and thickness $d = 15$ mm, annular slot antenna diameter $2b = 10.5$ cm and width $s = 5$ mm at a wave frequency $\omega/2\pi = 2.45$ GHz, which corresponds to the experiment reported in [20]. The collision frequency was set to $0.02\omega$ and the density profile was modeled as

$$n_e(z) = \begin{cases} n_0\left[1-(z/\Lambda-1)^2\right] & \text{for} \quad z < \Lambda, \\ n_0 & \text{for} \quad z > \Lambda. \end{cases} \tag{11}$$

Here we present the results for the density dependence of the power reflection coefficient (Fig. 6) and for the power—density dependence (Fig. 7) for two values (2 and 20 mm) of the density profile scale-length $\Lambda$, compared with results for homogeneous plasma ($\Lambda = 0$).

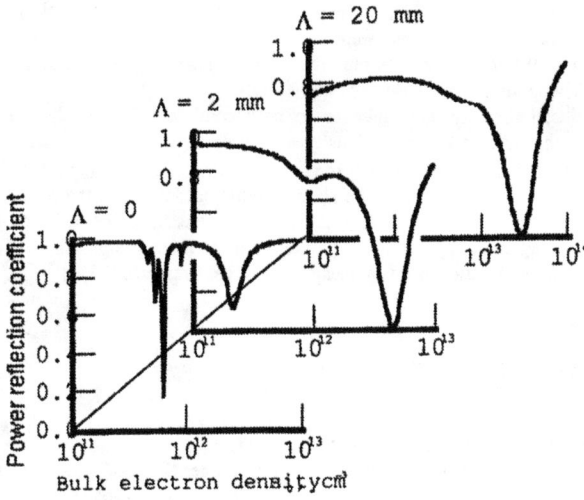

**Figure 6.** Dependence of the power reflection coefficient on the bulk plasma density $n_0$ in homogeneous ($\Lambda = 0$) and inhomogeneous ($\Lambda = 2$ and 20 mm) planar surface-wave plasma; $\Lambda$ is the lengthscale of the plasma density profile.

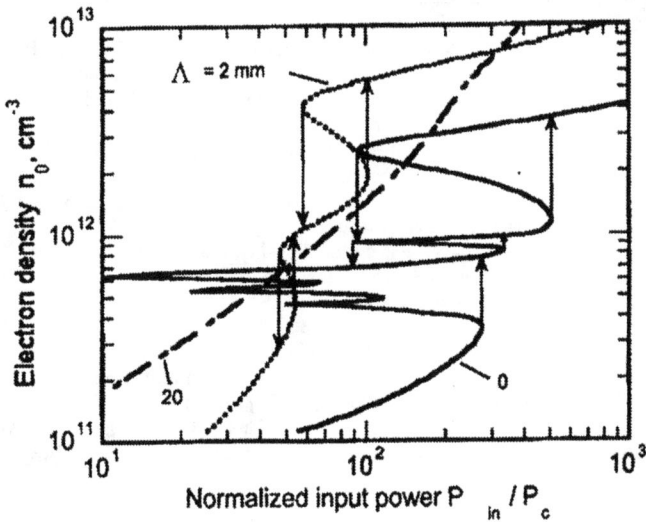

**Figure 7.** Dependence of the bulk plasma density $n_0$ on the normalized incident wave power $P_{in}$ in homogeneous ($\Lambda = 0$) and inhomogeneous ($\Lambda = 2$ and 20 mm) planar surface-wave plasma; $\Lambda$ is the lengthscale of the plasma density profile.

## 5. SUMMARY AND CONCLUSION

In this work we present experimental evidence for the local excitation of electron plasma oscillations in planar surface wave plasma. The oscillations were identified close to the plasma boundary. The electron density at the position of the resonance was measured and found to correspond to the cut-off density corresponding to the driving microwave frequency of 2.45 GHz.

The surface wave excitation by slot antennas in such conditions was studied theoreticaly. The numerical results show that when the plasma density profile becomes smoother (increasing the density profile scalelength $\Lambda$) the individual eigenmode resonances appear at higher bulk plasma densities. This is not surprising, since the surface wave, due to its short penetration depth into tha plamsa, sees only the lower density plasma layer close to the boundary. At the same time, the eigenmode resonances become wider and eventually overlap and disappear leading to fewer or no mode jumps, which is consistent with the observation of (almost) mode-jump-free SW plasma at low gas pressures [23].

## References

1. Sugai H., Ghanashev I., Nagatsu M. Plasma Sources Sci. Technol., 1998, **7**, 192.
2. Ghanashev I., Nagatsu M., Xu G. and Sugai H. Jpn. J. Appl. Phys., 1997, **36**, 4704.
3. Ghanashev I., Nagatsu M. and Sugai H. Jpn. J. Appl. Phys., 1997, **36**, 337.
4. Ghanashev I and Sugai H. Physics of Plasmas, 2000, **7**, 3051.
5. Ghanashev I., Sugai H., Morita S. and Toyoda N. Plasma Sources Sci. Technol., 1999, **8**, 363.
6. Stepanov K.N. Zh. Tekh. Fiz., 1965, **35**, 1002 [Sov. Phys. Tech. Phys., 1965, **10**, 773].
7. Zethoff M. and Kortshagen U. J. Phys. D: Appl. Phys., 1992, **25**, 1574.
8. Aliev Yu.M., Maximov A.V., Kortshagen U., Schlüter H. and Shivarova A. Phys. Rev., 1995, **E51**, 6091.
9. Kudela J., Terebessy T. and Kando M. Appl. Phys. Lett., 2000, **76**, 1249.
10. Stenzel R.L., Wong A.Y. and Kim H.C. Phys. Rev. Lett., 1974, **32**, 654.
11. Durandet A., Arnal Y., Margot-Chaker J. and Moisan M. J. Phys. D: Appl. Phys., 1989, **22**, 1288 (in French).
12. Grosse S. Proceedings of NATO ASI Advanced Technologies Based on Wave and Beam Generated Plasmas, Sozopol, Bulgaria, May 22—June 1, 1998 (NATO ASI Series, Partnership Subseries 3 High Technology), Eds. H. Schlüter and A. Shivarova (Amsterdam: Kluwer, 1999) 517.
13. Räuchle E Journal de Physique IV, 1998, **8**, Pr7-99.
14. Nagatsu M., Xu G., Ghanashev I., Kanoh M. and Sugai H. Proc. 13th Symp. Plasma Processing, Tokyo, August 29--31, 1996 (Tokyo: Jpn. Soc. Appl. Phys., 1996) 9.
15. Nagatsu M., Xu G., Yamage M., Kanoh M. and Sugai H. Jpn. J. Appl. Phys., 1996, **35**, L341.
16. Kokura H., Nakamura K., Ghanashev I. and Sugai H. Jpn. J. Appl. Phys., 1999, **38**, 5262.
17. Ghanashev I., Morita S., Toyoda N., Nagatsu M. and Sugai H. Jpn. J. Appl. Phys., 1999, **38**, 4313.
18. Ghanashev I., Sugai H. 52nd Anuual Gaseous Electronics Conference, 5-8 Oct. 1999, Norfolk, Virginia, USA [Bulletin of the Am. Phys. Soc., 1999, **44**, No. 4] p. 70.
19. Essam A.F., Ghanashev I., Sugai H. Jpn. J. Appl. Phys., 2000, **39**, 4181.
20. Odrobina I., Kúdela J. and Kando M. Plasma Sources Sci. Technol., 1998, **7**, 238.
21. Ghanashev I., Abdel-Fattah E. and Sugai H. Proc. 17th Symposium on Plasma Processing, Nagasaki, 26-28. Jan. 2000, Ed. H. Sugai (Tokyo: Japan Society of Applied Physics, 2000) 291.
22. Glaude V. M. M., Moisan M., Pantel R., Leprince P. and Marec J. J. Appl. Phys., 1980, **51**, 5693.
23. Nagatsu M., Morita S., Ghanashev I., Ito A., Toyoda N. and Sugai H. J. Phys. D: Appl. Phys., 2000, **33**, 1143.

# INFLUENCE OF THE PLASMA INHOMOGENEITY
# ON THE RF POWER DEPOSITION IN HELICON DISCHARGES

**M. Krämer\* and B. Lorenz**

Experimentalphysik II, Ruhr-Universität Bochum
D-44780 Bochum, Germany
\*email: mk@plas.ep2.ruhr-uni-bochum.de

**Abstract.** A highly ionised pulsed helicon discharge produced through a $m = 1$ or $m = 2$ helical antenna is investigated. Various diagnostics are applied to measure the discharge parameters and the radio frequency (rf) field distribution. Special attention is paid to the axial asymmetry of the discharge which is characteristic for helicon devices with helical antenna coupling. The axial profiles of the rf wave fields as well as the energy deposition profiles reveal that the rf power is predominantly transferred and absorbed via the $m = +1$ helicon mode travelling in positive magnetic field direction. The experimental findings are compared with numerical results obtained from a helicon wave guide model and a fully electromagnetic model taking into account the rf current distribution of the launching antenna as well as the finite size of the plasma column. The axial asymmetry of the helicon discharge can be understood in terms of the radial inhomogeneity of the helicon plasma. Furthermore, the calculations show that the small-scale Trivelpiece-Gould (TG) waves are excited near the plasma edge. In the present discharge, these waves play, most likely, only a minor role for the rf power absorption. We studied also the break-down of the helicon discharge and found that due to the existence of hot electrons, a discharge of moderate density is formed under the antenna. In a second step, the hot electrons relax to a thermal energy distribution, the density increases further and the helicon modes start propagating.

## 1. INTRODUCTION

Helicon discharges are very attractive sources for plasma applications because high electron densities can be achieved at relatively low gas pressures [1]. Many investigations have been carried out in the last years, however, the basic physical understanding of these discharges has still to be improved. One of the characteristic features of helicon discharges is the predominance of the helicon modes with positive azimuthal mode number $m$ [2,3,4,5]. This leads to an axial discharge asymmetry with respect to the antenna if the rf power is launched to the plasma via helical antennas. The rf power is then mainly coupled to the $m > 0$ modes and, thus, deposited on one side of the antenna so that the plasma is generated mainly on this side. The resulting asymmetry can be attributed to the different propagation behaviours of the helicon modes in non-uniform plasma as recent theoretical-numerical studies have shown [4,6,7].

Along with the *fast* helicon mode which is decisive for the power absorption and plasma production in the core of the discharge column, the *slow* wave, also denoted as Trivelpiece-Gould (TG) wave, can simultaneously be excited [8].Contrary to the *global* helicon modes, the TG waves are essentially quasi-static waves which propagate mainly at the plasma edge. They may therefore play a role for the rf power deposition in the outer zones of the plasma column. Moreover, they can affect the antenna-plasma coupling and, thus, the total rf power deposited in the plasma.

In the present paper we investigate the helicon discharges with helical antenna coupling treating the points addressed above in more detail. We aim at studying the axial asymmetry as well as the temporal evolution of pulsed $m = 1$ and $m = 2$ helicon discharges to obtain some more insight in the discharge physics. In particular, we focussed on the formation of the helicon discharge, that is, the evolution of the rf fields and the plasma in the early discharge.

## 2. THEORY

### 2.1. Wave propagation in the helicon regime

The wave frequencies in the helicon regime are well below the electron cyclotron frequency $\omega_{ce}$ and well above the lower-hybrid (LH) frequency $\omega_{LH}$ so that the ion dynamics can be neglected. In general, the plasma density in helicon sources is so high that the frequency is much lower than the plasma frequency, and the helicon wave frequencies are typically in the range $\omega_{LH} \ll \omega \ll \omega_{ce}, \omega_{pe}$. In linear theory, the wave dispersion relations for a collisional magnetised plasma (magnetic field in z-direction) can be derived from Maxwell's equations in combination with the electron equation of motion. Applying the wave *ansatz* $\exp[i(\mathbf{k} \cdot \mathbf{r} - \omega t)]$ for the electromagnetic fields and the plasma currents, we finally arrive at the wave equation

$$\mathbf{N} \times (\mathbf{N} \times \mathbf{E}) + \varepsilon \cdot \mathbf{E} = 0 \tag{1}$$

where N is the vector of the refractive index and $\varepsilon$ is the permittivity tensor having the elements $\varepsilon_{xx} = \varepsilon_{yy} = S$, $\varepsilon_{zz} = P$, $\varepsilon_{xy} = -\varepsilon_{yx} = -iD$, $\varepsilon_{xz} = \varepsilon_{zx} = \varepsilon_{zy} = \varepsilon_{yz} = 0$. For the helicon regime, the tensor elements become approximately $S \approx 1 + (\omega_{pe}^2 / \omega_{ce}^2)(1 + i v_c / \omega)$, $D \approx \omega_{pe}^2 / (\omega \omega_{ce})$, $P = -\omega_{pe}^2 / (\omega^2 (1 + i v_c / \omega))$ where $v_c$ is a phenomenological collisional frequency.

From the wave equation (1) we obtain immediately the dispersion relation which can be written as biquadratic equation in $N_\perp$, the perpendicular component of N, yielding

$$A N_\perp^4 + B N_\perp^2 + C = 0 \quad \text{with the solutions} \quad N_{\perp s, f} = \left[ \left( B \pm \sqrt{B^2 - 4AC} \right) / 2A \right]^{\frac{1}{2}}. \tag{2}$$

This form accounts for the fact that in a bounded magnetised plasma column the parallel wave number ($k_z$) spectrum of the antenna current is well defined by the antenna geometry. The coefficients A, B, C are functions of the plasma parameters as well as the parallel component of the refractive index $N_z = k_z \omega / c$. The two solutions can be attributed to the *slow* (s,+) and *fast* (f,-) waves, respectively, according to their phase velocities. The *slow* wave is also denoted as the Trivelpiece-Gould mode in bounded plasma while the second solution describes the *fast* helicon mode. For the parameters of most helicon experiments, the two wave solutions are well separated, and one obtains approximate dispersion relations reading

$$N_{\perp s} = N_z \sqrt{-P/S} = \frac{\omega_{pe} \omega_{ce}}{\omega \left( \omega_{pe}^2 + \omega_{ce}^2 \right)^{1/2}} \left( 1 - i \frac{v_c}{2\omega} \left( 1 + \frac{\omega_{pe}^2}{\omega_{pe}^2 + \omega_{ce}^2} \right) \right) \tag{3}$$

and

$$N = \frac{\omega_{pe}^2}{\omega \omega_{ce} N_z} \left( 1 + i \frac{v_c N}{\omega_{ce} N_z} \right), \quad N = \left( N_{\perp f}^2 + N_z^2 \right)^{1/2} \tag{4}$$

The slow wave dispersion relation (3) is identical with the dispersion relation for quasi-static waves in cold plasma, i.e., the *slow* waves are potential waves. The *rf* power is transferred along the resonance cone at an angle $\psi = 90° - \vartheta$ that is formed by the group velocity with respect to the magnetic field. As $\psi$ is of the order of $\omega / \omega_{ce}$, the group velocity and, thus, the *rf* power flow are nearly parallel to the magnetic field.

The *fast* waves are mainly electromagnetic with an electrostatic fraction increasing with the propagation angle $\vartheta$. Their dispersion relation (eq. (4)) is identical with the dispersion relation for oblique whistler waves or the helicon modes in bounded plasma.

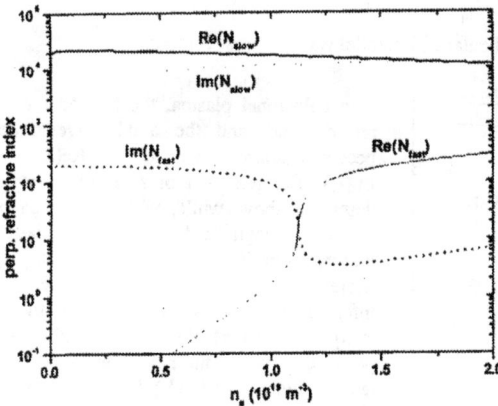

Figure 1. Fast and slow wave dispersion relation.-
$f = 13.56$ MHZ, $B_0 = 60$ mT, $N_z = 200$.

In Fig.1 we have plotted the dispersion curves for typical parameters of the present experiment assuming a parallel wave number given by the antenna length $L_a \approx 20$ cm, i.e., $k_z = \pi/L_a$. (Their parallel wavenumber is roughly given by the antenna spectrum with typical values of the order $\pi L_a$.) It can be seen that the slow wave has a perpendicular refractive index with large real and imaginary parts, that is, the slow waves have a short wavelength and damping length. As a consequence, they can propagate only near the plasma edge when they are excited by an antenna outside the plasma. The fast wave is however only weakly damped, and, for the present conditions where the perpendicular wavelength is of the order

of the plasma diameter, they are able to form helicon *modes* standing across the plasma column and travelling along the plasma axis. Moreover, the fast wave reveals a cut off at a certain density so that a minimum density is necessary for exciting these modes.

## 2.2. Helicon wave guide model

When we consider only the helicon mode propagation, it is convenient to apply the EMHD equations which are obtained from the basic equations (Maxwell's equations and the equation of motion) by neglecting the displacement current in Ampere's law and the time-dependence in the equation of motion. We first study the propagation of *undamped* helicon modes in a radially non-uniform plasma column [7]. Applying again the wave *ansatz* for the varying quantities and combining the resulting equations, we obtain the wave equation

$$\frac{\partial^2}{\partial r^2} B_z + f(r) \frac{\partial}{\partial r} B_z + g(r) B_z = 0 \qquad (5)$$

where $f(r) = \dfrac{1}{r} - \dfrac{2\alpha^2 \beta}{\alpha^2 - k_z^2}$, $\quad g(r) = \alpha^2 - k_z^2 - \dfrac{m^2}{r^2} - \dfrac{m\alpha\beta(\alpha^2 + k_z^2)}{k_z r(\alpha^2 - k_z^2)}$, $\quad \alpha = \dfrac{\omega}{\omega_{ce}} \dfrac{\omega_{pe}^2}{c^2 k_z}$, $\quad \beta \equiv L_n^{-1} = \dfrac{n_e'}{n_e}$,

and the prime denotes the derivative with respect to $r$.

For helicon modes travelling along a plasma cylinder, the boundary conditions have to be taken into account. Assuming that the plasma is surrounded by a perfectly conducting tube (radius $r_w = r_p$), the radial component of the *rf* magnetic field vanishes due to $\nabla \cdot \mathbf{B} = 0$. This condition along with the wave equation can be solved simultaneously by standard procedures to compute the axial wavenumber.

For a homogeneous plasma, the wave equation is reduced to Bessel's equation of *m*-th order. which is independent of the sign of *m*. However, in a non-uniform plasma the propagation behaviours of the helicon modes with positive and negative azimuthal mode numbers *m* may differ considerably. For, $g(r)$ (eq.(5)) may be negative for the $m < 0$ modes if the radial density gradient is steep enough so that these modes become evanescent. Assuming a Gaussian density profile of width $w$, one can deduce the approximate condition for evanescence

$$q = \frac{\omega_{ce}}{\omega} \frac{c^2}{\omega_{pe}^2} \frac{2|m|}{w^2} > p = O(1) \qquad (6)$$

where the $p$ depends on the ratio of the perpendicular and parallel wave number components.

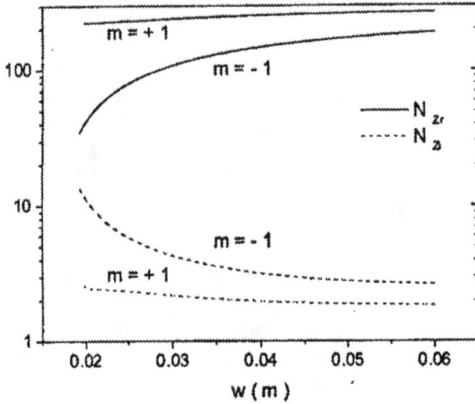

**Figure 2.** Dispersion relations of m = +1 and m = -1 helicon modes

In a collisional plasma, the helicon modes are damped, and the axial wavenumber becomes complex, i.e., $k_z = \text{Re}(k_z) + i \,\text{Im}(k_z)$. The real part of $k_z$ is obtained as described above while, with knowledge of the electromagnetic fields and the plasma current density, the spatial damping decrement $\text{Im}(k_z)$ [$\text{Im}(k_z) \ll \text{Re}(k_z)$] can be inferred from the relation between the $rf$ energy (Poynting) flux and the absorbed power density. As the power decay per unit length, $dP/dz = -2\,\text{Im}(k_z)\,P$, is equal to the absorbed power density $Q$ integrated over the cross section $A$, we simply have

$$\int_A Q(r,\vartheta,z)\,dA = \frac{dP}{dz} = -2\,\text{Im}(k_z)\,P(z) \qquad (7)$$

from which the damping decrement can be derived. Fig.2 shows the dispersion curves of the $m = +1$ and $m = -1$ modes for typical discharge parameters ($n_e(r = 0) = 4.2 \times 10^{18}$ m$^{-3}$, $B_0 = 33$ mT, $p = 1$ Pa).

In Fig.2 we demonstrate the effect of the radial plasma inhomogeneity on the propagation of the helicon modes by varying the width of the (Gaussian) density profile. It turns out that the $m = +1$ mode propagates independently of the profile width whereas the $m = -1$ mode can only propagate if the profile is not too narrow. The cut off of the $m = -1$ mode can easily be seen from the merging of the real and imaginary parts of the axial refractive index.

## 2.3 Antenna-plasma model

A fully electromagnetic model based on the linear theory for a collisional magnetised plasma was applied to describe the antenna-plasma system [6]. The treatment takes into account the current distribution of the antenna as well as the finite radial and axial dimensions of the plasma assumed to be uniform in z-direction, but non-uniform in radial direction. The magnetised plasma column is coaxially contained in a cylindrical cavity with perfectly conducting walls. The rf current paths of the antenna are located on a cylindrical surface, however, the $rf$ current distribution can be nearly arbitrary.

The electromagnetic fields are represented by a Fourier decomposition taking into account the boundary conditions for the electric fields at the conducting end plates. A corresponding decomposition has to be made for the antenna current. We finally obtain a set of coupled differential equations for each Fourier component which is solved numerically to give the electromagnetic fields in the whole system. Applying Poynting's theorem, the $rf$ power deposited in the plasma column as well as the plasma resistance can be determined. The code also permits to compute the energy density of the electromagnetic field as well as the absorbed power density.

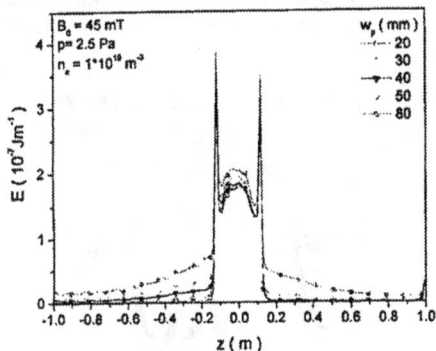

Figure 3a. Wave energy

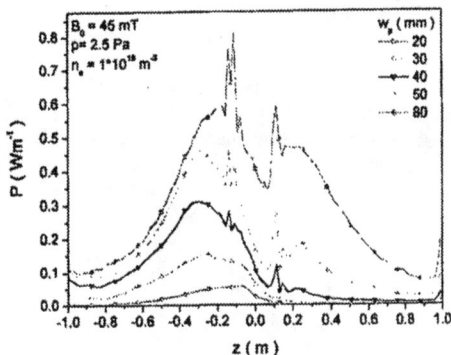

Figure 3b. Deposited rf power

In Fig.3 we show the axial distribution of the rf energy density integrated over the plasma cross section for Gaussian density profiles of different width w. As is seen in Fig.3a, the axial profiles of the wave energy become more asymmetric with narrowing of the density profile. In case of a (nearly) uniform plasma density, however, the profiles are nearly symmetric. The rf power deposited in the plasma reveals a very similar behavior (Fig.3b). Due to the better antenna-plasma coupling, the rf power is largest and is mainly absorbed under the antenna if the density profile is nearly uniform. However, a pronounced asymmetry arises if the density profile becomes sufficiently narrow. The axial absorption profile then exhibits a maximum outside of the antenna on the positive magnetic field ($m = +1$) side. Apparently the radial gradient of the electron density gives rise to this asymmetry.

## 3. EXPERIMENTAL

The investigations have been carried out on the pulsed large-volume helicon discharge HE-L in argon gas [5]. The plasma is produced by rf power pulses ($P_{RF} < 4$ kW, $f_{RF} = 13.56$ MHz, $\tau_{pulse} = 2$ –3 ms, $f_{pulse} = 25$ Hz, typically) alternatively through $m = 1$ and $m = 2$ helical antennas surrounding the quartz tube (Fig.4 and 5). The $m = 1$ antenna is a Shoji type antenna consisting of two helical current paths with opposite directions of the rf current (180° turn over the antenna length $L_a = 22$ cm) while the $m = 2$ antenna has 4 helical current paths with alternate rf current directions. The experimental parameters measured by standard methods (electric probe diagnostics and 1 mm and 8 mm interferometry) were $n_e < 2 \times 10^{19}$ m$^{-3}$, $T_e = 3 - 4$ eV, $B_0 < 0.1$ T, $p = 0.1 - 2$ Pa, $r_p = 7.4$ cm and $L_p = 200$ cm.

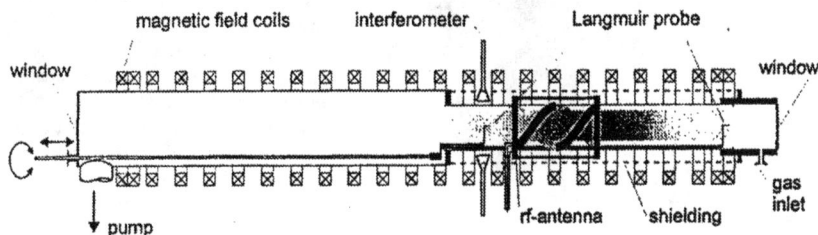

Figure 4. Experimental set-up of HE-L

Due to the radial *rf* power deposition profile peaking on the axis, the $m = 1$ discharge can be sustained in a wider pressure range, $p = 0.1 - 3$ Pa, while stable conditions for the $m = 2$ discharge were achieved for $p = 2 - 4$ Pa.

The helicon wave field was measured by a movable array of magnetic (B-dot) probes picking up all the three components of the *rf* magnetic field. An *rf* double probe (tip distance 1.1 mm) was employed to sense the small-scale fluctuations of the *rf* electric field.

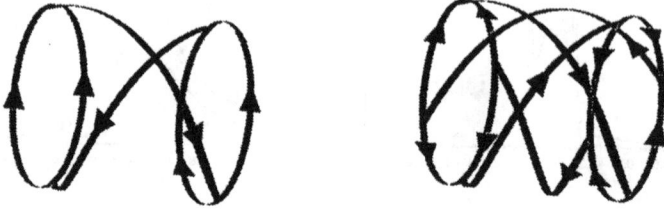

Figure 5. Helical m = 1 (left) and m = 2 (right) antennas

# 4. RESULTS

## 4.1. Discharge asymmetry and helicon wave propagation and damping

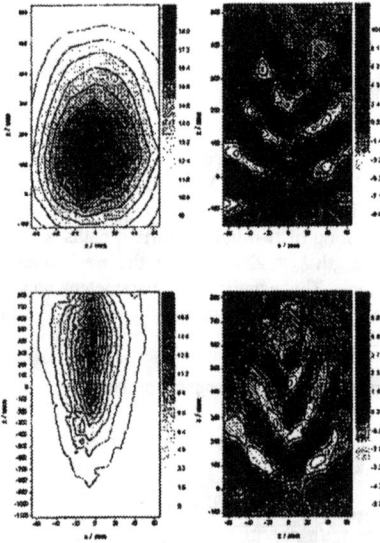

Figure 6. Contour plots of the rf magnetic field and the density for the m = 1 (bottom) and m = 2 mode (top). Centre of the antenna at z = 0

A characteristic feature of helicon discharges with helical antenna coupling is the pronounced asymmetry with respect to the centre of the antenna (Fig.6, left diagrams). The density maximum is shifted to the positive magnetic field side of the antenna. When the direction of the stationary magnetic field is reversed, the density maximum shifts to the opposite side of the antenna. Furthermore, the degree of asymmetry grows with increasing magnetic field and/or decreasing density. The cause for the discharge asymmetry is the directionality of helical antennas exciting predominantly helicon modes that travel in positive magnetic field direction. To be specific, the Shoji antenna excites mainly $m = +1$ modes and the $m = 2$ helical antenna $m = +2$ modes. In fact, as is seen in Fig.6 (right), the wave pattern of the perpendicular *rf* magnetic field strength exhibits wave fronts moving in positive magnetic field direction at small angle with respect to the field. As expected from the radial *rf* power deposition profiles, a plasma with a relatively broad radial profile is generated for $m = +2$ mode excitation while the $m = +1$ discharge has a significantly narrower profile, particularly at low pressures. From the two contour plots we calculated the axial wavelengths and compared them with the whistler wave dispersion relation as well as with the computations from the helicon wave guide model (Sec.2.2). Reasonable agreement has been achieved between the computational and experimental results for both discharges in a wide parameter range (Fig.7).

It is obvious that plasma production is closely related to the helicon wave absorption. To estimate the damping rate, we integrated the rf field energy over the plasma cross section and found that the wave field is nearly exponentially damped outside the antenna. In case of the $m = 2$ discharge, the damping decrements deduced from the decay of the wave fields fits reasonably with the values expected for electron-neutral and electron-ion collisions. However, in the $m = 1$ discharge at low gas pressures, the wave damping is too strong to be accounted for collisions so that, most likely, the major fraction of the helicon wave energy is lost by some non-collisional mechanism.

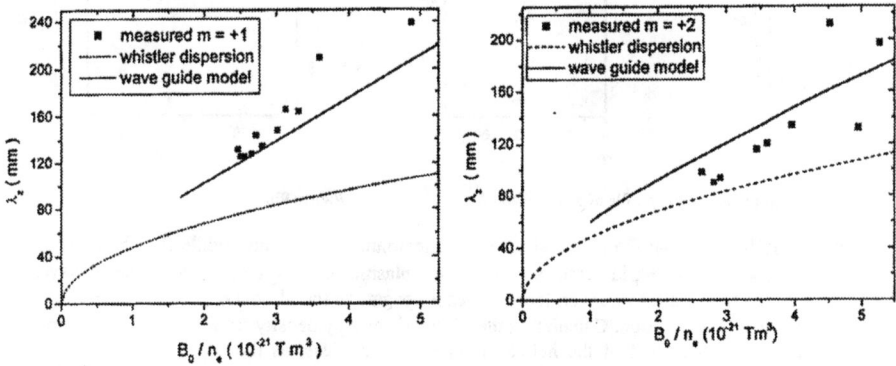

Figure 7. Dispersion of $m = +1$ and $m = +2$ helicon modes

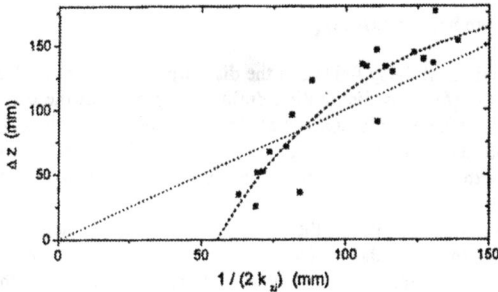

Figure 8. $\Delta z$ versus calculated power decay length

In Fig.8 we demonstrate that the plasma production is closely related with the helicon absorption. Assuming that the total rf power is only transferred to damped helicon modes with positive mode numbers, one would expect that the absorbed power scales as $\exp(-2k_z z)$ where $k_{zi}$ is the axial damping decrement. If one further assumes that the density has the same dependence, the distance $\Delta z$ of the centre of mass of the axial density distribution and the middle of the antenna is approximately $1/(2k_{zi})$. Of course, due to axial diffusion the density profile is broader in axial direction

than that of the rf energy. However, this should only weakly affect the position of the centre of mass of the profile. In the diagram, we plotted the dependence of $\Delta z$ on the power decay length $1/(2k_{zi})$ for the $m = 2$ helicon discharge taking the damping values from the helicon wave guide model. It turns out that $\Delta z$, i.e., the degree of asymmetry, scales roughly with $1/(2k_{zi})$ which has been computed from the wave guide model for a collisional plasma.

124

## 4.2 Measurements of the rf electric fields and TG excitation

The *rf* electric probes measure essentially the *rf* potentials associated with the TG (potential) waves and/or the electric part of the helicon modes. The TG waves are clearly more interesting inasmuch as they cannot be diagnosed by magnetic probes. In Fig.9a we plotted the radial profiles of the *rf* potential amplitude and the (mean) phase difference obtained from the *rf* double probe measurements.

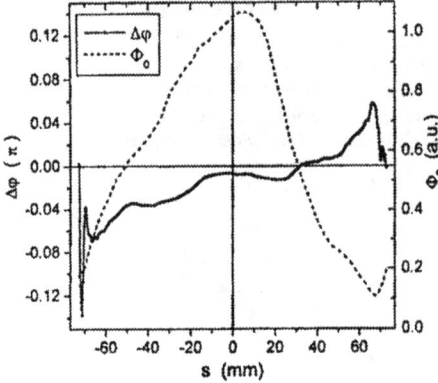

| Figure 9a. Radial profile of potential and Δφ | Figure 9b. Profile of (averaged) |E|² |

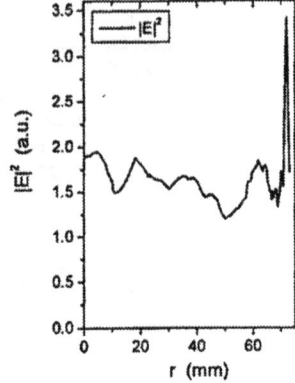

It can clearly be seen that the potential reveals a maximum in the centre while the phase difference between the probe signals, $\Delta\phi$, increases strongly at the plasma edge. Apparently small-scale waves are excited close to the tube wall. Moreover, these waves propagate to the plasma edge as it is expected from the slow wave dispersion relation. Calculating the rf electric energy density from $\Phi_{rf}$ and $\Delta\phi$ it turns out that it is much larger than that of the helicon modes. Nevertheless, as the edge region, where the *rf* electric energy is relatively high, is thin, the major faction of the *rf* power is transferred to the helicon modes.

### 4.3. Evolution of the helicon wave fields and the helicon discharge

It is instructive to study how the spatial distributions of the *rf* fields and the discharge parameters evolve. In Fig.10, we show the temporal evolution of the *rf* energy distribution during the *rf* pulse of the $m = 1$ discharge. In the first half millisecond the *rf* fields are mainly concentrated near the helical windings of the antenna. After about 0.6 ms a maximum originates on the tube axis which steadily increases. At the end of the pulse, the profile forms a largely extended crest associated with the $m = +1$ helicon mode travelling in positive magnetic field direction.

Most interesting is the evolution of the axial *rf* profiles on the axis (Fig.11), because the power deposition has there a maximum. It can be seen that both the helicon modes are excited, however, the $m = -1$ mode has a much smaller amplitude. Furthermore, the $m = +1$ wave field appears first while the $m = -1$ field originates with a significant time delay. For smaller magnetic field strengths and smaller (!) values of $n_e/B_0$ ($n_e$ decreases more strongly than $B_0$) the $m = -1$ modes propagate no longer in accordance with the condition (6) for evanescence. The *rf* measurements in the afterglow of the helicon discharge performed with small *rf* power reveal the same behaviour. An explanation for the weaker excitation of the $m = -1$ modes is, most likely, the considerable damping of the $m = -1$ modes near the cut off (see Sec. 2.2).

Valuable information on the plasma formation was obtained from measurements with a small energy analyser. In Fig.12 (left) we have plotted the temperature of the bulk electrons as well as the energy of a hot electron group and the plasma potential.

$\tau = 0.55$ ms

$\tau = 0.60$ ms

$\tau = 0.75$ ms

$\tau = 1.95$ ms

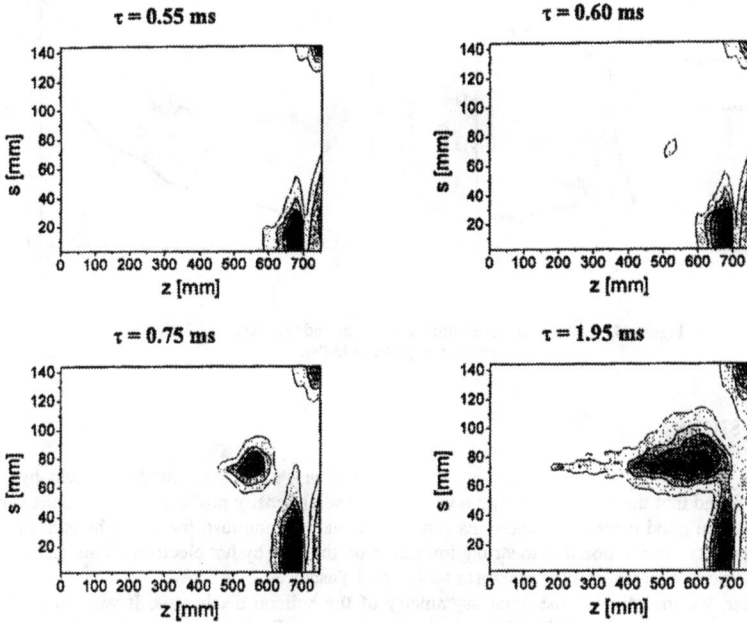

Figure 10. 2D-Plots of rf energy density ($|B|^2$)

It turns out that in the early discharge with very low density there is a group of electrons with energies of $11 - 12$ eV. After about 200 μs the energy rises up to about 20 eV at 300 μs and then drops strongly to the energy of the bulk electrons, that is, the electron group relaxes to a thermal distribution. Simultaneously with the energy increase of the electron group, the electron density rises strongly. After reaching a small maximum, $n_e$ increases again, however, with a longer rise time. Correlated with this second step, the *rf* fields strongly steepen and the helicon mode starts propagating.

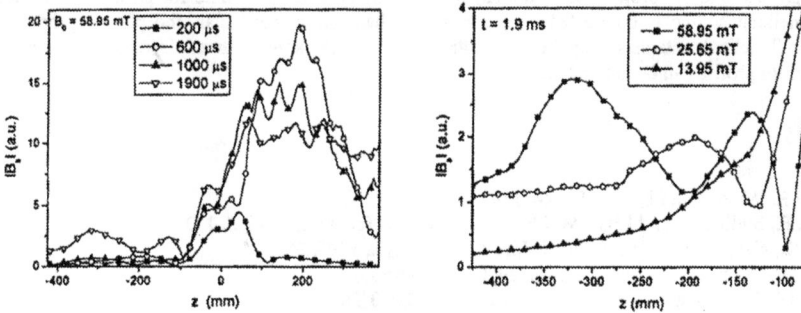

Figure 11. Axial profiles of the rf amplitude ($B_\vartheta$)

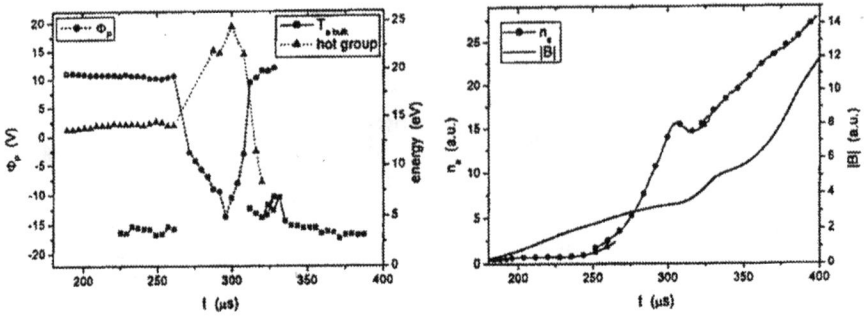

**Figure 12.** Evolution of the electron temperature and plasmapotential (right), density and magnetic field (left)

## 5. CONCLUSIONS

We studied in detail the plasma production by $m = +1$ and $m = +2$ helicon modes excited by helical antennas. We found that the $m = 1$ discharge reveals broad radial density profiles in the pressure range of a few Pascal where good discharge conditions can be achieved. In contrast, the $m = 1$ helicon discharge has a peaked density distribution due to strong ionisation on the axis by hot electrons. This discharge has a high ionisation degree, especially at pressures far below 1 Pascal.

In particular, we investigated the axial asymmetry of the helicon discharges. It was found that the degree of asymmetry depends on the plasma parameters $n_e$ and $B_0$. It can partly be explained by the helicon mode propagation in non-uniform plasmas, as it was confirmed by the results obtained from two numerical models. We found that the formation of the helicon discharge is mainly determined by helicon modes having positive azimuthal mode numbers.

The investigation of the formation of the helicon discharge with $m = 1$ helical antenna reveal that the $m = +1$ helicon modes originate just after a base plasma, located under the antenna, has been established. Finally, the observations of short-scale waves with high electric fields give evidence that Trivelpiece-Gould modes are exited at the plasma edge. However, these modes play probably only a minor role for the rf power deposition and, hence, for the plasma production.

We emphasise that the pronounced axial asymmetry may also be favourable for plasma applications where the helicon source is connected with a reaction chamber. In particular, the plasma production through $m = 2$ helical antennas may be very attractive because large-volume plasmas with a nearly uniform density profile can be generated.

## References

1. Boswell R. W. Plasma Phys., 1984, **26**, 1147
2. Chen F. F., Sudit I. D. and Light M. Plasma Sources Sci.Technol., 1996, **5**, 173.
3. Suzuki K., Nakamura K. and Sugai H. Jpn. J. Appl. Phys., 1996, **35**, 4044.
4. Enk Th. and Krämer M. Phys. Plasmas, 2000, **37**, (in press).
5. Krämer M., Enk Th. and Lorenz B. Physica Scripta, 2000,**T84**, 132.
6. Fischer B., Krämer M. and Enk Th. Plasma Phys. Control. Fusion, 1994, **36**, 2003.
7. Krämer M. Phys. Plasmas, 1999, **36**, 1052.
8. Shamrai K. P. and Taranov V. B. Plasma Phys. Control. Fusion, 1994, **36**, 1714.

# ELECTRODE MICROWAVE DISCHARGES IN MOLECULAR GASES.

Yu.A. Lebedev, M.V. Mokeev, A.V. Tatarinov and I.L.Epstein

A.V. Topchiev Institute of Petrochemical Synthesis RAS, Leninsky Prospect, 29, Moscow, 117071, Russia

Abstract. Results of experimental study and numerical simulation of nonequilibrium electrode microwave discharge in $H_2$ and $N_2$ at pressures 0.5-15 Torr are summarized. Discharge has nonuniform structure with bright near electrode region and outer ball-shape region with sharp boundary. Plasma is characterized by monotonous decrease of electron temperature, flat distributions of electron density and DC plasma potential inside the ball structure. Self-consistent modelling in quasi-static approximation showed that results of calculations are in partial qualitative agreement with results of experiments. It was concluded that superposition of microwave and DC fields are necessary to explain the sharp discharge boundary. Picture of physical processes in electrode microwave discharges is discussed.

## 1. INTRODUCTION

Investigations of microwave discharges are now aimed primarily at studying electrodeless discharges excited in various devices and even in free space [1]. This line of investigations are also motivated by multiple applications, e.g., plasma chemistry. The absence of electrodes is considered to be one of the important advantages of such discharges, because, in this case, it is possible to avoid the contamination of a plasma and plasma-treated surfaces by the products of electrode erosion. At the same time, electrode discharges have some merits, e.g. these discharges permit the plasma to be created at a low microwave power.

Electrode discharges can be classified as initiated discharges with one peculiarity: electromagnetic energy is transmitted to the system along the initiator–antenna. The plasma can be created in a given region, which can be located close to the treated surface, thus allowing to avoid difficulties connected with the transport of active particles to this surface. As a rule, the electrode discharge plasma is inhomogeneous. On the one hand, the non-homogeneity facilitates the transportation of active particles to the treated surface, and, on the other hand, it is an additional source of the plasma non-equilibrium.

Recent experiments indicate that the risk of contamination might be overestimated [2]. In experiments on the diamond film deposition on a silicon substrate in a microwave electrode discharge under fairly severe conditions (the electrode was heated to the red-heat temperature, and the temperature of the substrate located at a distance of 1 cm from it was 1000°C), ESCA analysis showed no traces of the products of electrode erosion in plasma-treated substrates.

The erosion problem has been studied extensively for RF discharges (see, e.g., [3]). As the field frequency increases, the thickness of the electrode-to-plasma sheath and the energy of the ions bombarding the electrode decrease. In the microwave range the current continuity is provided by the displacement current through the sheath. In this case, the role of the electrode in generation of charged particles is negligible ($\gamma$-processes associated with the secondary electron emission from the electrodes can be neglected), and the discharge exists in the form of an $\alpha$-discharge, in which the processes of volume ionization are of importance. Presumably, the electrode only determines the structure of the electromagnetic field. The lack of data on the parameters of electrode discharges substantially restricts the possibility of studying the physical-chemical processes occurring in plasma. Some results of plasma chemical applications of such discharges are presented in [2,4,5].

During last years a lot of experimental results were obtained ([4-11]) but the physical processes occurred in the electrode microwave discharges are still far from a complete understanding. This paper is the attempt to summarize some results of experiments and modelling in the light of design of the picture of physical processes which can explain the observed discharge phenomena in the case when the plasma dimensions were much less then the diameter of the discharge chamber.

128

## 2. EXPERIMENTAL

Two types of similar discharge cameras which differed in the cylinder diameters of discharge vessels ($R_1$=7 cm and $R_2$=4.2.cm) have been used for experiments [6-10]. Plasmas in both arrangements revealed similar features, so the properties of the described plasmas seems to be typical for the electrode microwave discharge systems.

The discharge chambers were the stainless steel cylinders excited by an antenna at the butt-end of a chamber (Fig. 1). The antenna was an element of a coaxial-to-waveguide converter, which could be tuned with the shorting pistons in the waveguide transmitted lines. The lower end of one of discharge chambers was a movable shorting piston that provided an additional possibility of tuning the system and changing the position of substrates immersed to plasma (this piston was used as a heated substrate holder). Another chamber had unchangeable internal dimensions.

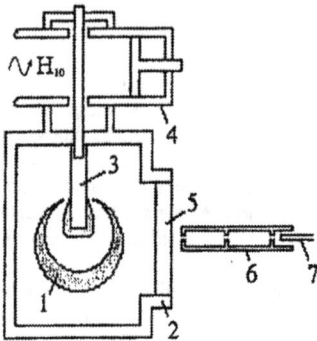

Figure 1. Experimental device.
1 – plasma, 2 – discharge chamber, 3 – antenna, 4– waveguide-to-coaxial converter, 5 – optical window, 6 – collimator, 7 – optical fiber

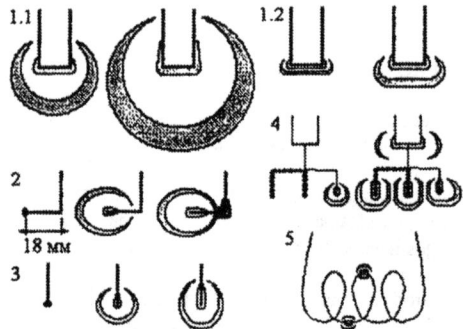

Figure 2. Shapes of electrodes and dependencies of discharge shapes on geometry of electrodes and plasma conditions. 1– thick cylinder electrode at low (1.1) and high (1.2) pressure, 2 – thin bent electrode, 3 – thin direct electrode, 4 – trident electrode, 5 – spiral electrode.

A cylindrical tube and solid electrodes of different shapes with diameters of 0.5-6 mm manufactured from stainless steel and copper and have been used as the antennas (Fig. 2).

Most of results has been obtained when the discharge was ignited at the end of cylindrical electrode (antenna). Plasma dimensions were much less than that of the discharge camera so these discharge systems can be attributed to single-electrode systems (the second one is the metal discharge camera). This is one of fundamental differences from conventional discharges where the plasma volume is strongly defined by the discharge vessel. This discharge property defines peculiarity both the electrodynamics of discharge and dependencies of plasma parameters on microwave power. Some information on the electrode microwave discharge initiated between the electrode and close placed metal plane or metal lug has been presented in [6].

The sizes of discharge chambers were not specially chosen. The system can be considered as a non-regular coaxial line (regularity is disturbed by a break of the central conductor) loaded with the plasma produced at the end of the central electrode. Since the radii of the outer electrodes (camera) were high enough to satisfy the inequality $\lambda(r_{el}+R)>1$ (where $r_{el}$ and $R$ are the inner and outer radii of the coaxial line conductors and $\lambda$= 12.4 cm is the of the electromagnetic radiation wavelength in vacuum), not only the transverse TEM mode, but also the other spatial modes which could exist in such systems. The transformation of the electromagnetic field structure occurred exactly inside in the region occupied by plasma due to both the presence of plasma and break up of the electrode. And, vice versa, plasma is created as a result of the action of the set of electromagnetic field modes.

Microwave power (2.45 GHz) was transmitted from the magnetron generators with output powers of 2.5 kW or 150 W. In last case the stabilized DC power supply was used to feed the magnetron. The power absorbed in the system was measured by a directional couplers by subtracting the reflected power from the incident one. It should be noted that the absorbed power can not be attributed completely to the plasma due to power losses in the discharge camera, the coaxial-to-waveguide converter and feeders. The share of these losses is unclear in general case [12].

The set-up was equipped with the system of optical measurements. Integrated over the spectrum plasma emission with spatial resolution was used for studying discharge structure. Spatially integrated or with spatial resolution signals were used to characterize the emission properties of plasma in the spectral range 400-800 nm. The spatial resolution of measurements was provided by a collimator (spatial resolution was not worst than 1 mm) which could be moved parallel to the antenna axis or in the perpendicular directions. The optical plasma emission was directed through the optical fiber to the photomultiplier or to the input slit of the monochromator, according to the needs of experiment.

Parameters of electron component of plasma and distributions of DC voltage in the plasma were measured with the help of a double electric probe. The details of these measurements were described in [13]. The tungsten probes of a diameter of 100 μm in the quartz capillaries were introduced through the orifice in the lateral surface of the chamber and could be moved along the radius of the discharge column. The length of the open part of the probes was of 1 mm, and the distance between them was 3-4 mm. The measurements were performed in the plane perpendicular to the axis of the antenna and located at distance, which exceed the thickness of high luminosity of the sheath observed near electrode. Resistive filters (20 kΩ) were used to decrease the possible influence of a microwave field on the voltage-current characteristic of the probe.

Plasma gases were Ar, Ne, $H_2$, $N_2$, $O_2$, $CH_4$, $C_2H_2$, air, and their mixtures. Gas flow systems were used. The total gas flow rate was less than 1000 sccm. Results for $H_2$ and $N_2$ are presented in this paper.

The aspects of plasma chemical applications of electrode microwave discharge for surface processing were studied in cooperation with the Angström Laboratory, Uppsala University (Sweden). This discharge device was used for the diamond growth [2] in hydrogen and hydrogen-methane mixtures (1-8%) at 1-15 Torr and to deposit $CN_x$ films in $N_2+C_2H_2$ mixtures [5]. The composite of nanotubes oriented perpendicular to the substrate surface was also obtained [5].

## 3. MODELLING

The radial distributions of the hydrogen microwave plasma density and electric field inside coaxial and spherical system of electrodes has been studied in quasi-static approximation. Self-consistent solution of electron balance equation and the electric field distribution was numerically defined with the plasma absorbed power as the parameter.

The radial distribution of quasi-static electric field was given in accordance with chosen geometry of the system and the dielectric constant of plasma [14]. Only component of the electric field $E_n$ was taken into account which is normal to the surface of inner electrode:

$$E_n = \frac{\overline{E}}{\varepsilon} = E_0 \left(\frac{r_{el}}{r}\right)^k \left\{(1-n)^2 + \frac{v_{el}^2}{\omega^2} \cdot n^2\right\}^{-\frac{1}{2}}, \tag{1}$$

where k = 1 and 2 for the case of cylinder and sphere geometry respectively, $n = n_e/n_c$ is the reduced electron density and $n_c$ is the critical plasma density $n_c = \left(v_{en}^2 + \omega^2\right)m/4\pi e^2$, $\omega$ is the field frequency, $v_{en}$ is the electron-heavy particle collision frequency, $E_0$ is the microwave field strength on the surface of the electrode, $r_{el}$ is the radius of the inner electrode.

The electron balance equation with zero boundary conditions includes processes of the direct ionization, ambipolar diffusion and volume dissociative electron-ion recombination. The ionization frequency and the coefficients of diffusion and recombination were a function of the electric field. All the coefficients were calculated using the Boltzmann equation [15]. The gas temperature was supposed to be constant and equal to 300 K. The input parameters for calculations were the gas pressure, absorbed power and geometry of the system.

The balance equations for hydrogen atoms and molecules were also solved to study the process of hydrogen dissociation.

## 4. RESULTS AND DISCUSSION

### 4.1. Phenomenology of the discharge

Electrode microwave discharges are the localized plasma structures which are generated in the presence of exciting electrode. The external dimensions of plasma region are defined by the pressure, microwave power and diameter of the inner electrode. All described phenomena were observed when the plasma dimensions were less than the dimensions of the discharge chamber and did not depend on the latter. The general feature of the electrode microwave discharges is the presence of bright plasma sheath near the electrode and less luminous plasma region with sharp external boundary. As usual this region is ball-shaped in molecular gases or in the mixtures with large content of it. One of distinctive features of the electrode microwave discharges is that they exist only at high levels of reflected power (the standing wave ratio is changed between 5 and 15). At low incident microwave power the discharge can be ignited by external initiators, e.g. the Tesla-coil. Phenomenology of the discharges with different electrodes is presented below:

a. Tube and rod antennas

- The discharge is ignited at the end of antenna and it partially covers it. The less the antenna diameter the less is the ignition and maintenance microwave field. The discharge could be ignited at incident powers of 2 W with 0.5 mm diameter antenna. Increase in the incident power leads to growth up of the discharge sheath along the antenna. From the butt-end of the antenna the sheath looks like a ring-shaped bright region along antenna periphery. When pressure or incident power are such that the size of plasma region and electrode are comparable, the shape of plasma follows the shape of the electrode (the case of a thin plasma region). Slight increase in the plasma emission in the plasma periphery was observed at some conditions.

- As microwave power is increased up to certain threshold level, which is defined by the pressure and electrode diameter, the discharge runs away towards the microwave generator along the antenna. It can be moved back by matching the adjusting elements in microwave system but this leads to the decrease of absorbed power. Thus in the particular configuration of the gas discharge system it is impossible to generate the plasma at the top of antenna with the absorbed power, which exceeds the threshold value. In this case the discharge can simultaneously exist at the end of antenna, along it and at the place of an antenna input of the discharge chamber. The last one is the "main" discharge and it strongly reduce the electromagnetic wave penetration into the camera. The reflected power is decreased in the feeder and in almost all incident power is absorbed. These regimes were off our interest because in those cases the pure electrode discharge placed at the end of antenna is only small supplementary discharge.

- The dumbbell-like structures of the discharge were obtained in $H_2$ and $N_2$: two balls stringed the antenna, one of which was placed at the end of antenna and the other placed at distance of $\lambda/4$ upward the antenna. Both balls were connected by weak plasma slab which covers the antenna. The observed structure is differed from the conventional standing-wave shape: balls had the sharp boundary from both the side of microwave generator and the opposite side. The whole plasma system had the symmetry plane placed between the balls centers.

- The plasma ball exhibits properties of medium with the elastic boundary: when the probes touch the boundary and moves further, the boundary layer was bent inside. At certain moment the plasma covers the probes and the boundary returns to its initial position. In the reverse motion of the probes the boundary was bent outside moving together with the probes. Finally the plasma loses contact with the probes and several oscillations in shape and diameter of the plasma occur before it comes to the state of equilibrium with an initial diameter of the ball.

- If the plasma gas is introduced through the channel in the electrode, the ball shape is distorted and plasma flux expands along the axis of the channel for the distance exceeded the ball dimensions. The length of the plasma flux depends on the flow rate of the plasma gas.

b. Trident electrode.

Firstly the discharge is ignited near the top of one end of the trident. The structure of the plasma region is the same as described above. When the incident power is increased the other ends are covered by the same plasma. The diameters of plasma balls are increased with increase of the incident power but the balls never overlap and the dark region exists between plasma regions at any power. The shapes of the balls are changed to ellipse with increased power.

c. Spiral electrode.

The ball-shaped plasma formations are generated at several points of the spiral. Positions of this points are changed with the incident power. The structure of all discharges is the same as described above: the surface of the spiral was partly covered by a bright sheath surrounded by the plasma ball.

## 4.2. Probe measurements

a) Space outside luminous plasma region.

The space outside the plasma region is spatially uniform in electric field and double probe voltage-current characteristics pass through zero current at zero voltage difference between the probes $\Delta U_p(I_p=0)=0$. Figures 3,4 show some characteristics of this region [7].

Figure 4 shows the electromagnetic field E distributions. This picture was obtained from the measured values of electron temperatures with the assumption that the local value of electron temperature corresponds the local value of the heating field (the characteristic length of electron energy losses in collisions with heavy particles was much less than the observed temperature decay length). The E values were determined with the help of the Boltzmann equation and measured values of electron temperatures.

It is seen that the electric field strength exponentially decays with radius. The only wave mode with such a structure is a surface wave [16]. This wave spreads along the plasma boundary and can cause the increase of intensity of plasma emission as it was observed in experiments. The role of the surface wave in maintenance of the electrode discharge is unclear at present.

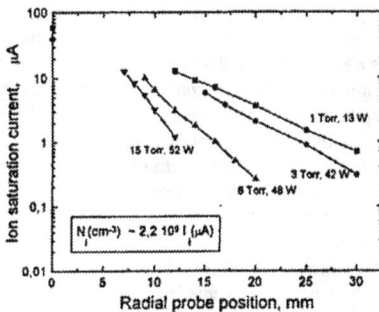

Figure 3. Ion saturation current for double probe system in hydrogen in the space outside the plasma ball structure.

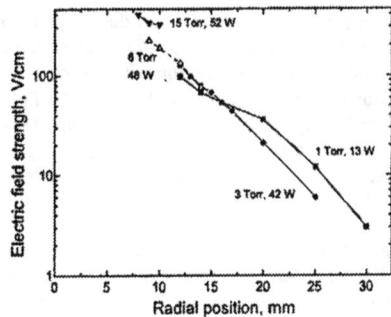

Figure 4. Electric field strength in the space outside of the hydrogen plasma ball structure.

b) Internal region of the luminous discharge.

Measurements showed that the voltage-current characteristics of double probe system passed through the zero current at a certain DC voltage difference $\Delta U_p(I_p=0)$ between the probes [9, 10]. The electron temperature was monotonously decreased with radius while the ion saturation current and $\Delta U_p(I_p=0)$ were slightly changed in the plasma ball and decreased abruptly at the boundary (Fig. 5,6). The last one fell down to zero in the external region of the discharge. The estimated electron density exceeded the critical value inside the plasma sphere.

At present a simple interpretation of these results is hardly possible. Some comments on the nature of $\Delta U_p(I_p=0)$ was done in [9,10]. Nonlinear interaction of electromagnetic waves with the plasma accompanied by the frequency transformation can be the origin of appearance of $\Delta U_p(I_p=0)\neq0$ [17]. Such a transformation can be effective in the region of maximal values of electric fields namely in the bright near electrode sheath. The DC field can produce the structure of electrostatic trap and double sheath can exist between the plasma ball and surrounding providing the sharp boundary of the plasma ball.

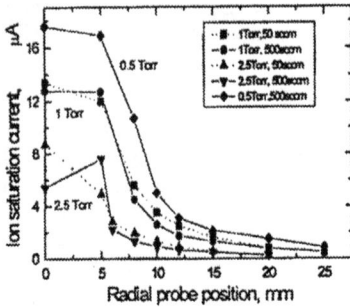

Figure 5 Ion saturation current for double probe system inside the nitrogen plasma ball structure.

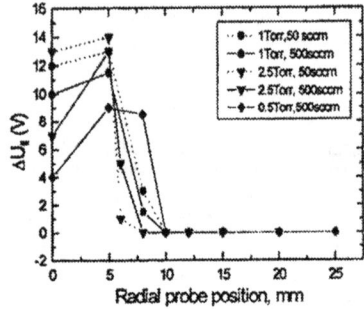

Figure 6. Difference of floating potentials of probes inside the nitrogen plasma ball structure.

## 4.3. Optical measurements

Figures 7,8 show distributions of the integrated over the spectrum intensity of plasma emission along the axis of the rod antenna with diameter of 6 mm. One can see a bright near-electrode plasma sheath surrounded by a plasma region with weaker emission. As a rule, the total discharge emission is the linear function of microwave power. The growth of plasma brightness near the external discharge boundary was observed in hydrogen discharge. It is important that in spite of small thickness of the sheath, the major part of plasma emission is attributed exactly to this region thus the total plasma emission defines the properties near electrode region. Measurements showed that the intensities of atomic lines of H and Ar (5% of Ar was added to $H_2$ for diagnostics) and molecular bands in $H_2$ plasma present linear dependence on microwave power. This means that emissive states of heavy particles are excited by the direct electron impact. Analysis also showed that the degree of hydrogen dissociation in near electrode sheath is of 1% and H atom emission is caused by the process of dissociative excitation of $H_2$ molecules.

One feature can be noted in comparing $H_2$ and $N_2$ discharges which seems indicate the differences in mechanisms of physical processes in both plasmas: the discharge in $N_2$ is more symmetric with respect to the end of antenna, i.e. the antenna is covered by the plasma ball in $N_2$ in more extent than in $H_2$ discharge. This difference can be related to differences in mechanisms of plasma emission excitation and ionization in both gases: in the $H_2$ plasma the ionization and excitation of heavy particles occurs due to direct electron impact whereas in $N_2$ plasma these processes go along with a participation of excited particles and by means of secondary processes. Thus plasma of $N_2$ can exist in the regions with smaller electric fields than necessary for $H_2$ plasma. As high energy part of the electron energy distribution function is sensitive to the electric field strength, the visible structure of discharge is defined by spatial distributions of the electron density and field.

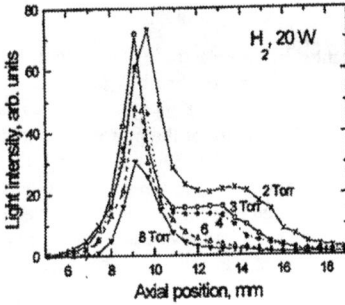

Figure 7. Axial distributions of the intensity of hydrogen plasma emission

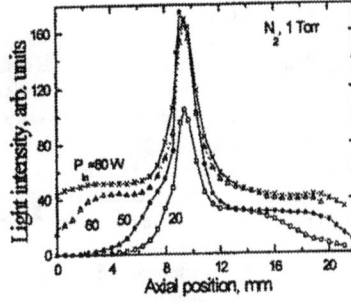

Figure 8. Axial distributions of the intensity of nitrogen plasma emission

## 4.4. Gas temperature

Gas temperature defines the peculiarities of interaction of electromagnetic field with a plasma through the parameter $\nu/\omega$, where $\nu$ is the effective frequency of electron-heavy particle collisions, as well as the kinetics of plasma processes. The gas temperature was defined in hydrogen plasma through the relative intensities of rotational lines of the electron excited molecules $H_2(d^2\Pi_u)$ [18]. Calculations were based on intensities of Q and R-branches emission for diagonal ($v'=v''=0,1,2$) bands of Fulcher $\alpha$-system $H_2(d^2\Pi_u \rightarrow a^3\Sigma_g)$:

$$\ln\left(\frac{I_{\cdot\rightarrow\cdot\cdot}}{v_{\cdot\rightarrow\cdot\cdot}^4 S_{j',j''}}\right) = -\frac{hc}{kT_{rot}^*}F(j') + const , \qquad (2)$$

where $I_{\cdot\rightarrow\cdot\cdot}$ is the intensity of emission of rotational spectra, $v_{\cdot\rightarrow\cdot\cdot}$ is the frequency of rotational transition, $S_{j',j''}$ is the Hönl-London factor, $F(j')$ is rotational energy of higher state, Rotational temperature of the ground state of $H_2$ was calculated as:

$$T_{rot}^0 = T_{rot}^* \frac{B^0}{B'} , \qquad (3)$$

where $B^0$ and $B'$ are rotational constants of the ground and excited states. The rotational temperature of the ground state was assumed to equal the gas temperature.

The total intensity of the discharge emission was used for calculations. The main input in this emission give the bright thin near-electrode sheath. Thus the defined temperatures should be attributed to this region of the discharge. The estimated values for rotational temperature of excited state were of 300-350 K at pressure 1 Torr. Taking into account the difference in the rotational constants for excited and ground states, the gas temperature was not exceed 700 K. This value slightly changed with gas pressure.

This result can be directly related with clarifying the processes of absorption of microwaves in a plasma, namely with study of the role of plasma resonance. The role of this mechanism decreases with increase of the ratio $\nu/\omega$. As it was shown by the modelling, increase of the electromagnetic field strength in the region of plasma resonance (n=1) become negligibly small at pressures 3-4 Torr and radial field profile is defined by the electrodynamics of discharge device. As the gas heating is small this process can not considerably increase the range of pressures where the role of resonance is high. From the other hand, the results are known where the same plasma structures were generated in microwave discharge at pressure of 100 Torr [19]. Thus it can be concluded that resonance processes can not be responsible for existence of the discharge at all pressures at least in near-electrode plasma region.

## 4.5. Results of Modelling

It was shown that for low absorbed power the discharge is presented by thin electrode plasma layer with the electron density everywhere below the critical value. As the level of the absorbed power is high enough for the electron density to reach the critical value, at pressures when the plasma resonance can exist (small values of $v/\omega$), the structure of the discharge changes. Further rise of the power leads to the radial expansion of the discharge with electron density just above the critical value (Fig.10). Thus it is impossible to achieve densities essentially higher than the critical value by rising the input (and hence the absorbed) power. Such a behavior seems to be general for unbounded plasma with $n>1$ sustained by microwaves when plasma is not limited in one or more directions by a discharge vessel. As an example we refer to well known phenomena in the surface wave sustained plasma in the long tubes where the length of plasma slab increases with the input power.

In the cases of plasma resonance the radial dependencies of the absorbed power were similar to that observed in experiments for plasma emission (Fig.10). Thus the results of modelling has revealed a qualitative agreement with part of the experimental data (it should be noted that the same plasma behavior was also observed in experiments at pressures when the role of resonance was negligible).

The results of simulations occurred to be insensitive to the change of the radius of the chamber. The decrease of the radius of the internal electrode lead to increase of the peak value of the electric field. This result agrees with the experimental evidence of the fact that the discharge sustained at thin internal electrode exists at lower level of microwave power than that at the thick electrode.

Calculations showed that sharp boundary in electron density distribution, as it was observed in experiments for diffusion controlled regimes, can not be realized at any microwave field radial distributions including the model stepwise shape.

Simulations of the dissociation of hydrogen under conditions of plasma resonance in the range of the absorbed power 1-20 W showed that maximum of the degree of dissociation did not exceed 25 %. It was shown that the maximum of the dissociation degree was mainly determined by the fields inside the plasma formation and not by a thin peak of the electric field situated nearby the internal electrode.

Qualitatively the results for cylindrical and spherical systems of electrodes are not considerably different. In spherical system all spatial distributions are more driven towards the internal electrodes.

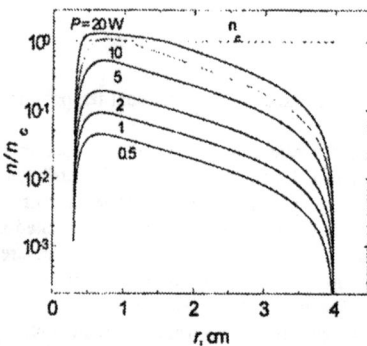

Figure 9. Normalized radial distributions of electron density in the cylinder chamber with R=4 cm, $r_{el}$=0.3cm in hydrogen at pressure 1 Torr.

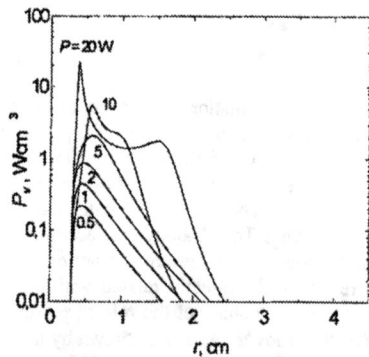

Figure 10. Radial distributions of absorbed power per unit of volume in the cylinder chamber with R=4 cm, $r_{el}$=0.3cm in hydrogen at pressure 1 Torr.

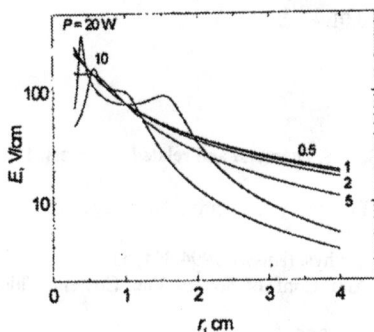

Figure 11. Radial distributions of electric field strength in the cylinder chamber with R=4 cm, $r_{ef}$=0.3cm in hydrogen at pressure 1 Torr.

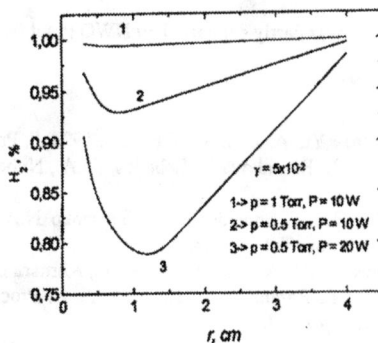

Figure 12. Radial distributions of relative concentrations of hydrogen molecules in the chamber with coaxial system of electrodes. γ is the probability of hydrogen atoms recombination on the stainless steel surface.

## 5. CONCLUSION

This paper is devoted to phenomena observed in microwave electrode discharges in molecular gases when microwave energy is introduced to the discharge chamber by means of antennas of different shapes and when the size of plasma is much less than the discharge chamber dimensions. Electrode microwave discharge appears near the exciter/antenna and has nonuniform structure: the exciter is surrounded by the bright narrow plasma sheath which covered by outward less emissive plasma region with sharp external boundary. The latter is ball-shaped when the diameter of exciter is less than that of the plasma and details of the structure depend on the mechanism of ionization.

The luminous discharge region is surrounded by dark space characterized by exponential spatial decay of the electron density. Some results of experiments can be interpreted by the existence of the surface wave spreads along the sharp spherical boundary of the discharge.

The discharge seems to consist of two coupled regions which play different roles.

The bright electrode region is characterized by high level of plasma emission and electromagnetic field strength, low degree of gas dissociation. In this region the direct processes of excitation and ionization by electron impact prevail over the stepwise processes. The field profile in this region has small influence on the processes in the main body of the discharge (the ball region).

The ball plasma region is characterized by monotonous decrease of the electron temperature, of the plateau profile of plasma density and DC voltage between the double floating probes. The last values are sharply decreased in the boundary layer and DC voltage falls down to zero level.

Quasi-static modelling showed the partial qualitative agreement with experimental results (increase of plasma diameter with absorbed power, slight change of electron density inside the plasma ball) only in the presence of plasma resonance ($v<\omega$). It was shown also that microwave field with any radial profile can not alone provide the observed abrupt fall down of electron density in the boundary region in the case of diffusion controlled discharge. This means that additional DC field should be superimposed with microwave field in a plasma which traps electrons inside the plasma ball. Such a field was indicated by probe measurements. The near electrode plasma sheath, where the highest electromagnetic field exists, can be the source of DC field. Thus the role of this region seems to be high in the formation of plasma structure.

136

**Acknowledgments**

This study was partly supported by NWO grant 047.011.000.01.

**References**

1. Lebedev Yu. A. J. Phys. IV France. 1998, **8**, Pr7-369.
2. Bardos L, Barankova H, Lebedev Yu.A., Nyberg T., Berg S. Diamond and related materials, 1997, **6**, 224.
3. Raizer Yu.P., Shneider M.N., Yatsenko N.A. Radio Frequency Capacitive Discharge. CRC Press. Boca Raton, Tokyo, London, 1995.
4. Brovkin V.G., Kolesnichenko Yu.F., Khmara D.V. Appl. Phys. (Russ.), 1994, N4, 5.
5. Bardos L, Barankova H, Lebedev Yu.A. Proc. 42-nd Ann. Conf. of Soc. of Vac. Coaters, Chicago, 1999, paper E-7.
6. Bardos L., Lebedev Yu.A. Plasma Phys. Reports, 1998, **24**, 956.
7. Bardos L., Lebedev Yu.A. Technical Physics, 1998, **43**, 1428.
8. Lebedev Yu.A., Mokeev M.V. Tatarinov A.V. Plasma Phys. Reports, 2000, **26**, 272.
9. Lebedev Yu.A., Mokeev M.V. High Temperature, 2000, **38**, 358.
10. Bardos L., Lebedev Yu. High Temperature, 2000, **38**, 552.
11. Räuchle E. . J. Phys. IV France 1998, **8**, Pr7-99.
12. Lebedev Yu.A. Plasma Sources Sci.&Technol. 1995, **4**, 4740.
13. Lebedev Yu.A. High Temperature, 1995, **33**, 846.
14. Gildenburg V.B.,Gol'tsman V.L., Semenov V.E. News of high school, Radiophysics (Rus.), 1974, **17**, 1718.
15. Lebedev Yu.A., Epstein I.L. J. Moscow Phys. Soc., 1995, **5**, 103.
16. Moisan M., Shivarova A., and Trivelpiece A.W. Plasma Phys., 1982, **24**, 1331.
17. Brandt A.A., Tihomirov Yu.V. Plasma multiplicators of frequency. Publ. "Nauka" Moscow, 1974.
18. Zhou Qing, D. K. Otorbaev, G. J. H. Brussaard, M. C. M. Van de Sanden, D. C. Schram J. Appl. Phys., 1996. **80**, 1312
19. Brovkin V.G., Kolesnichenko Yu.F., Khmara D.V. In "Ball lightning in the laboratory", Publ. "Chemistry" (Moscow, Russia) 1994, 119.

# MODE STRUCTURES OF SURFACE WAVE PLASMA EXCITED AT 915 MHz

**M. Nagatsu, A. Ito, N.Toyoda[*], H. Sugai**

Graduate School of Engineering, Nagoya University,
Furo-cho, Nagoya 464-8603, Japan
* Nissin Inc., 10-7 Kamei-cho, Takarazuka 665-0047, Japan

**Abstract.** Mode characteristics in the surface wave plasmas (SWPs) excited at 915 MHz have been studied for Ar discharges in a chamber with a square cross section of 40x40 cm². Various types of slot antennas are tested at pressures of 20-100 mTorr. The density jumps were observed in the planar SWP at 915 MHz similarly to the previous observation at 2.45 GHz SWP, when two-longitudinal slots or crossed slots were used as a launching antenna. With these antennas, the various optical emission patterns were observed just below the square quartz window, when varying the incident power or pressure. From space-resolved electric field measurements, some of them were distinctly identified as transverse magnetic modes $TM_{mn}$ having the mode numbers $(m,n)=(3,2)$ and $(1,4)$, especially in the cases of two-longitudinal slots and crossed slots. The electron density measurements and theoretical SW dispersion relations support the modes identified in the experiment.

## 1. INTRODUCTION

High electron density surface wave plasmas (SWPs) produced by microwave discharge conventionally at 2.45 GHz have been recently developed for large area plasma processing, such as etching, ashing and CVD processing. Large-area planar SWPs are easily produced in a wide range of pressures from a few mTorr to several Torr [1-8]. Furthermore, the SWP source is attractive particularly in manufacturing of liquid crystal display which needs a plasma with square cross section larger than 100cmx100cm.

So far, there has been no report on the production of square planar SWPs of large dimension. In addition, it is of importance to lower the discharge frequency from 2.45 Ghz to 915 MHz, in order to enlarge the plasma dimension: the free space wavelength in vacuum at 915 MHz is a factor of 2,7 larger than at 2.45 GHz, which makes the antenna dimension large and the excitation of long-wavelength SWs easy. Furthermore, it is noted that 915 MHz SWP can be sustained at a low value of electron density, say $1 \times 10^{10}$ cm$^{-3}$, since it has the lower critical electron density for SWP than that in 2.45 GHz SWP by factor of 7 [9].

In this paper we studied the plasma production characteristics by examining a different slot structures and also investigated the mode structures of 915 MHz SWP in detail by performing the spatially resolved electric field measurements. Observed electric field distributions are compared with the optical emission patterns and those obtained from the theoretical analysis of SW modes.

## 2. EXPERIMENTAL SETUP

A schematic drawing of the experimental setup is shown in Fig.1. The 915 MHz UHF wave guided by a rectangular waveguide was fed into an aluminum chamber with cross section of 40x40 cm² and a height of 30 cm filled with Ar gas through slot antennas cut in the board waveguide face. For vacuum sealing, a quartz plate with a cross section of 25x25 cm² at lower surface and thickness of 35 mm was used. Thus, the plasma has an interface partly with the squared quartz plate (central region) and partly with the surrounding metal plate. The output power of 915 MGz magnetron is

138

variable from zero to 5 kW by an attenuator. For slot antennas, we have tested here four types of slot antenna structures: (a)crossed slots, (b)two-transverse slots, (c)two-longitudinal slots and (d) a combination of 2-transverse and 2-longitudinal slots, which are described in ref. 10, except for crossed slots. The vacuum chamber was pumped down to the order of $10^{-6}$ Torr by a turbo-molecular pump. For the spatial distribution measurements of the plasma parameters, we used a Langmuir probe with a 0.7 mm–diameter, 7 mm–long platinum wire tip. The local electric field components at 915 MHz were also measured using a wire tip of Langmuir probe, which picks up the electric field component mainly along the wire tip. By bending the wire tip along $x-$, $y-$ or $z$–direction, each component of of the electric field was measured, although small amount of other components might be simultaneously received. The wave amplitudes were directly measured using a spectrum analyzer.

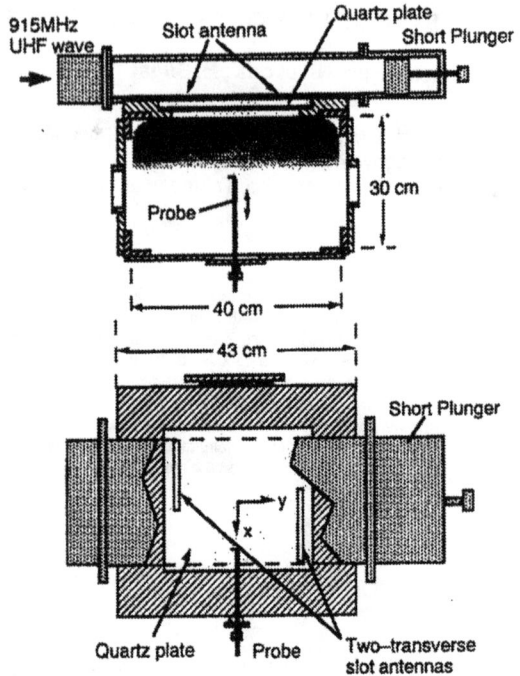

**Figure 1.** Schematic drawing of the experimental setup of 915 MHz SWP.

## 3. RESULTS AND DISCUSSION

First, we show the experimental results of plasma production efficiency by plotting the ion saturation currents of the Langmuir probe versus the net incident power for four slot structures in Figs. 2(a) and 2 (b) at pressures of 28 mTorr and 80 mTorr, respectively. At 28 mTorr, 4–combined and 2–transverse slots show the very similar plasma production efficiency, where both ion saturation currents tend to saturate when the net incident power exceeds ~500 W. As for 2–longitudinal and crossed slots, we clearly see the density jumps at ~850 W and ~550 W, respectively, similar to those observed in the previous cylindrical SWP experiment.[10,11] At a higher pressure, 80 mTorr, the ion saturation currents for 4–combined slots increase linearly with the net incident power up to 1 kW. From the optimum antenna design, we found that the 4-combined slots is most efficient for the plasma production among the present slot structures. These results are almost the same as the previous results of 2.45 GHz SWP reported in ref. 11. In Fig. 2, we also showed some photographs of typical optical emission patterns observed just below the wave launcher.

In the cases of the 4–combined slots and 2–transverse slots, a similar optical emission with complicated mode patterns was observed when $P_i$=700 W at 28 mTorr. In the 2–longitudinal and crossed slots antenna, where density jumps are clearly observed, we found a regularly spaced optical emission patterns below the quartz plate: one is the 4–lines shape of emission pattern in the case of crossed slots, and the other is the 2×3 multiple emission pattern in the case of 2–longitudinal slots. When the pressure was increased, the optical emission patterns changed. When the net incident power was 700W, we observed the same ion saturation current and the same optical emission pattern in the cases of the 2–longitudinal and crossed slots.

The present results suggest that the optical emission patterns, that is, the electric field structures near the quartz–plasma interface are determined from the boundary conditions (the geometry and the dielectric constants of quartz and plasma) even when the slot structures were different. These behaviors of mode characteristics in the present square-shape 915 MHz SWP were very analogous to the 2.45 GHz SW mode characteristics observed in a 22 cm–diameter cylindrical vacuum chamber.[11]

It is a new finding that a linear plasma production performance is available in the 915 MHz SWP when appropriate slot structures were employed. However, it is not understood well why the plasma production efficiencies in the 4-combined slots and 2-transverse slots were linear against the net incident power, differently from those in the 2-longitudinal and crossed slots.

In other words, it is still an open quation why the density jump occurs in the other slot structures, such as 2-longitudinal or crossed slots. These could be figured out by the rigorous treatment based on the electro-magnetic wave analysis taking the coupling of waveguide applicator–dielectric–plasma in a given chamber geometry into account. Now, another question is how the SW field structures were determined

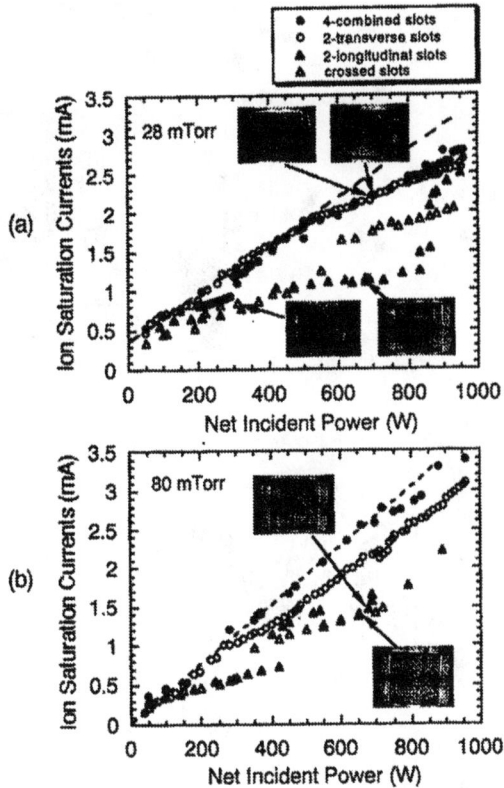

**Figure 2.** Experimental results of the plasma production efficiencies for four slot structures at pressures of (a) 28 mTorr and (b) 80 mTorr, respectively. The inserted photos show slot structures and optical emission patterns.

under the present quartz-metal combined boundary condition. To make it clear, we have carried out measurements of two–dimensional distributions of electric field intensities, by scanning the wire probe along x-axis (perpendicular to the waveguide axis) or y–axis (parallel to the axis) just below the quartz surface, as shown in Fig. 3.

Figures 4 and 5 show the electric field distributions of x–, y– and z–components measured with a wire probe scanning along x-axis in the 2 longitudinal slots antenna when $P_i$=700 W at 80 mTorr, and those along y-axis in the crossed slots antenna when $P_i$=300 W at 28 mTorr, respectively. As shown in Figs. 4 and 5, the inner region is occupied by the quartz from $x$=-12.5 cm to $x$=12.5 cm and the outer region is occupied by the metal plate. First, it is seen from Fig. 4 that the field components vary periodically, that is, x–components vary as $\sin k_x x$, while y– and z–components vary as $\cos k_x x$ in the quartz region. Here, $k_x$ is x–component of wave number and given by $m\pi/a$, where a=24.8 cm is the size of square quartz window. As for y–axis distribution of electric field intensity, we similarly found that they varies as $\sin k_y y$ for x– and z–components and as $\cos k_y y$ for y–components, where $k_y$=$n\pi/a$. From the field measurements, the boundary conditions at the quartz edge are given by

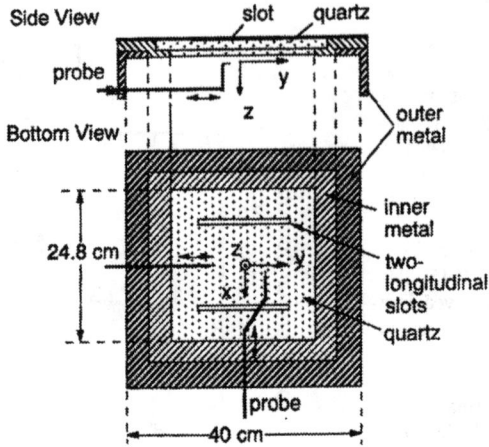

**Figure 3.** Illustration of the electric field measurements.

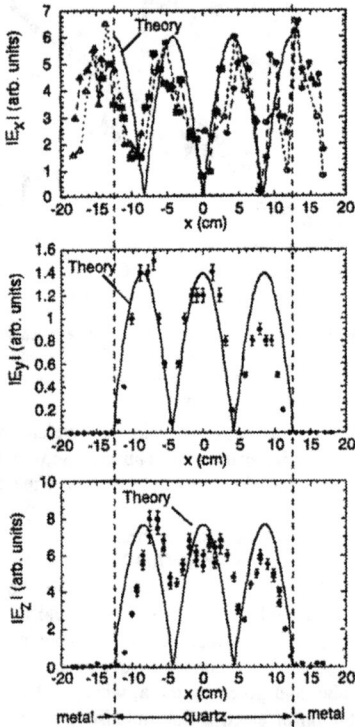

**Figure 4.** Electric field distributions of $x$-, $y$- and $z$-components measured with a wire probe scanning along $x$-axis in the 2 longitudinal slots antenna when $P_i$=700 W at 80 mTorr.

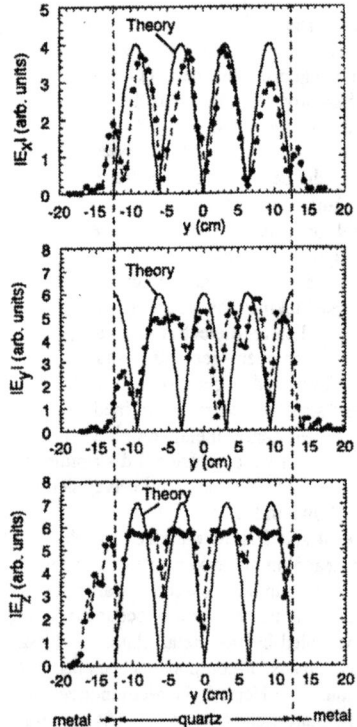

**Figure 5.** Electric field distributions of $x$-, $y$- and $z$-components measured with a wire probe scanning along $x$-axis in the crossed slots antenna when $P_i$=300 W at 28 mTorr.

$$E_y = E_z = 0 \quad \text{at} \quad x = \pm\frac{a}{2}$$

$$E_x = E_z = 0 \quad \text{at} \quad y = \pm\frac{a}{2} \tag{1}$$

Then, assuming an evanescent wave structure along the $z$-direction with the damping constant of $\gamma$, we can express the spatial distributions of electric field components for the transverse magnetic (TM) waves as follows.[12]

$$E_x = Ak_x\gamma \sin k_x x \cos k_y y \exp(-\gamma z)$$

$$E_y = Ak_y\gamma \cos k_x x \sin k_y y \exp(-\gamma z) \tag{2}$$

$$E_z = A(k_x{}^2 + k_y{}^2) \cos k_x x \cos k_y y \exp(-\gamma z)$$

From the spatial field distributions along $x$ and $y$, we can identify the observed electric field distributions in Fig. 4 as m=3 and n=2 ( that is, $TM_{32}$ mode) and those in Fig. 5 as m=1 and n=4 ($TM_{14}$ mode), respectively.        According to the boundary conditions of the observed fields given by eq. (1), we can regard as if the plasma were confined in a rectangular metal chamber having the same cross section as the square quartz.        Thus, we can reduce the troublesome analysis under the complicated boundary to a simple one under a rectangular quartz–plasma boundary.

Using two interface-model[13], we can analyze the SW dispersion relations for 915 MHz SWPs. With a=24.8 cm, numerical calculations show that the electron densities are given by $n_e$=2. 9×10$^{11}$ cm$^{-3}$ for $TM_{32}$ mode and $n_e$=1.3×10$^{11}$ cm$^{-3}$ for $TM_{14}$ mode, as shown in Fig. 6.        To compare them with the experiments, we also carried out the measurements of electron densities in the corresponding SWP discharges.

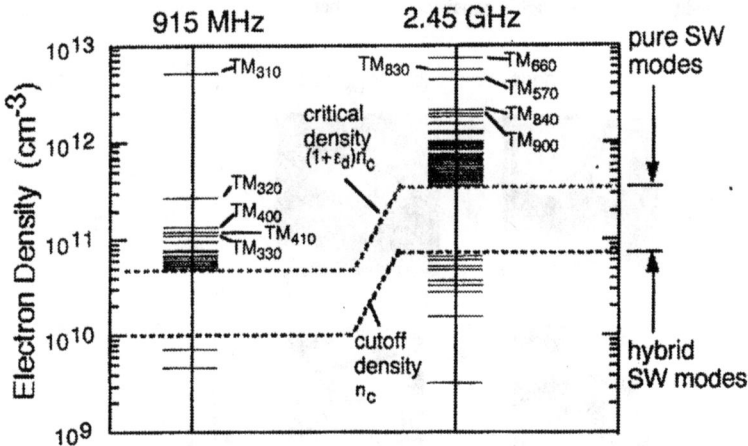

**Figure 6.** Numerical calculation results of theoretical dispersion relations of SW using wave frequency 915 MHz and quartz size of 24.8 cm.

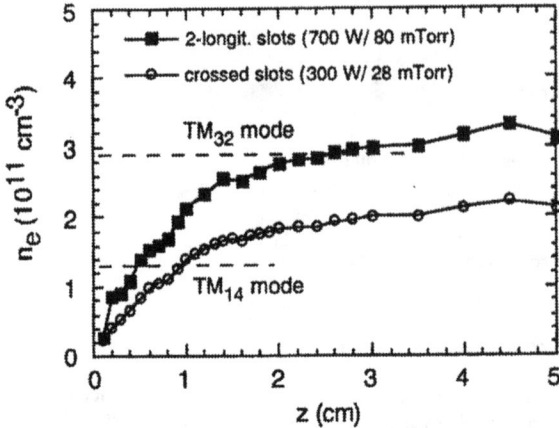

**Figure 7.** Axial distributions of the electron densities measured in the 2 longitudinal slots antenna when $P_i$=700 W at 80 mTorr and in the crossed slots antenna when $P_i$=300 W at 28 mTorr.

Figures 7 show the results of axial distributions of the electron densities. Theoretical values are fairly consistent with the electron densities at the axial positions where the electric fields exponentially drop. The damping constant $\gamma$ of the evanescent waves are measured by scanning the short-wire probe and we found that $\gamma \sim 1.4$–$1.6$ cm$^{-1}$ for TM$_{32}$ mode and $\gamma \sim 0.72$–$0.82$ cm$^{-1}$ for TM$_{32}$ mode. On the other hand, the damping constant $\gamma$ is theoretically given by the dispersion relation calculations; that is, $\gamma \sim 1.1$ cm$^{-1}$ for TM$_{32}$ mode and $\gamma \sim 0.83$ cm$^{-1}$ for TM$_{14}$ mode. Comparing them with experimental values, we found that they are roughly consistent each other.

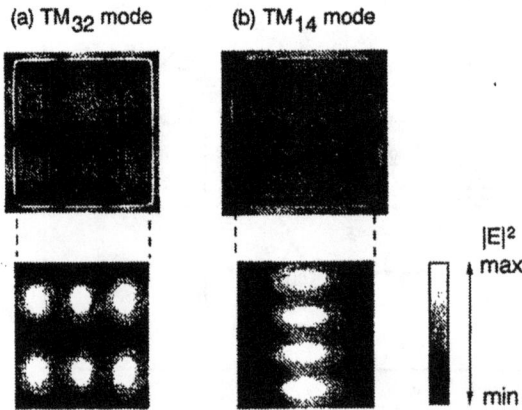

**Figure 8.** Two-dimensional plots of total electric field intensities(lower) numerically calculated using eq. (2). The corresponding optical emission photographs (upper) are shown for comparison.

The total electric field intensities of these modes are numerically calculated from the analytical expression of each field component, represented by eq. (2). Substituting the values of damping constants and wavenumbers into eq. (2), we can obtain the two-dimemsional plots of total electric field intensities, $|E|^2$. The results are shown in Fig. 8 together with corresponding optical emission photographs, from which we found a good agreement between the optical emission patterns in the quartz region and numerical calculation. At the quartz boundary, the $x$– and $y$–field components should continuously connect with the field lines which terminate perpendicularly to the metal top plate. Hence, these field components in the metal surface may relate with the strong optical emission, as shown in Fig. 8.

## 4. CONCLUSIONS

In conclusion, we studied the mode characteristics in the SWPs excited by 915 MHz UHF waves for Ar discharges in the vacuum chamber with a square cross section of 40×40 cm$^2$ by testing various slot antenna shapes. Results show that the density jumps occur in the planar SWPs at 915 MHz similarly to previous results of 2.45 GHz SWP, when two-longitudinal slots or crossed slots were used as a launching antenna. With these antennas, the various optical emission patterns were observed just below the square quartz window plate, when varying the incident power or pressure. With the spatially resolved electric field measurement, we found that some of them were distinctly identified as the transverse magnetic modes TM$_{mn}$ having the mode numbers of (m,n)=(3,2) and (1,4), especially in the cases of two-longitudinal slots and crossed slots. The electron density measurements under these TM modes support the experimental verification in the mode identification derived from theoretical SW dispersion relations. Lastly, we conclude that the TM modes in square SWPs can be similarly sustained even when the wave frequency changed from 2.45 GHz to 915 MHz and the square quartz plate was not fully covered the interface with the plasma.

## ACKNOWLEDGMENTS

The authors would like to thank Nissin Inc. for supporting the present work. This work was also supported by a Grant–in–Aid for Scientific Research from the Ministry of Education, Science, Sports and Culture in Japan, and by Creative and Fundamental R&D Program for SMEs from Japan Small Business Corporation (JSBC) and New Energy and Industrial Technology Development Organization (NEDO) in Japan.

## REFERENCES

1. Moisan M, Zakrzewski Z, J. Phys. D: Appl. Phys. 1999l **24** 1025.

2. Komachi K, Kobayashi S , J. Microwave Power & Electromagnetic Energy 1990 **25** 236.

3. Werner F, Korzec D, Engemann J, Plasma Sources Sci. Technol. 1994 **3** 473.

4. Bluem E, Bechu S, Boisse-Laporte C, Leprince P, Marec J, J. Phys. D: Appl. Phys. 1995 **28** 1529.

5. Nagatsu M, Xu G, Yamage M, Kanoh M, Sugai H, Jpn. J. Appl. Phys. 1996 **35** L341.

6. Odrobina I, Kudera J, Kando M, Plasma Sources Sci. Technol. 1998 **7** 238.

7. Yasaka Y, Nozaki D, Koga K, Yamamoto et al, Plasma Sources Sci. Technol. 1999 **8** 530.

8. Sugai H, Ghanashev I, Nagatsu M, Plasma Sources Sci. Technol. 1998 **7** 427.

9. Nagatsu M, Ito A, Toyoda N, Sugai H, Jpn. J. Appl. Phys. 1999 **37** L468.

10. Nagatsu M, Xu G, Ghanashev I, Kanoh M, Sugai H, Plasma Sources Sci. Technol. 1997 **6** 427.

11. Nagatsu M, Morita S, Ghanashev I, et al, J. Phys D: Appl. Phys. 2000 **33** 1143.

12. For example, Jackson J D : Classical Electrodyanamics, John Wiley & Sons, Inc., NewYork, 1962 Chap. 8 p.235.

13. Ghanashev I, Nagatsu M, Sugai H, Jpn. J. Appl. Phys. 1997 **36** 337.

# SURFACE MICROWAVE DISCHARGE ON DIELECTRIC BODY IN A SUPERSONIC FLOW OF AIR

V.M.Shibkov, A.V.Chernikov, V.A.Chernikov, A.P.Ershov, L.V.Shibkova, I.B.Timofeev, D.A.Vinogradov, A.V.Voskanyan

Department of Physics, Moscow State University, 119899, Moscow, Russia

**Abstract.** A new way of creation of the stable and constantly reproduced at various experiments surface microwave discharge into a boundary layer near a flat dielectrical body streamlined a supersonic flow of air is developed and experimentally investigated at a wide range of air pressure $p=10^{-3} \div 10^{3}$ torr. The breakdown characteristics determining a threshold of appearance of a surface microwave discharge are studied. Longitudinal size and velocity of spreading of a surface microwave discharge in air are determined at a wide range of gas pressure, power of microwave pulse and its duration. The spatial-temporary evolution of gas and vibrational temperatures is investigated. It is shown that the supersonic flow does not render strong influence on a common view of the surface microwave discharge on a dielectrical body, threshold of its appearance and gas heating. The surface microwave discharge can be used as a plasma source for microelectronics, material processing, and supersonic aerodynamics.

## 1. INTRODUCTION

Nonequilibrium low-temperature plasma generated by microwave and radio-high-frequency discharges is extensively used for various applications such as aerodynamics and material processing (etching, surface cleaning, sputtering, deposition, ion implantation, and so on). So, for example, a new direction of aerodynamics - so-called plasma supersonic aerodynamics was arisen [1,2]. In this case the various type of the gas discharges are applied with the purpose of influence on the characteristics of a gas flow near a surface of the flying bodies. However a physics of the discharge in a supersonic flow of gas until the present time is in a phase of development. There are many unsolved questions, among which are a problem of a gas breakdown in a flow, a creation and maintenance of the stable discharge in a supersonic flow of air, influence of a flow on parameters of plasma of the gas discharge and influence of the discharge on the characteristics of a supersonic flow.

The first laboratory experiments have shown an opportunity of a drag reduction at creation of the discharges of direct and alternative currents before a body, streamlined supersonic airflow. However electrode discharges in a flow are unstable and spatially non-uniform. Such discharges result in strong erosion of electrodes and model surface and reliably are not reproduced in various realizations. There was a task of search of optimum ways of creation of nonequilibrium plasma in a supersonic flow. One of such ways offered in our laboratory is the new version of a surface microwave discharge, namely, a microwave discharge on an external surface of a dielectric body streamlined a supersonic flow of air.

The low-temperature plasma widely also finds application in microelectronics. In this case a large diameter of plasma ($\varnothing \sim 300$ mm), high uniformity of plasma density ($\sim 1 \div 2$ %), high ion plasma density ($n_i \sim 10^{11} \div 10^{12}$ cm$^{-3}$), and low ion energy ($\varepsilon_i \sim 1 \div 10$ eV) are required for modern plasma sources. Nowadays the most perspective sources of plasma are the microwave sources on a basis of a electron cyclotron resonance (ECR source), helicon sources of high-frequency plasma and high-frequency sources with inductive and transformer coupled plasma. The direct transfer of an electromagnetic field energy to electrons occurs for the account of an electron cyclotron resonance. It allows to achieve a high density of the charged particles at low gas pressure that is the major advantage of this method. However complex configuration of ring magnetic traps and multipolar systems of magnets are used in ECR sources for creation of dense and homogeneous plasma of large volume. Moreover, with increase of a diameter of a plate it becomes more and more difficult to support a homogeneous plasma by magnetic fields. The complex configurations of a magnetic field used at practice essentially increase cost and considerably worsen convenience of operation and reliability of *ECR* sources. This concerns to helicon plasma sources too.

The microwave surfatron is the other direction at development of plasma chemical reactor of low pressure plasma. In a microwave surfatron the discharge represents a plasma waveguide on which the surface wave supporting ionization of gas is distributed. Thus for a wide range of frequencies it is possible to receive a plasma column of significant extent and forms determined by a dielectric wall limiting the discharge. A high electron concentration (exceeding critical density) and its appreciable unhomogeneity in a direction of a surface wave distribution are main feature of such discharge. It is known [3,4] that at creation of the high-frequency and microwave discharges inside dielectric tube filled by gas at a low pressure, the electromagnetic energy delivered to system is transformed to a surface wave. Thus, there is a self-sustaining system when a plasma medium created by the surface wave is necessary for a surface wave existence, i.e. the presence of plasma is a necessary condition for distribution of a surface wave. The surface wave is travelled in space so long as its energy is sufficient for creation of plasma with an electron density no less then a critical value. This way of plasma production and device for its creation refer to as a surfatron. This way is enough detailed investigated and widely is used. In this case we have the plasma - dielectric - free space system, i.e. plasma, created by a surface wave, exists inside discharge tube, filled the gas with the lowered pressure. Plasma is limited to walls of dielectric tube, which separate plasma from a free space.

This paper deals with a surface microwave discharge outside a dielectric body into a low pressure chamber. In this case we offer as though to turn out surveyed above system inside out. In this case plasma supported by a surface microwave is formed on an exterior surface of a dielectric. The similar experiments are carried out in [5-7]. In papers [5,6] the microwave generator at a frequency of *2,45 GHz* excited a rectangular resonator connected to a coaxial line by means of a loop for magnetic excitation. A dielectric quartz tube was placed between the outer and inner metal coaxial electrodes. The linear surface wave discharge formed the outer electrode. In [7] the planar microwave plasma applicator at a frequency of *2,45 GHz* was used. The design of this applicator is based on a principle of distributed tunable coupling of microwave power in conjunction with the use of an additional smoothing waveguide which supports homogeneous plasma excitation. The technical realization of this principle uses two rectangular waveguides in *T*-shape configuration. This gives the possibility to generate plane plasmas.

## 2. EXPERIMENTAL INSTALLATION

In the report the experimental data on study of properties of a surface microwave discharge created in a supersonic flow of air with Mach number *M=2* are discussed. The magnetron generator had the following characteristics: a wavelength $\lambda=2,4$ cm; a pulsed microwave power $W<600$ kW; a pulse duration $\tau=1\div300$ $\mu s$; a pulse period-to-pulse duration ratio $Q=1000$. The microwave energy introduced into the vacuum discharge chamber by a waveguide system of rectangular section $9,5x19$ mm through a directional coupler. All waveguide system was hermetic. It was filled by $SF_6$ at pressures up to 5 atmospheres for avoidance of an electrical breakdown. The discharge was formed in a cylindrical chamber. The inside diameter of the chamber is equal *1 m*, its length is equal *3 m*. The vacuum system allowed to investigate the surface microwave discharge in a diapason of pressures from $10^{-3}$ *Torr* up to *760 Torr*. The waveguide entered into the discharge chamber ended the special antenna, on which the surface microwave discharge in a supersonic airflow was created. The discharge was formed on an external surface of a dielectric body of rectangular section *S=1x2 cm* and length *L=15 cm* and on a dielectric flat plate by *1 cm* in thickness, *14 cm* in length and *20 cm* in width. The direction of a supersonic flow was opposite to the direction of a surface microwave discharge spreading.

## 3. DIAGNOSTICS METHODS

For obtaining of the threshold characteristics of a surface microwave discharge the dependencies of the minimum input power, at which the discharge on an exterior surface of a dielectric body starts to be formed, on at different values of air pressure in a vacuum chamber and at different duration of microwave pulses were measured. In experiments the moment of formation of the discharge was registered visually or on appearance of a signal on the screen of a double-beam oscilloscope from the collimated photoelectric multiplying tube, tuned on area of the antenna at an edge of a waveguide. Thus, on the second beam of an oscilloscope the signal from a microwave detector was applied. The amplitude of this signal was proportional to a pulsed microwave power.

The common view of a surface microwave discharge was registered on a film and a video camera in two projections (side view and top view). It has allowed to fix the longitudinal size of a surface discharge and to measure longitudinal velocity of its distribution at different air pressures, pulse duration and microwave powers.

For an experimental investigation of a surface microwave discharge in supersonic flow it is necessary to use contactless methods and the optical diagnostics of plasma fully satisfy this requirement. The vibrational temperature $T_v$ was measured by the relative intensity of the nitrogen and $CN$ bands. The gas temperature $T_g$ was determined by the distribution of the intensity of rotational lines of molecular bands of a second positive system of nitrogen.

## 4. EXPERIMENTAL RESULTS

Before to study the parameters of a surface microwave discharge it was necessary to determine conditions of its appearance and degree of influence of a supersonic flow on a common view of the discharge. *Fig.1* represents the dependence on an air pressure of minimal microwave power which is necessary for a breakdown and beginning of formation of a surface discharge on a dielectric body. One can see that the power of the used generator is sufficient for creation of the surface microwave discharge in a range of air pressures $p=10^{-3} \div 10^3 Torr$. The minimum power which is necessary for formation of a surface discharge decreases at first at increasing of an air pressure and then grows. The obtained dependence is analogue of a Pashen curve. Such behaviour of power is explained by decrease of diffusion losses of electrons at increasing of an air pressure and growth of inelastic losses at it. For compensation of the electrons losses it is needed a large ionization frequency. The ionization frequency is growing function of electron temperature $v_i=f(T_e)$, i.e. reduced electrical field $E/n$. From here it is required to make a large power for creation of surface microwave discharge at small ($p<10^1 Torr$) and large ($p>10^1 Torr$) pressure, as it is observed at experiment (Fig.1).

On Fig.2 the threshold characteristics of a surface discharge created at different pulse durations are submitted. It is shown that at a fixed air pressure the power necessary for formation of a surface discharge sharply decreases with growth of a pulse duration from $1,5$ up to $10 \mu s$, whereas for $\tau \geq 50 \mu s$ the threshold does not depend almost on pulse duration.

Let's admit that the external field is included quickly in comparison with characteristic time of creation of the charged particles and remains constant during time of development of a discharge. In this assumption $v_i(t)$, $v_a(t)=const$ after the moment $t=0$ switches on of a electric field and the equation for full number of electrons in discharge area can be written down as

$$\frac{dn_e}{dt} = \left(v_i - v_a - v_D\right)n_e.$$ (1)

At these assumption the equation (1) has the exponential solution:

$$n_e = n_{eo}\, exp\left\{\left(v_i - v_a - v_D\right)t\right\},$$ (2)

where $n_{eo}$ is a number of backgrounded electrons, $v_i$, $v_a$, $v_D$ are the frequencies of ionization, attachment and diffusion.

From equation (2) it is possible to receive a frequency of ionization

$$v_i = \frac{1}{\tau} ln\frac{n_e}{n_{eo}} + v_a + v_D.$$ (3)

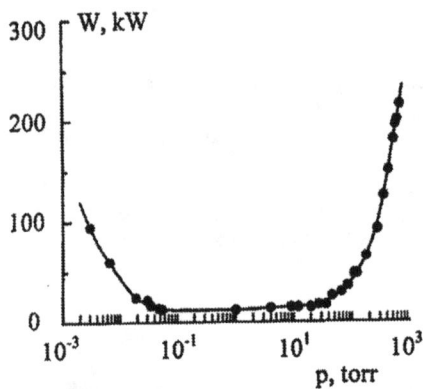

Figure 1. Dependence of a minimum pulsed microwave power which is necessary for creation of a surface discharge on air pressure at $\tau=50\ \mu s$ and $f=40\ Hz$.

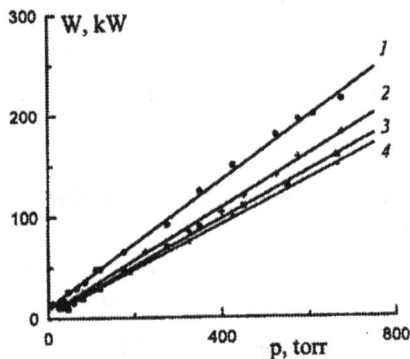

Figure 2. The threshold characteristics of a surface discharge created at different microwave pulses $\tau$, $\mu s$: 1 - 1,5; 2 - 5; 3 - 10; 4 - 50-100.

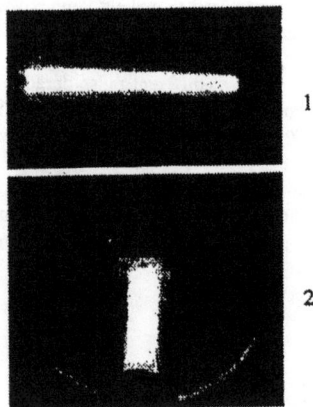

Figure 3. The common view of a surface microwave discharge on an exterior surface of a dielectric antenna at $M=2$, $p=40\ Torr$, $\tau=50\ \mu s$ and $f=100\ Hz$ (1 - side view, 2 - top view under $45^0$ to a vertical).

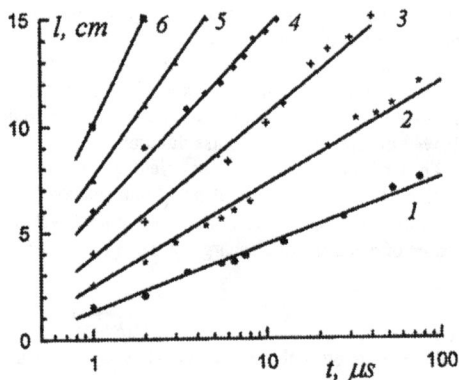

Figure 4. Dependencies of the longitudinal sizes of a surface discharge on duration of a microwave pulse at $p=10\ Torr$, $f=20\ Hz$ and microwave power $W$, $kW$: 1 -25; 2 - 35; 3 - 55; 4 -75; 5 - 100; 6 -175.

In case of low gas pressure ($p<1$ Torr) the diffusion coefficient is great ($D\sim1/p$), and the losses of electrons due to diffusion are significant $v_D>v_a$. The large ionization rate is necessary for their compensation, that is a strong electric field is necessary. Thus the threshold field is proportional to frequency of a microwave radiation and in inverse proportion to pressure of gas and sizes of discharge volume.

In case of high pressure the diffusion losses of electrons are insignificant and even not too large ionization rate provides gas breakdown. The losses of electrons for the account of attachment become main process. They also limit frequency of ionization. At pressure $p>10 Torr$ $v_a>v_D$ and the ionization frequency is defined by expression

$$v_i = \frac{1}{\tau} ln \frac{n_e}{n_{eo}} + v_a. \tag{4}$$

Let's estimate a frequency of ionization for conditions of our experiment: $\tau=50$ $\mu s$, $n_{eo}=10^2$ $cm^{-3}$, $n_e=10^{12}$ $cm^{-3}$. In this case first term in the right part of expression (4) is approximately equal $4\cdot10^5$ $s^{-1}$, and second - $6\cdot10^5$ $s^{-1}$, i.e. the ionization frequency is equal $10^6$ $s^{-1}$. From here one can see that the ionization frequency at $p=10$ Torr should be more than $10^6$ $s^{-1}$. Thus at $\tau\leq50$ $\mu s$ for definition of a threshold of gas breakdown it is necessary to take into account duration of a microwave pulse. It is well coordinated with experimental data (Fig.2).

During existence of the discharge number of processes, such as a gas heating, an excitation of vibrational of freedom degrees of molecules, an accumulation of long living metastable molecules, an accumulation of charged particles (electrons, positive and negative ions), a change of chemical structure of gas and number of other processes, proceeds in plasma. All these processes can result to change of conditions of secondary breakdown of gas. For a surface microwave discharge in a supersonic flow of air the time of replacement of gas near a body is equal $t=L/v$. This time is determined by the longitudinal size $L$ of body and velocity $v$ of supersonic airflow. Let's estimate this time for our case. The size $L=15$ $cm$ and flow velocity $v=500$ $cm/s$, then $t=300$ $\mu s$. A period of repetition of pulses in experiment is equal $T=25-50$ $ms$. From here all changes which have arisen during action of the previous microwave pulse will disappear to arrival of the following pulse and the breakdown power should not depend on frequency of repetition of pulses, as it is observed in experiment.

The common view of a surface microwave discharge on an exterior surface of a dielectric antenna of rectangular section $1x2$ $cm^2$ and length $15$ $cm$ in a supersonic airflow at Mach number $M=2$ is given on Fig.3. One can see that the discharge represents uniformly luminous plasma coating all surface of a dielectric body and a supersonic airflow does not destroy the surface discharge. The direction of propagation of a supersonic airflow is opposite to a direction of a surface wave traveling, i.e. direction of propagation of the discharge.

The longitudinal sizes of a surface discharge and its longitudinal velocities at $p=10$ Torr as a function of a microwave pulse duration are submitted on Fig.4 and Fig.5. Parameter of this curves is the microwave power. It is shown that the longitudinal size of a surface discharge and its velocity at fixed pulse duration grow with an increasing of a microwave power. Thus, on initial stages of a surface discharge existence this velocity is large and reaches of value $v=10^7$ $cm/s$ at $W=175$ $kW$, whereas at late stages the velocity of the discharge propagation decreases up to $v=10^4$ $cm/s$ at $W=25$ $kW$. From Fig.5 one can also see that all curves have the identical slope. The similar results are obtained also at other air pressures ($p=40$, $62$ and $100$ Torr). The processing of the received data has shown, that the time dependence of longitudinal velocity of the surface microwave discharges can be described by the law:

$$v = A\cdot t^{-(0.875\pm0.035)}, \tag{5}$$

where $A$ is a coefficient dependent on microwave power, $t$ is a time of microwave discharge existence.

Experimentally received dependence $A=f(W)$ for various pressure of air is given on Fig.6. One can see that at air pressure $p=10\div100$ Torr coefficient $A$ is directly proportional to $W/W_o$, where $W_o$ is a threshold

150

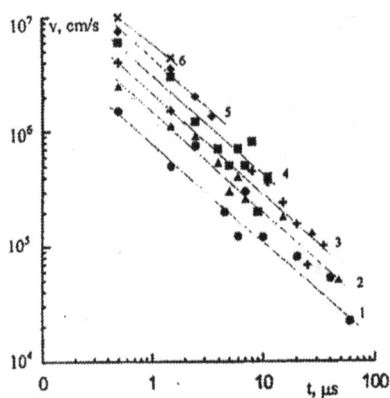

Figure 5. Longitudinal velocity of the surface microwave discharge on the antenna of rectangular section at p=10 torr, f=20 Hz, and microwave power W, kW: 1 - 25; 2 - 35; 3 - 55; 4 - 75; 5 - 100; 6 - 175.

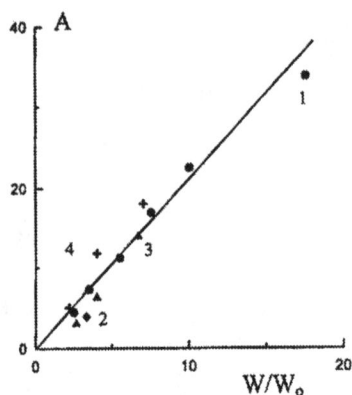

Figure 6. Coefficient A as a function of a microwave power at different air pressure p, torr: 1 - 10; 2 - 40; 3 - 62; 4 -100.

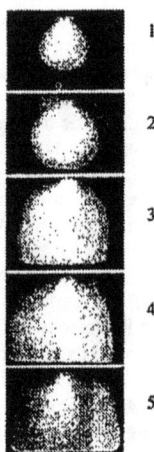

Figure 7. The common view of a surface microwave discharge (top view) on a dielectric plate at p=40 torr, τ=50 μs, f=20 Hz, and W, kW: 1-50; 2-100; 3-150; 4-200; 5-250.

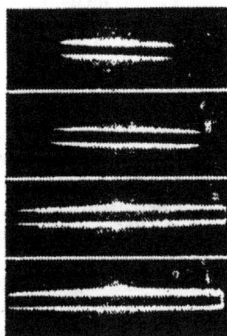

Figure 8. The common view of a surface microwave discharge (front view) on a dielectric plate at p=40 torr, τ=50 μs, f=20 Hz, and W, kW: 1-50; 2-100; 3-200; 4-250.

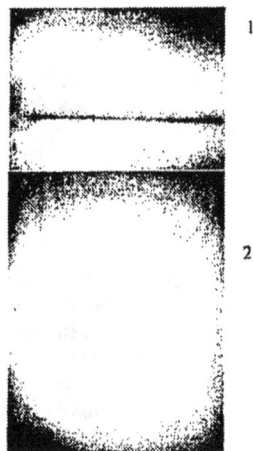

Figure 9. The common view of a surface microwave discharge (1 - side view, 2 - top view) on a dielectric plate at p=7·10⁻³ torr, τ=50 μs, f=20 Hz, and W=100 kW.

power for the given pressure of gas. Then using dependence $A=f(W)$ from the equation (5) it is possible to receive, that the longitudinal size of the surface microwave discharge is described by the law

$$L = 17\frac{W}{W_o} \cdot t^{(0.125\pm0.035)}, \tag{6}$$

here a length $L$ of surface microwave discharge has a dimension [$cm$], and a time $t$ has a dimension [$s$].

From here one can see that our experimental installation allows to receive the surface microwave discharges with length up to $1\ m$. It was shown also that the supersonic flow of air at a Mach number $M=2$ does not influence on common view of a surface discharge on a dielectric antenna of rectangular section and on values of microwave power which are necessary for its creation.

The experiments on creation of a surface microwave discharge on a flat plate were also carried out. For this purpose the raylon plate by $1\ cm$ thickness, $14\ cm$ length and $20\ cm$ width was used instead of the antenna. On Fig.7-9 the common views of a surface discharges on a plate are submitted at different values of input power and air pressure. At large pressure ($p>1\ Torr$) in a place of plasma creation the microwave energy is transformed in a surface wave, which spreads in all directions, creating thin ($\sim1\ mm$) uniformly luminous layer of plasma on a surface of a dielectric body. The part of a surface of a plate is coated by plasma increases when a microwave power growths. There is a reflection of a surface wave from edge of a plate when surface wave reaches to plate edge, that leads to a mode of a standing wave. With reduction of pressure the radial sizes of the discharge are increased and achieved a values $L_e\sim10\ cm$ for a pulse duration $\tau=100\ \mu s$ and air pressure $p\sim10^{-3}\ Torr$. This is explained by that fact, that at reduction of pressure the role of diffusion of the charged particles is increased. Spatial scale of nonuniformity of electron concentration $L_e$ is possible to estimate from expression

$$L_e = \sqrt{D_a\tau}, \tag{7}$$

where $D_a$ is a coefficient of ambipolar diffusion. Then for air pressure $p=10^{-3}\div10^{-2}\ Torr$ value $L_e\sim10\ cm$, that is well coordinated with experiment (Fig.9).

The velocity of propagation of a surface discharge on a plate also depends on power and duration of a microwave pulse as a velocity of the discharge propagation on the antenna. At power $W\approx350\ kW$ the surface discharge coats all plate ($S=14\times20\ cm^2$) for time $t\sim1\ \mu s$.

In various sections of a surface microwave discharge the dependencies of gas and vibrational temperatures on a pulsed microwave power were measured. The gas temperatures in section $z=2,5\ cm$ as a function of a microwave power is submitted in Fig.10. The dependence of vibrational temperature on a microwave power measured by molecular bands of the second positive system of nitrogen (point) and on molecular bands of $CN$ (daggers) is submitted in Fig.11.

It is shown, that the gas temperature increases from $\sim500\ K$ at $W=35\ kW$ up to $\sim1700\ K$ at $W=175\ kW$, whereas the vibrational temperature remains practically constant under these conditions, insignificantly decreasing with increase of microwave power. At this the maximal gas heating is observed in a place of excitation of a surface microwave discharge and gas temperature decreases by the end of the discharge.

The time evolution of gas temperature under conditions of surface microwave discharge is submitted on Fig.12. It was shown that on initial stage of existence of the surface microwave discharge the fast gas heating with rate $dT_g/dt>50\ K/\mu s$ is observed.

The similar result was received by us at research of kinetics of gas heating in conditions of the freely localized microwave discharge in air in the focused beam of electromagnetic radiation [8-11]. Various mechanisms are known which might lead to the heating of molecular gas. The contribution of these mechanisms to heating of molecular gas is detailed analysed in works [8-11].

In [8-11] it was shown that the mechanism connected with effective excitation of electron-exited states of nitrogen molecules at large values of the reduced electrical field $E/n\geq100\ Td$ and their subsequent quenching is responsible for the fast heating. At this the part of excitation energy of these states is transferred in the heat of air.

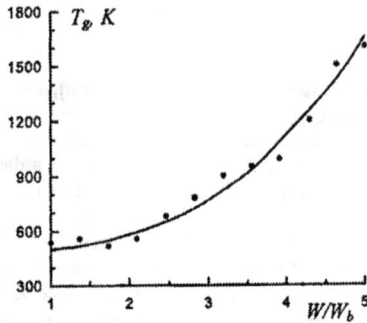

**Figure 10.** The gas temperature of a surface discharge as a function of a microwave power at $p=40$ Torr, $\tau=50$ $\mu s$ and $f=40$ Hz in section $z=2,5$ cm. ($W_b$ is breakdown power at air pressure $p=40$ Torr).

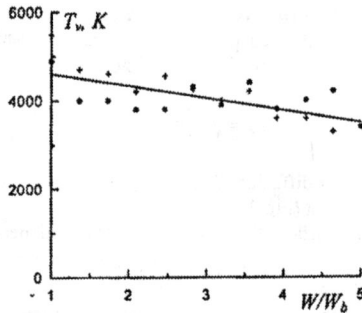

**Figure 11.** The vibrational temperature of a surface discharge, measured by molecular bands of the second positive system of nitrogen (point) and on molecular bands of $CN$ (daggers), as a function of a microwave power at $p=40$ Torr, $\tau=50$ $\mu s$ and $f=40$ Hz in section $z=2,5$ cm. ($W_b$ is breakdown power at air pressure $p=40$ Torr).

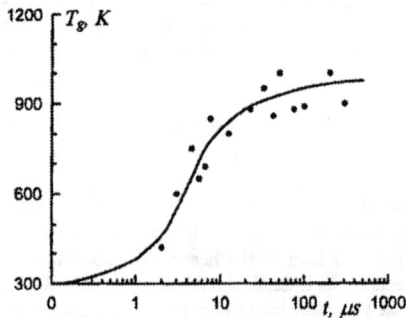

**Figure 12.** Time evolution of gas temperature under conditions of surface microwave discharge at $p=10$ Torr, $\tau=100$ $\mu s$, $f=10$ Hz and $W=100$ kW.

Our estimations show that only the quenching of electron-exited long-living states of the nitrogen molecules which are effectively created in conditions of a surface microwave discharge in air provides the observed gas heating rate.

## Acknowledgments

The work was supported by the EOARD (ISTC grant # 1866p).

## References

1. Intern. Space Planes and Hypersonic System and Technologies Conf. Workshop on Weakly Ionized Gases: Proceedings Colorado, USA, AIAA-1997; Norfolk, USA, AIAA-1998; Norfolk, USA, AIAA-99.
2. The First and Second Workshops on Magneto- and Plasma Aerodynamics for Aerospace Applications: Proceedings Mocsow, Russia, IVTAN-1999; Mocsow, Russia, IVTAN-2000.
3. Moisan M., Zakrzewski Z. J. Phys. D: Appl. Phys., 1991, 24, 1025.
4. Daviaud S., Boisse-Laporte C., Leprince P., Marec J. J.Phys.D: Appl.Phys. 1989, 22, 770.
5. Räuchle E. Journal de Physique IV, 1998, 8-Pr7, 99.
6. Gritsinin S.I., Kossyi I.A., Malykh N.I., Misakyan M.A., Temchin S.M., Bark Y.B. Preprint No1, Russian Academy of Sciences, General Physics Institute. Moscow, Russia, 1999.
7. Ohl A. Journal de Physique IV, 1998, 8-Pr7, 83.
8. Zarin A.S., Kuzovnikov A.A., Shibkov V.M.. Freely localized microwave discharge in air. -Moscow: Oil and gas, 1996, 204p (Rus).
9. Shibkov V.M. High Temperature. 1997, 35, 681.
10. Shibkov V.M. High Temperature. 1997, 35, 858.
11. Shibkov V.M. Kinetics of gas heating in plasma created in supersonic airflow. 9th Intern. Space Planes and Hypersonic Systems and Technologies Conf. Norfolk, Virginia, USA, 1999, AIAA-99-4965.

# CW LARGE VOLUME MICROWAVE DISCHARGE AT LOW PRESSURE

## A. A. Skovoroda, V. A. Zhil'tsov

NFI, RRC "Kurchatov Institute", 123182, Kurchatov sq. 1, Moscow, Russia

**Abstract.** The experiment on installation PNX-U is described. PNX-U is the prototype of plasma neutralizer of 1MeV negative $H$ ions in injector developed for ITER. The plasma neutralizer on a basis of ECR microwave discharge of low pressure in multipole magnetic trap (three-dimensional magnetic wall) is used. The basic purpose of the present experiment consists in check of the basic principles: CW and high power efficiency of plasma production. The experiments was curried out at gas ($H_2$, $Ar$) pressure ~$10^{-4}$ *torr* at stationary magnetic field in slits 0.35T at quasi CW (0.3s) input of 40κW microwave power on frequency 7GHz. The linear density $nl = 2\ 10^{18}\text{m}^{-2}$ at plasma length $l = 2$m (volume of plasma ~ 0.5m³) is obtained. The electron temperature 20-30eV is maximal on periphery of installation (on distance ~10cm from a wall) in ECR area and falls up to 5-10eV at the center (radius of plasma 0.3m), where practically there is no magnetic field. The total confinement time is ~1ms. The energy confinement time is determined not only by magnetic confinement of the charged particles, but also by radiation losses. The confinement of particles is essential different for peripheral and central areas. At microwave power switch off fast decay of the peripheral hot electrons (in time ~0.03ms) and slow decay of the central cold plasma (in time ~3ms = classical cusp life time) were observed. At microwave power switch on the potential barrier (~50V) was formed because of peripheral ECR heating. Potential confinement increases confinement time of peripheral plasma up to ~ 0.3 ms and central plasma up to ~0.15s. The average ionization degree 0.25 is obtained at average input power density of 80kWm⁻³ in 0.5m³ plasma volume.

## 1. INTRODUCTION

To provide an effective (80%) stripping of $D^-$ beam PN plasma linear density have to be at a level of $n_e l$~2-3 $10^{19}\text{m}^{-2}$. As the plasma neutralizer (PN) must be able to replace the gas neutralizer (the maximum efficiency of gas neutralization is 60%) in the ITER Neutral Beam Injector (NBI) of the reference concept its length, $l$, is defined and equal to 3-4m [1]. So the necessary plasma density in neutralizer should be $n_e$~5-7·$10^{18}\text{m}^{-3}$ and the degree of ionization >0.3. The total beam cross-section is large ~$1\text{m}^2$, hence the plasma volume is large ~$10\text{m}^3$. To keep final high PN efficiency, the plasma should be created very economically with power density ~100kWm⁻³.

The 3D multi-cusp magnetic wall configuration has been chosen for PN plasma confinement system [2]. An electrodeless microwave discharge at a low gas pressure (<$10^{-2}$Pa, $H_2$, $Ar$) has been used to generate a cold high ionized plasma in the multi-cusp configuration [3]. After some preliminary experiments the large volume (0.5m³) model PNX-U was constructed to proof of principle of the proposed PN scheme. The first results of the experiments have been presented in [4,5].

## 2. PNX-U INSTALLATION AND DIAGNOSTIC

Figure 1 shows PNX-U magnetic field configuration and the coils arrangement. The magnetic system consists of cylindrical side part and two end parts: a front one and a back one. The side part is arranged with ring coils connected in pairs with interchanging current direction from pair to pair. The inner diameter of the side part coil is 0.6m. The ends closing of the formatted multi-cusp magnetic system is fulfilled with transition to a ring coil of reduced diameter and finally on the axis to a coil of a racetrack form. The large part of experiments was fulfilled at dc coil current in the range of 2.5-2.7 kA. Two tangential microwave inputs are located at gaps between coils of a co-current pair to prevent large plasma flow to a launcher [4,5]. The plasma density exceeds the cut-off and the surface plasma - waveguide discharge is realized [6].

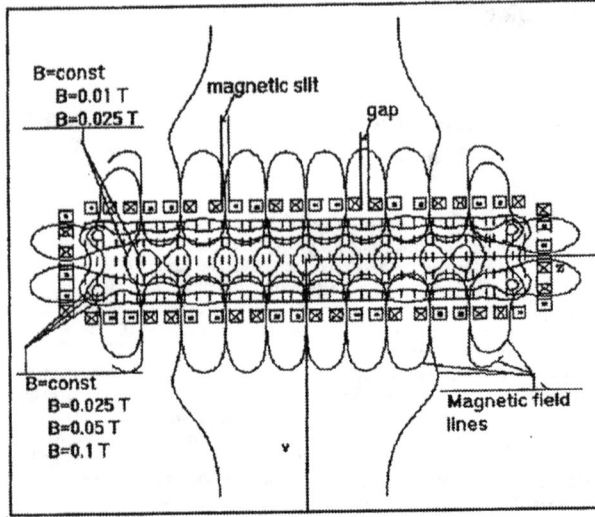

**Figure 1.** PNX-U magnetic field configuration, 2.5kA

The set of diagnostic systems is installed. Plasma linear density is measured with 4-mm microwave interferometer. The horns are oriented along the installation axis. The plasma density, potential and electron temperature profiles were measured locally with movable Langmuir probes. The arrangement of immovable Langmuir probes was situated behind the side magnetic slit. Ten probes are displaced across the slit plasma outflow to measure profiles. The set of multi-grid end loss analyzers (ELA) gives a possibility to get an information about electron or ion mean energy and plasma potential in different parts of the installation. The outgoing plasma particles reach ELA along magnetic field lines. The ELA-1 "sees" the particle flow coming from the periphery of the magnetic system. The ELA-2 "sees" the central region.

The PNX-U vacuum chamber is divided into three parts with two partitions. The central part and the right one have their own independent vacuum pumping with turbo-molecular pumps. The left part is evacuated via edge side windows and axial hole. There are two collectors of axial plasma flows at the front and the back ends. Two circular collectors accepts the end ring slits plasma flows. Each of the ring collectors is divided into two independent parts to estimate a vertical uniformity of the plasma flows. All collectors are water-cooled and are used for calorimetric measurements and are electrically insulated. Each of the collectors can be biased with different potentials. This allows to get their probe-like characteristics and in particular to measure a full ion current to each of the collectors. All collectors form the gas-boxes with gauges. This gives a possibility to measure gas pressure variation in the collectors during plasma flowing at different conditions.

The laser irradiation system and monochromator were used for laser-induced fluorescence local measurements of ion temperature and ionization degree in Ar discharge. The local emission spectroscopy is used for density profile measurements. The total light intensity is measured by photo-multiplier; gas composition is analyzed by mass-spectrometer; microwave directional couplers measure the incident and reflected microwave power.

## 3. EXPERIMENTAL RESULTS

The use of piezo-valve for an additional gas puffing gave a possibility to get a steady-state "H-mode" of operation with average plasma density $n_e \sim 1.5 n_{cut-off}$. An Ar plasma with $n_e l = 2 \cdot 10^{18} m^{-2}$ ($n_e = 10^{18} m^{-3}$) was generated at average injected microwave power density 80 kW/m³. Fig.2 shows the oscillogramms of such discharge. One can see the good microwave matching at large density. The another important observation consists in difference of density and Ar line decay. The plasma density disappears much longer, than light intensity.

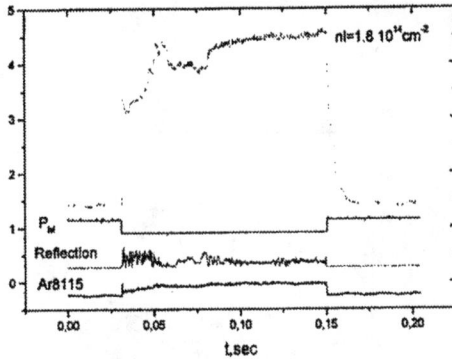

Figure 2. The H-regime signal oscillogramms: $n_e l$ interferometer, $P_M$ the injected microwave power, Reflected power, Ar line intensity

Such a behavior of signals is connected with existence of two groups of electrons: hot electrons on the periphery and cold electrons in the center. Figure 3 shows the probes measured space profiles of electron temperature and potential. One can see that the ECR heating and electron magnetic confinement on the periphery lead to the large electron temperature and plasma potential increment.

The ELA and optics measurements confirm the data obtained by probes. Figure 4 shows the ion flow energy analysis. The electrons are total stopped by –300V potential on one grid and ion energy function is analyzed by variable positive potential from 0V to +82V on the second grid. One can see that all ions have the minimal energy 53V, determined by maximal positive plasma potential along magnetic field lines. This value is coincides with probe data on periphery (see Fig. 3). The ion temperature, determined from sharp decrease of ions current after 53V, is estimated as 6V. This value of ion temperature coincides with value measured by laser-induced fluorescence diagnostic. The parabolic electron density profile on Fig. 3 coincides with profile, obtained by Ar line intensity local measurements.

Fig. 5 shows the electron flow energy analysis. The ions are total stopped by +60V potential on one grid and electron energy distribution is analyzed by variable negative potential from -50V to 0V on the second grid. The obtained value of electron temperature $T_e \sim 12$eV coincides with probe data in the center (see Fig. 3).

## 3.1. Potential confinement

The potential distribution on Fig. 3 is formed only at electron heating on the periphery by microwave. When the heating is switched off, hot electrons on the periphery decay very quickly with time ~30µs. Such time coincides with estimation for classic mirror confinement and is observed experimentally, as decay time of Ar line intensity (see Fig. 2). The decay time of cold electrons density from installation is large ~3ms (see Fig. 2) and coincides with classic cusp confinement time scaling

$$\tau[s] = 3\frac{B[T](V/LN)[m^2]}{T_e[eV]}$$ (B is magnetic field in the slit, V is plasma volume, LN is the total slits length).

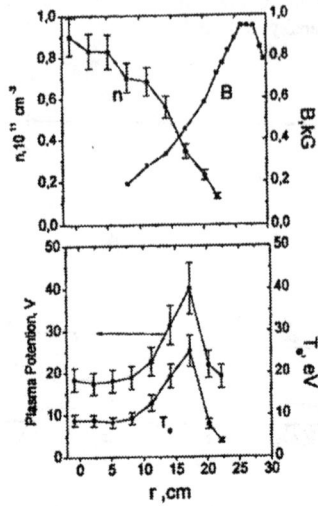

**Figure 3.** Magnetic field, density, electron temperature and potential profiles in Ar discharge at microwave power 5kW, $p=10^{-2}$Pa. The probe is moved in gap, axis is on $r=0$m, coils wall is on $r=0.3$m.

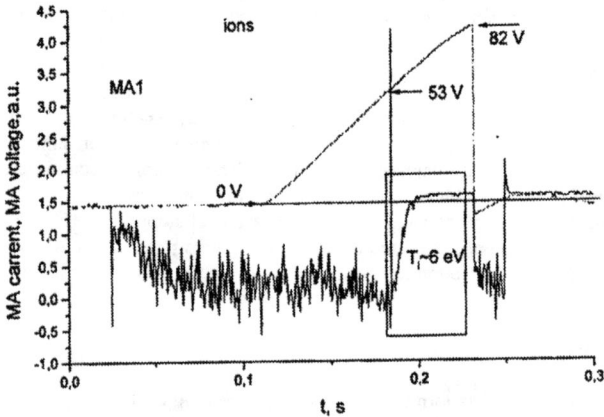

**Figure 4.** Oscillogramms of ELA-1(periphery) ion current and positive retarding potential. Ar discharge at microwave power 15kW, $p=10^{-2}$Pa, discharge duration 0.25s.

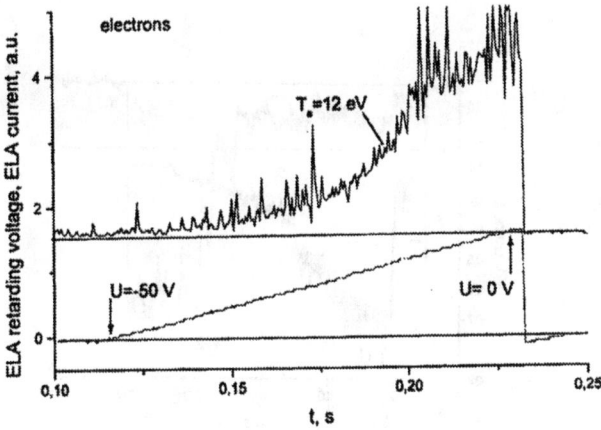

Figure 5. Oscillogramms of the ELA-2 (center) electron current and negative retarding potential. Ar discharge at microwave power 15kW, $p=10^{-2}$Pa, discharge duration 0.25s.

When the heating is switched on, the enhancement of plasma confinement is observed. Figure 6 shows the oscillogramms of outgoing ion flux into axial gas box. One can see the decreasing of ion flux, while the density in the center is increased (see Fig. 2).

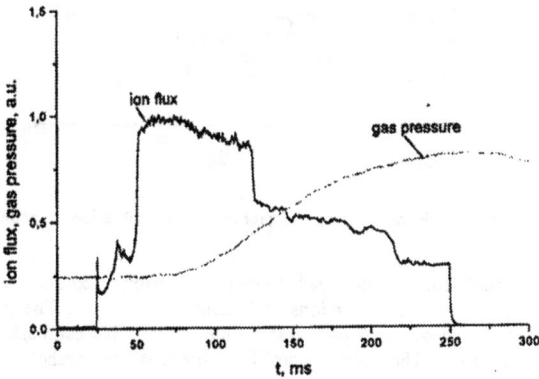

Figure 6. Oscillogramms of ion flux and gas pressure in axial plasma collector.
Ar, $10^{-4}$, 40 kW

The another signals show the analogous behavior. Fig. 7 shows the ELA-1 (periphery) and ELA-2 (center) ion flows. Fig. 8 shows the probes ion saturation currents. One can see that the periphery plasma flow decreases and the center flow increases due to plasma density increment. We should note that the change of local parameters with time is much larger, then the change of average along axis parameters (compare Fig. 2 and 7).

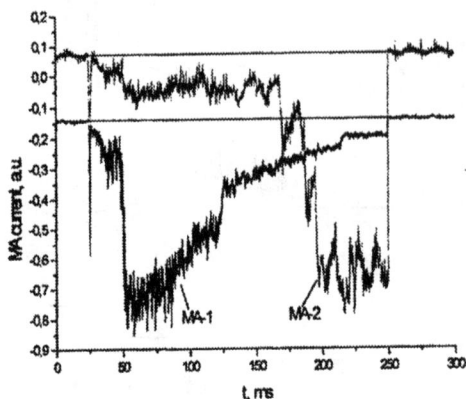

Figure 7. Oscillogramms of ELA-1 and ELA-2 ion fluxes. Ar, $10^{-4}$, 40 kW

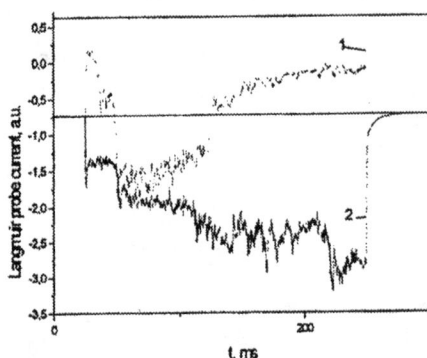

Figure 8. Oscillograms of Langmuir probes: 1- peripheral region, 2- central plasma region.

The measured potential profile should leads to large increasing of cold ion confinement in the center. The depth of potential well is ~20V for ions with temperature ~5eV. The large enhancement factor exp(20/5)~50 appear in classical cusp confinement time scaling. We estimate the particle confinement time for central ions as 0.15s. The potential profile determines the parabolic density profile, which is surprising in the low-pressure discharge in multi-cusp magnetic system.

The potential barrier ~50V exists for periphery hot electrons too. At hot electron temperature ~20eV enhancement factor exp(50/20)~10 appear in classical mirror confinement time. We estimate the particle confinement time for hot electrons as 0.3ms. Else the volume of hot electron plasma is ~3 times smaller as the total plasma volume, we estimate the total confinement time as ~1ms.

What is the reason of cold electron formation in the center? This region is surrounded by hot electron periphery and the particle confinement time is large here. The radiation cooling explains this phenomena. In Table 1 the power balance in PNX-U is presented.

Table 1. Power balance

| Power | kW | Measured or estimated | Estimation formula |
|---|---|---|---|
| Klystron power | +40 | measured | |
| Reflected into waveguide | -8 | measured | |
| Charged particles inside insides the trap | -8 | estimation | $P = \dfrac{nV(T_{eh} + U_{ioniz.} + \varphi_{pl})}{\tau}$  $V=0.5m^3$, $n=10^{12}cm^{-3}$,  $\tau=1ms$, ( )=100eV |
| Radiation | -8 | Estimation | $W = \dfrac{dP_r}{dV}V_h$  $V_h=0.15m^3$,  $dP_r/dV\sim0{,}05 Wcm^{-3}$ |
| Scattered microwaves and plasma production outside the magnets | -16 | Estimation | 40-8-8-8 |

In the case of PNX-U operation with hydrogen the plasma density was 1.5-2 times less than with argon. The density profiles were flat and the potential confinement was no so observable. It is planned to continue experiments with hydrogen.

## 4. PN-ITER DESIGN

The status of experimental achievements at PNX-U in compare with the designed and with the necessary for PN ITER parameter meanings is demonstrated Tab. 2 below. The super-conducting PNX-SU installation is the proposed intermediate step. The requisite for PN-ITER plasma parameters should be obtained on PNX-SU. The calculations, based on PNX-U results, have predicted the microwave power level of 700kW for PN-ITER with deuterium. The use of Ar as processing gas can give the additional possibilities (radiation cooling of electrons, multi-charged ions, and high ionization degree) for power decrement and plasma density increment. The complementary investigation of such "radiation discharge" on PNX-SU installation is proposed.

Table 2. Main parameters of plasma neutralizers

| | PNX-U | | PNX-SU | PN ITER |
|---|---|---|---|---|
| | Experiment | Design | | |
| Plasma length, m | 2 | 2.2 | 2.5 | 3-4 |
| Plasma volume, $m^3$ | 0.5 | 0.5 | 1 | 6 |
| Linear density, $m^{-2}$ | ~$2\cdot10^{18}$ | $1.4\cdot10^{18}$ | $1.3\cdot10^{19}$ | $2\cdot10^{19}$ |
| Plasma density, $m^{-3}$ | $1\cdot10^{18}$ | $0.7\cdot10^{18}$ | $5\cdot10^{18}$ | ~$7\cdot10^{18}$ |
| Ionization degree | 0.4 | ~0.2 | 0.4 | 0.4 |
| Microwave power, kW | 50 | 50 | 150 | 700 |
| Frequency, GHz | 7 (klystron) | | 24 (gyrotron) | |
| Magnetic system | copper | | super-conducting (NbTi) | |
| Magnetic field in slits, T | 0.36 | 0.5 | 1 | 1 |

## 5. CONCLUSION

The experiments have shown that the cut off plasma density can be easily provided in the multi-cusp configuration with microwave near wall ECR discharge at a low pressure. The potential improvement of the magnetic cusp confinement allows obtain the plasma density more than the cut off one at smaller input microwave power. The high ionization degree can be achieved at a low plasma electron temperature and acceptable microwave power. The radiation type argon discharge in multi-cusp magnetic configuration was investigated. PN could be integrated into existed ITER NBI design.

## References

1. ITER Design Description Document, N 53 DDD 15 97-11-30 W 0.2 (1998).
2. Skovoroda A. A., "Plasma neutralizer for ITER injector", IAE-5544/6, Moscow (1992).
3. Kulygin V. M., Skovoroda A. A., Zhil'tsov V. A., Plasma Devices and Operations, 1998, 6, 13
4. Zhil'tsov V.A., Klimenko E.YU, Kosarev P.M. et al. Nuclear Fusion 2000, 40, 509; Fusion Energy 1998 (Proc.17th Int. Conf. Yokohama, 1998), IAEA, Vienna (1999) (CD-ROM file ITERP2/04).
5. Skovoroda A.A. and OGRA Team, Proc. Intern.Workshop "Strong Microwaves in Plasmas", Ed. by A.Litvak, Nizhny Novgorod, 2000, V.1, 371.
6. Gildenburg V.B., Skovoroda A.A., Plasma Phys. Reports, 1996, 22, 246.

# LIGHT EMISSION IN LOW-PRESSURE LARGE-AREA MICROWAVE DISCHARGES

T. Terebessy, J. Kudela* and M. Kando

Graduate School of Electronic Science and Technology, Shizuoka University,
Johoku 3-5-1, Hamamatsu 432-8561, Japan
*Satellite Venture Business Laboratory, Shizuoka University,
Johoku 3-5-1, Hamamatsu 432-8561, Japan

**Abstract.** There are many works reporting on periodic structures in electromagnetic field distribution and light emission patterns in bounded microwave plasmas. These structures are often associated with standing wave modes formed due to wave reflection from the boundaries. However, in microwave discharges, a rich spectrum of the field distribution and light emission patterns can also exist because of plasma instabilities. In this work we present discharge light emission patterns observed in argon at low-pressures (< 50 mTorr). The patters exhibit small-scale wavelength periodic structures, which clearly do not correspond with the field distribution of electromagnetic standing wave modes. The phenomenon is briefly discussed on the basis of plasma instabilities.

## 1. INTRODUCTION

The research and development of large-area microwave plasma sources working without the use of static magnetic fields have undergone a significant progress in past few years [1-8]. Considering plasma density and plasma production area as the main criterion, these sources appear to be a promising candidate for the next-generation of high-density plasma sources for plasma processing. The overdense plasma is produced over large areas by guided electromagnetic waves propagating along the plasma-dielectric interface. When the power transferred to the wave is sufficient, wave-reflection occurs at the plasma source boundary resulting in standing-wave formation. Under such condition the plasma-dielectric waveguide can be treated as a resonator and the discharge light-emission often coincides with the electromagnetic standing-wave modes of this resonator. The agreement between the light emission and field distribution of the standing-wave mode in the large-area plasmas has been confirmed in Refs. 4, 9. However, in most cases the light-emission patterns in large-area microwave plasmas exhibit complicated structures, the interpretation of which is not so straightforward. The complexity of the patterns suggests that the light emission is the result of self-organization of the discharge. As it is known, in microwave discharges in general, a rich spectrum of the field-distributions and light-emission patterns can also exist because of plasma instabilities, see e.g. Refs. 10-14. The instabilities can contribute to the power absorption, therefore, the study of the light emission may be useful also in the connection with the discharge heating mechanism, which is still not well understood.

In this work we will present light-emission patterns of microwave argon discharges sustained at low gas pressures (<50 mTorr). The patterns exhibit self-organized structures and are clearly not consistent with the field distribution of any electromagnetic standing-wave mode in the plasma-dielectric resonator. The phenomenon is briefly discussed on the basis of plasma instabilities.

## 2. EXPERIMENTAL APPARATUS

The measurements were carried out in argon in a large area planar plasma source powered by 2.45-GHz microwaves. The plasma source, in details described in the Refs. 5,15, or elsewhere in this proceeding [16], is schematically shown in Fig. 1. Vacuum part consists of a stainless steel chamber, cylindrical in shape, with 312-mm diameter and 350-mm height. Microwaves are coupled to the plasma chamber through a 15-mm-thick quartz window by a fully azimuthally symmetrical microwave applicator. The applicator is a tunable cylindrical cavity, designed for the $TM_{011}$-mode, with an annular slot at its bottom wall. The cavity is made of copper cylinder with the inner diameter of 110 mm and the wall thickness of

5 mm. The bottom wall of the cavity is a 5-mm-thick and 90-mm-diameter disk, which together with the cylindrical body of the cavity forms an annular slot with 10-mm width and 110-mm outer diameter.

The diagnostic tools used in the experiment were CCD-camera and a Langmuir probe. The CCD-camera was mounted at the main view-port on the bottom flange of plasma chamber and the Langmuir probe was inserted into the chamber through a bottom-port along the side-wall as shown in Fig.1. The probe was movable in the axial and azimuthal directions allowing the measurement at any radial and axial position in the chamber. The probe characteristics were recorded by using a PC-based data acquisition system with a 12-bit resolution and 250-kHz sampling frequency. Each data point of the total 1000 points was averaged 4000 times in order to reduce the noise and to achieve the smooth probe characteristics. The characteristics were used for the evaluation of the ion densities and electron temperatures. The ion density was determined from the fit of ion saturation current [17] and the electron temperature from the linear part of $\ln J_e$-$V$ curves.

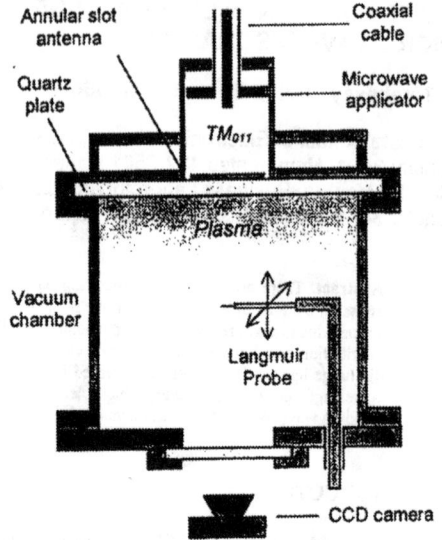

Figure 1. Schematic view of the plasma source.

## 3. RESULTS AND DISCUSSION

A large variety of the light-emission patterns, standing or rotating, can be observed in the plasma source, depending on the gas pressure, microwave power and tuning of the applicator [5, 15]. The tuning plays an essential role in the selection of a particular pattern, since it enables fine control of power transfer to the discharge along with the matching of the antenna-field with the self-consistent field of the plasma-dielectric waveguide. The effect of the tuning is demonstrated by Fig. 2a, where the power reflection coefficient along with the ion saturation current is shown as a function of the cavity height. The Langmuir probe was placed on the chamber axis 10 cm under the quartz plate. When the cavity is out of resonance, the discharge is sustained only under the cavity (Fig. 2b) and it exhibits increased power reflection coefficient and lower plasma densities. On the other hand, when the height of the cavity is near the resonance height for $TM_{011}$ mode, the condition for the efficient power transfer is fulfilled and a high-density discharge is sustained over the whole plasma source diameter (Fig. 2c). Such a large-area discharge is always overdense and is believed to be sustained by surface waves. The small discharges can be both, overdense or underdense. They remain localized because the condition for the self-consistent guiding of the wave is not fulfilled due to the mismatch antenna-field and surface wave field.

As mentioned above, here we focus on the discharges with light-emission patterns not corresponding to standing-wave modes. An example of such a discharge on large area is shown in Fig. 3a. In a standing-wave mode with a cylindrical geometry, the electric field varies in the azimuthal direction as $\exp(im\phi)$, where $m$ is the azimuthal mode number. Assuming the light emission intensity $I \sim |E|^2$, only even number of light-emission maxima can exist in the azimuthal direction. The light emission presented in Fig. 3a exhibits odd number of maxima (9). Moreover, Fig. 3b presents a fine structure with a scale-length much lower than the wavelength of the excitation field, which can exist within each light maximum. Light-emission patterns with odd number of maxima in the azimuthal direction can also exist in small discharges sustained under the cavity (Fig. 4a). Unlike the patterns in large-area discharge, which are

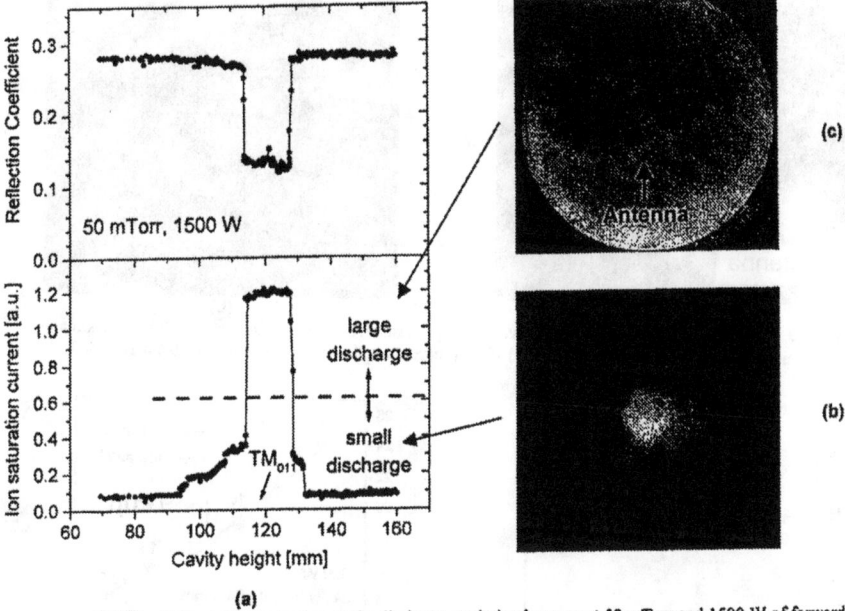

(a)

**Figure 2.** Effect of the applicator tuning on the discharge evolution in argon at 50 mTorr and 1500 W of forward microwave power: (a) cavity height vs. power reflection coefficient and ion saturation current measured on chamber axis 100 mm under the quartz; (b) light emission of a large discharge; (c) light emission of a small discharge.

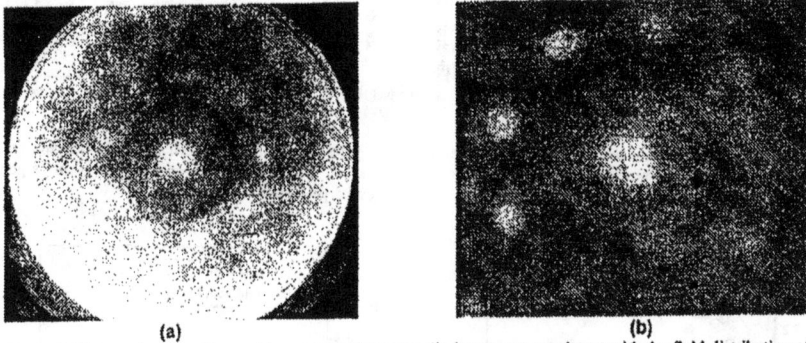

**Figure 3.** Light emission patterns of large-area microwave discharges not consistent with the field distribution of electromagnetic standing wave modes: (a) 9 light-maxima in the azimuthal direction in a 50 mTorr discharge sustained on 312-mm diameter; (b) small-scale structure inside the light maxima.

rather complicated to lead to a straightforward explanation of their origin, the small discharge patterns are much simpler and well reproducible. In the following sections, we will focus on small discharges, the large discharges will be the subject of future work.

The small discharges are typically less than 10 cm in diameter. In the pressure range ~10-50 mTorr and powers up to 1000 W, the emission has a form of a light-ring, sometimes with regular modulation in the azimuthal direction. The number of light maxima in the ring can reach both odd (Fig. 4a) and even values (Fig. 4b), and it is controllable by the gas pressure and microwave power. The observed number of

**(a)**                          **(b)**

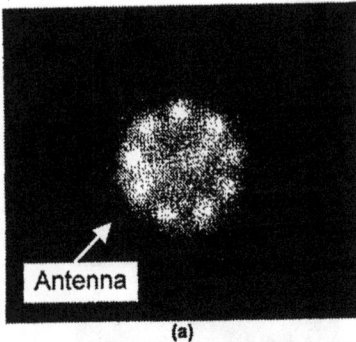

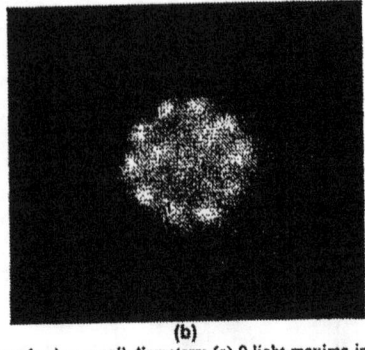

Figure 4. Light emission patterns of microwave discharges sustained on small diameters: (a) 9 light-maxima in the azimuthal direction, 19.4 mTorr, 600 W; (b) 10 light-maxima in the azimuthal direction, 20.4 mTorr, 600 W.

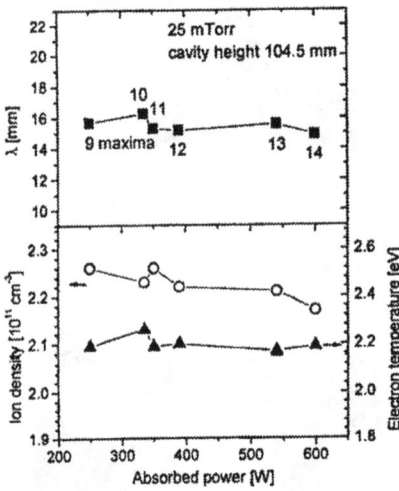

**Figure 5.** Distance between the light maxima and plasma parameters vs. absorbed microwave power in a 25 mTorr argon discharge.

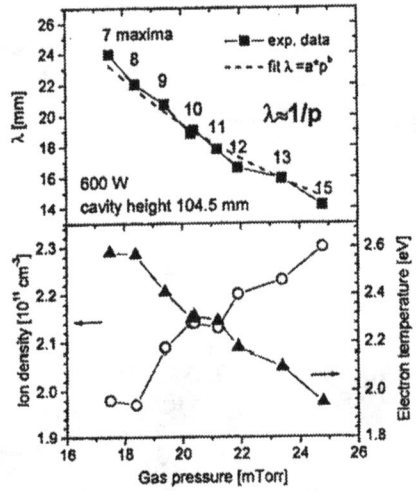

**Figure 6.** Distance between the light maxima and plasma parameters vs. gas pressure in argon discharge sustained by 600 W.

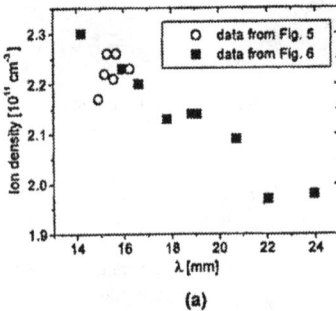

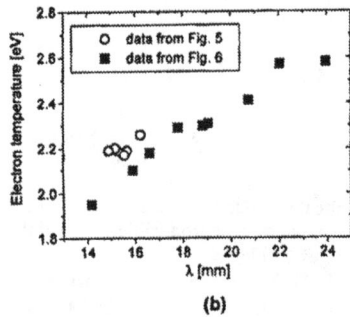

**(a)**                          **(b)**

Figure 7. Plasma parameters vs. distance between the maxima in small discharges.

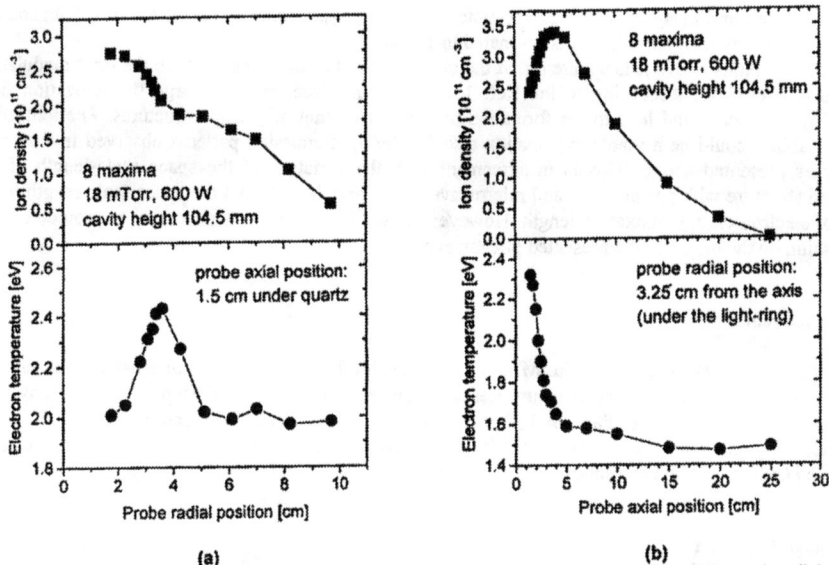

**Figure 8.** Spatial plasma parameter profiles in a small discharge: argon, 18 mTorr, 600 W; 8 light maxima; lightring diameter 6.5 cm; (a) radial profile measured 1.5 cm under the quartz; (b) axial profile measured under the light ring; probe radial and axial positions are the distances from the chamber axis and quartz plate, respectively.

maxima are between 3 and 16 with a distance 1-3 cm between them. Out of this range, the azimuthal light-modulation disappeares and only homogeneous light ring is observed by naked eye. The number of maxima is growing along with the diameter of the light-ring with increasing pressure or microwave power. When microwave power is changed and pressure is kept constant, the number of maxima and diameter of the light-ring change in such a way that the distance between the maxima $\lambda$ (estimated from the ratio: ring- circumference / number of maxima) remains constant (Fig. 5). Similarly, the plasma density and electron temperature measured in the light-ring 1.5 cm under the quartz are approximately constant and equal $2.2 \times 10^{11}$ cm$^{-3}$ and 2.2 eV, respectively (Fig. 5). However, when the gas pressure is changed and power is kept constant, $\lambda$ is changing and is approximately inversely proportional to the gas pressure (Fig. 6). As for the plasma parameters, the plasma density in the ring slightly increases and the electron temperature decreases with increasing gas pressure (Fig. 6). It should be noted that the plasma parameters 1.5 cm under the quartz do not change, when measured under the light maximum or minimum, respectively. The results above suggest that $\lambda$ is directly related to the internal parameters, plasma density and electron temperature, rather than to the external ones, gas pressure and microwave power. This is confirmed by combining all the results together in Fig. 7.

For more detailed characteristics, plasma parameter profiles were measured for a discharge with 8 maxima (Figs. 8a-b). The discharge was sustained at 18 mTorr and 800 W of absorbed microwave power, and the diameter of the light-ring was about 7.5 cm. The radial profiles of plasma density and electron temperature measured 1.5 cm under the quartz plate are shown in Fig. 8a. The density exhibits diffusion-like profile in the radial direction with the maximum value of about $2.8 \times 10^{11}$ cm$^{-3}$ on the chamber axis. The electron temperature, as expected, reaches its maximum under the ring and decays towards the wall and center. The value of temperature measured under the ring is about 2.4 eV, and its radial decrease saturates at approximately 2 eV. For complete information, the axial plasma parameter profiles of the discharge are shown in Fig. 8b. The profiles were measured under the ring, i.e. in a radial distance of 3.25 cm from the chamber axis. The plasma density reaches its maximum value $\approx 3.5 \times 10^{11}$ cm$^{-3}$ 4 cm under the quartz and it exhibits exponential-like decay towards the bottom wall of the chamber. This confirms that the plasma is produced in the vicinity of the quartz plate, which is supported also by decaying electron temperature away from the quartz plate. Assuming exponential decay of electron temperature, the electron energy relaxation length is estimated to 1.4 cm.

From the results above it is apparent that the discharge in the whole area, where the light modulation appears is overdense. According to the Ref. 13, in such a microwave discharge the ionization field instability may occur and lead to the formation of a rich spectrum of plasma structures. The ionization field instability could be a viable explanation also for the light-emission patterns observed in the small discharges presented above. This is in agreement with the variation of the space scale-length of the observed structure with gas pressure and microwave power and the fact that the space scale-length is the order of electron energy relaxation length. However, further investigation is necessary to provide a clear explanation on the origin of the presented phenomenon.

## Acknowledgement

The authors are indebted to Prof. Yu. M. Aliev of Lebedev Physical Institute of Russian Academy of Sciences, Dr. D. Korzec of Microstructure Research Center of University of Wuppertal (Germany) and Prof. J. Marec of University of Paris-Sud (France) for valuable comments and discussions. This work was supported by a Grant-in-Aid for scientific Research (B)(2) from the Ministry of Education, Science, Sports and Culture of Government of Japan.

## References

1. Komachi K. and Kobayashi S. J. Microwave Power Electromagn. Energy, 1989, 24, 140.
2. Kimura T., Yoshida Y., and Mizuguchi S.I. Jpn. J. Appl. Phys., Part 2, 1995, 34, 1076.
3. Werner F., Korzec D., and Engemman J. Plasma Sources Sci. Technol., 1994, 3, 473.
4. Nagatsu M., Xu G., Yamage M., Kanoh M., and Sugai H. Jpn. J. Appl. Phys., Part 2, 1996, 35, 341.
5. Odrobina I., Kudela J., and Kando M. Plasma Sources Sci. Technol., 1989, 7, 238.
6. Yasaka Y., Nozaki D., Koga K., Ando M., Yamamoto T., Goto N., Ishii N., and Morimoto T. Jpn. J. Appl. Phys., Part 1, 1999, 38, 4309.
7. Rauchle E. J. Phys. IV France, 1998, 8, 99.
8. Bluem E., Bechu S., Boisse-Laporte C., Leprince P., and Marec J. J. Phys. D: Appl. Phys., 1995, 28, 1529.
9. Nagatsu M., Xu G., Ghanashev I., Kanoh M., and Sugai H. Plasma Sources Sci. Technol., 1997, 6, 427.
10. Massey J.T. and Cannon S.M. J. Appl. Phys., 1965, 36, 361.
11. Massey J.T. J. Appl. Phys., 1965, 36, 373. ·
12. Brovkin V.G. and Kolesnichenko Yu.F. J. Moscow Phys. Soc,. 1995, 5, 23.
13. Sinkevich O.A. and Sosnin V.E. J. Phys. D: Appl. Phys., 1996, 29, 2609.
14. Vikharev A.L., Gorbachev A.M., O.A. Ivanov, Kolysko A.L., Kuznetsov O.Yu. JETP Letters, 1998, 67, 567.
15. Kudela J. Doctor thesis, Graduate School of Electronic Science and Technology, Shizuoka University, July 1999.
16. Kudela J., Terebessy T., and Kando M. *Hot electrons and EEDF anisotropy in large-area surface-wave discharges*, elsewhere in this proceeding.
17. Peterson E.W., Talbot L. AIAA J., 1970, 8, 2215.

# SURFACE WAVE PLASMA EXCITATION AT AN INNER TUBE OF COAXIAL QUARTZ GLASS TUBE

T. Yamamoto and M. Kando

Graduate School of Science and Engineering, Shizuoka University,
Johoku 3-5-1, Hamamatsu 432-8561, Japan

**Abstract.** To apply the low-pressure surface wave plasma to the light sources, the leakage of the microwave should be reduced to enough a low level not to disturb electronic instruments or not to affect on human bodies. This work is devoted to develop the method of low-pressure plasma production by surface wave with low level microwave leakage from the discharge tube. The coaxial quartz glass tube and 2.45 GHz microwave are used in the experiment. The surface wave plasma produced in argon at the pressures from 50 mTorr to 90 mTorr by slot antenna at the inner tube of the coaxial quartz glass discharge tube is examined by a Langmuir probe, a microwave leakage detector and a dipole antenna to detect the microwave electric field. As a result, it is found that 1) the plasma density measured near the outer glass tube exceeds the critical density, 2) the plasma density estimated from measured wave lengths of the surface wave agrees with that measured by the Langmuir probe, 3) microwave leakage decreases with increasing the pressure because the plasma with higher density at the higher pressure can shield out the microwave more efficiently. In order to suppress the microwave leakage still more, the regions that the plasma density is lower than the critical density should be completely excluded from the whole areas of the discharge tube.

## 1. Introduction

Plasma densities produced by surface wave plasmas are essentially higher than critical density. Even at the low pressure, it is not necessary to apply the magnetic field for the production of surface wave plasma. Furthermore, they can be excited in the discharge chamber with versatile structures under such experimental conditions as a wide range of filling gas pressure and electromagnetic wave frequency and various gas species[1],[2]. Therefore, it seems quite promising to apply the surface wave plasmas to electrodeless and mercury-free light sources. In general, surface wave plasmas are excited on the surface of the glass tube by some kinds of launcher and propagate near the surface of the glass tube along the tube axis. Therefore it is inevitable for the electromagnetic wave to radiate from the plasma if an effective electromagnetic wave shielding are not carried out. The leakage of the electromagnetic wave will disturb elaborate electronic instruments or will be a matter of concern for human body[3]. Thus, it is significant to establish the method to suppress the electromagnetic wave leakage from the viewpoint of the surface wave plasma application to wide industrial areas.

Fortunately the surface wave plasma can be used as an effective reflector of the electromagnetic wave because the plasma density is normally higher than the critical density, which will make the possibility to prevent the electromagnetic wave from radiating outside of the tube, if the launcher for the surface wave excitation were placed inside the plasma. In the present work, a coaxial quartz glass tube is used in the experiment to excite the surface wave plasma at the inner glass tube. The region where the surface wave propagates can be surrounded by high density plasmas in the present configuration of the discharge tube so that the electromagnetic wave is expected to be confined inside the column of the high density plasma. The present paper reports the properties of the present surface wave plasma produced by 2.45 GHz microwave from the view point of light source application together with the plasma processing, by using Langmuir probe and loop and dipole antennas to detect the microwave electric field. The plasma density measured by Langmuir probe coincided with that estimated from the surface wave dispersion measured by the interferometer. It is found that the microwave leakage from the outer glass tube to the atmosphere seems to be suppressed by the overdense plasma, depending on the plasma density, although the microwave leakage could not completely be excluded because the plasma density became dilute near the end of the discharge tube.

## 2. EXPERIMENTAL APPARATUS

The coaxial quartz glass discharge tube consists of the outer tube with an outer diameter of 45 mm and the inner tube with inner diameter of 13 mm, both are 300 mm long and 2.0 mm thick. It is set up into the coaxial waveguide as shown in Fig.1. A metal adapter is used to fix the discharge tube at the top of the discharge tube. The inner conductor of the coaxial waveguide is equipped with a slot antenna at 30.5 mm, corresponding to a quarter wavelength of 2.45 GHz microwave in the vacuum, from the end of the coaxial waveguide. The gap of the slot antenna is 1 mm and is used as a surface wave launcher. The distance between the adapter and the launcher which are connected to the ground is 300 mm. Figure 2 shows the expanded drawing of the slot antenna. Before the appearance of the plasma, a high frequency electric current flows across the slot antenna so that an intense electric field in the axial direction is induced there, which will excite the surface wave propagating along the surface of the inner glass tube. To avoid the occurrence of the breakdown between the slot antenna, a ceramic spacer made of alumina is placed. In Fig.3, the construction of the microwave circuit is depicted. Rectangular and coaxial movable plungers are adjusted to satisfy with the matching condition, by which the reflection of microwave could be reduced completely. The 2.45 GHz microwave is launched by the magnetron oscillator and is coupled with the discharge tube through the slot antenna. The microwave is continucus and has a power range from 200W to1000 W. Argon gas is flowing and is kept constant at the pressure from 50 mTorr to 100 mTorr throughout the experiment. The dipole antenna to measure the surface wave is inserted near the surface of the inner glass tube through the inner quartz glass tube and is moved along its axis. The Langmuir probe made of tungsten with 0.3 mm diameter and 1.5 mm length is placed near the surface of

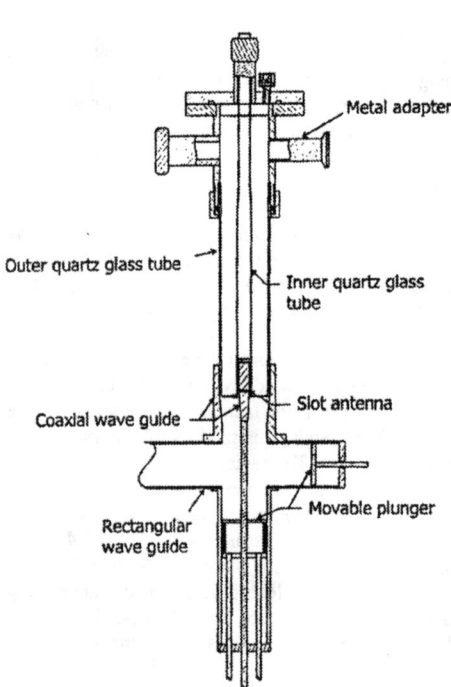

Figure 1. Coaxial quartz glass tube for the discharge

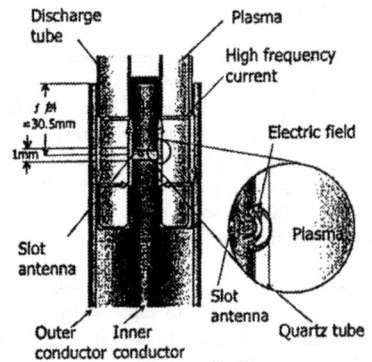

Figure 2. The expanded drawing of the slot antenna.

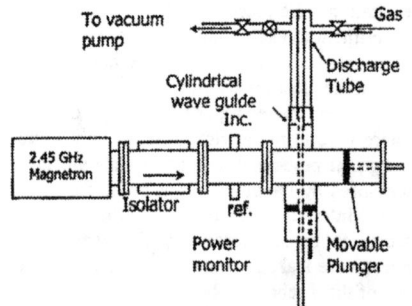

Figure 3. The schematic diagram of microwave circuit.

the outer glass tube and is moved along the axis of the discharge tube. The microwave leakage is measured in two dimensionally by the microwave leakage detector (Type:LD10M, Micro Denshi Co.Ltd).

## 3. EXPERIMENTAL RESULTS AND DISCUSSION

After evacuating the discharge chamber to the pressure lower than $10^{-6}$ Torr, argon gas is fed into the chamber. The plasma is produced throughout the discharge tube by applying the microwave. The microwave power reflected from the chamber is around a few % at the maximum.

### 3.1 Axial profiles of plasma density, electron temperature and floating potential.

In Fig.4, axial profiles of the plasma density, electron temperature and floating potential are shown as a function of a microwave incident power and a pressure. The position of z means the distance from the slot antenna. The plasma density decreases monotonically with z and is higher than the critical plasma density of $7x10^{10}$ cm$^{-3}$. It does not depend much on the incident microwave power and pressure. This may come from the reason that the plasma near the outer glass tube is transported by diffusion from the plasma production region near the inner glass tube. The floating potential and electron temperature are oscillating in the axial direction. Furthermore, at the point that the electron temperature is higher, the floating potential is lower. These features may be caused by the microwave leaked near the end of the discharge tube where the plasma density decreases lower than the critical density. Such a leaked microwave excites the standing wave on the surface of the outer glass tube and heat up the plasma. For the reference, the probe characteristics curves are shown in Fig.5.

### 3.2 Surface wave dispersion

The surface wave dispersion was measured by using the interferometer as shown in Fig 6. The signals from the dipole antenna and the magnetron oscillator are connected to the signal and reference inputs of the mixer. The interferometer wave patterns are obtained by moving the dipole antenna near the surface of the inner glass tube along the axis. Figure 7 shows the typical wave pattern. In general, the surface wave dispersion is the function of the plasma density so that the wave length of the surface wave can be uniquely related to the plasma density. In the present paper, the surface wave dispersion given by Eq.(1), derived under the plane geometry model[4], is used for the plasma density estimation from the wavelength.

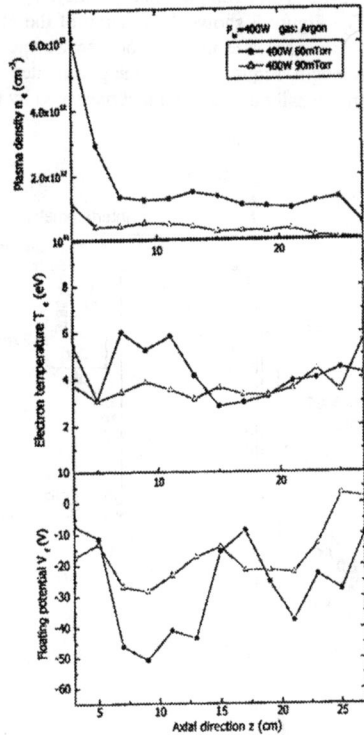

Figure 4. The axial profiles of plasma density, electron temperature and floating potential in argon measured by the Langmuir probe. z = 0 means the position of the slot antenna.

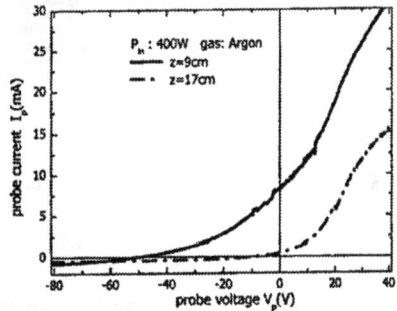

Figure 5. The probe characteristics measured at z=9 cm and 17 mm. Argon, 60 mTorr.

$$k_z = \kappa_d^{1/2} \frac{\omega}{c} \left[ \frac{\omega_{pe}^2 - \omega^2}{\omega_{pe}^2 - (1+\kappa_d)\omega^2} \right] \tag{1},$$

where $\omega_{pe}$ is electron plasma frequency, $\kappa_d$ is dielectric constant of quartz and $c$ is light velocity.

The Figure 8 shows the relation of the plasma density and the wavelength calculated by Eq.(1). The plasma densities estimated from the wavelength means the density averaged over the distance of a half wavelength. Figure 9 shows the plasma density obtained in such a way. It is recognized that the order of plasma density agrees with that measured by the Langmuir probe.

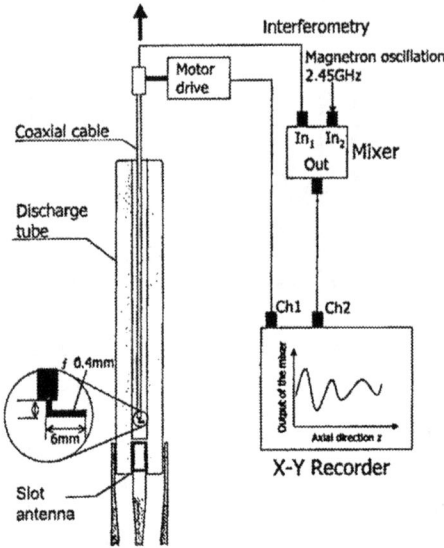

Figure 7. Measured interferometer wave pattern.

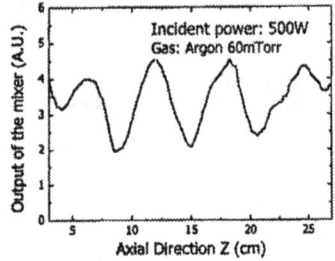

Figure 6. Interferometer and detail of the dipole antenna.

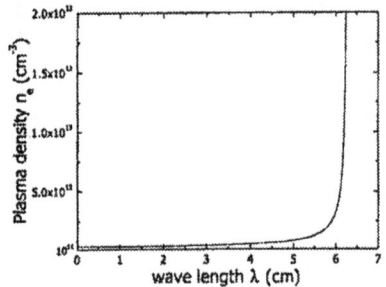

Figure 8. The relation of the plasma density and the wavelength of the surface wave dispersion.

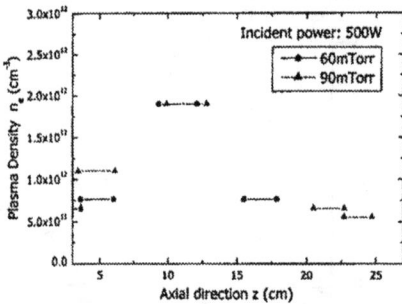

Figure 9. Plasma density estimated from the wavelength of the surface wave.

### 3.3 Microwave leakage measurement

Microwave radiated from the discharge tube is measured by the microwave leakage detector. It is made by a loop antenna and can measure a certain component of microwave electric field adjusting the direction of the loop antenna. At this moment, only the electric field parallel to the tube axis was measured in two dimensions. The axial profile of the microwave electric field intensity measured at 40 cm from the surface of the outer glass tube is shown in Fig.10. It is seen that the standing wave is excited on the surface of the outer glass tube. Figure 11 indicates the map of equi-electric field curves in the space near the discharge tube as a function of the pressure at the constant incident microwave power of 400 W. It is possible of deduce the following features from Figs.10 and 11;

1) the microwave leakage is not be reduced to 5 mW/cm$^2$ which is the limit of microwave leakage for the safety of the human body.
2) the microwave leakage increases with increasing the distance z from the slot antenna.
3) the microwave leakage decreases with increasing the pressure.
4) as is clearly shown in the map at the pressure of 90 mTorr, a maximum electric field exists around z =20 cm, which agrees with the results shown in Fig.10.

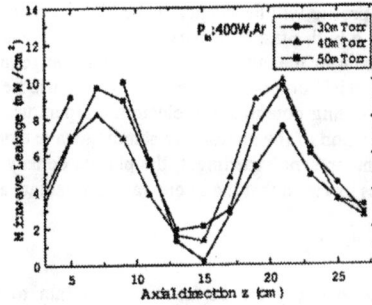

**Figure 10.** The axial profile of the microwave electric field intensity measured at 40 cm from the surface of the outer glass tube.

The plasma density decreases in the axial direction as shown in Fig.4. Therefore, the microwave leakage from the discharge tube will increase at a certain z where the plasma density becomes lower than the critical density and is supposed to be a large z close to the end of the discharge tube. However, the plasma density at large z seems not to depend on the pressure shown in Fig.4. The feature 3) suggests that the higher the pressure, the higher plasma density near the inner glass tube. In addition, the radial plasma

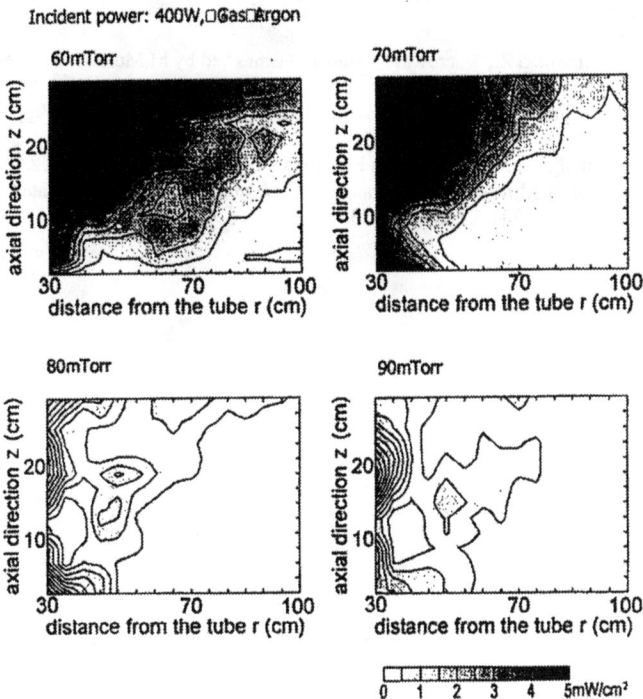

**Figure 11.** The map of equi-electric field curves in the space near the discharge tube as a function of the pressure.

density profile will be steeper as the higher pressure. Thus, the microwave electric field will be shield out by such a high plasma density near the inner glass tube at the higher pressure, although plasma density measured near the outer glass tube is not change much.

The axial component of microwave electric field has a maximum at $z = 20$ cm where the axial profiles of the floating potential and electron temperature also have a maximum or minimum value. This means the excitation of the microwave standing wave along the surface of the outer glass tube.

In the present experiment, the plasma density near the metal adapter at the top of the discharge tube becomes lower so that the microwave can radiate around the top region of the discharge tube.

## 4. Summary

In order to apply the surface wave plasma to the light sources, the microwave leakage should be decreased to around 5 mW/cm$^2$, which is the level regulated legally. In the present work, the plasma excited by the slot antenna at the inner tube of the coaxial quartz glass tube is examined. Measured wave dispersion and plasma density support that the plasma is produced by the surface wave. The microwave leakage decreased with increasing the pressure. However, it increased at the region apart from the slot antenna. These facts will be brought by the microwave shielding by the high density plasma. It is considered that the microwave may leak out from the region that the plasma density is lower than the critical density. In the present apparatus, plasma becomes so weak at a part of the connection of the discharge tube with the metal adapter for the microwave to leak out.

The standing wave excitation is measured on the outer glass tube. At the maximum of the standing wave, the electron temperature increases, which is caused by the microwave heating.

## References

1. Moisan M.and Zakrzewski Z.,"Microwave Excited Plasmas",ed by M.Moisan and J.Pelletier, Elsevier. 1992, Chap.5.
2. Boisse-Laporte C., Granier A., Dervisevic E., Leprince P. and Marec J. J.Phys.D:Appl.Phys.,1987, 20, 197.
3. Piejak R.B.Bulletin of 51st Annual Gaseous Electronics Conf.(1998, October) 1452.
4. M.A.Libermann and A.J.Lichtenmerg,"Principles of Plasma Discharges and Materials Processing", John Wiley & Sons, Inc.,1994, Chap.13, p.443.

# HOLEY-PLATE PLASMA SOURCE

Y.Yoshida

Toyo University, Kawagoe, Saitama 350-8585, Japan

**Abstract.** A surface-wave plasma source, with special focus on the holey-plate (HP) type source is briefly reviewed in due order of development. A partial-coaxial cavity-type HP source can produce high-density plasma under low gas pressures. This is designed to determine the most efficient method of coupling between the plasma and the surface wave. This source converts microwave power at 2.45 GHz into a surface wave as an evanescent electric field on the holey-plate, and can efficiently couple the field with the plasma. Another HP source, a rectangular waveguide type HP source for large wafers, has been developed. In this, the holey-plate is placed on an H-plane located at one end of the rectangular waveguide. A further developed version of the rectangular type, a parallel plate type with a holey top plate, is newly developed as a high-efficiency, lightweight, and low-cost plasma source.

## 1. INTRODUCTION

High-density, low-pressure plasma sources utilizing microwaves have recently been attracting much attention. We have first developed a coaxial-type open-ended cavity using both an ECR microwave and a dc planar magnetron plasma generator.[1] The ECR magnetic field is applied at the open end by the magnetron configuration magnets. The magnetron cathode is placed at the open end of the inner conductor of the coaxial-type cavity, which introduces the microwave power to the plasma. This cavity acts as a circle slot antenna. Using this cavity, we can produce a surface-wave enhanced plasma source.[2] This source can produce high-density and low electron temperature plasma. However, it does not have even plasma uniformity and does not produce good results under low gas pressure operation. In order to produce uniform and high-density microwave plasma in a large area without magnets, preliminary experiments have been performed with a new plasma source employing surface wave radiation from a holey plate.[3] This source is based on a parallel plate structure with a holey top plate. A major advantage is that much higher plasma densities can be expected because strong electric fields are obtainable on the surface of the holey plate. Another advantage is that uniform plasma can be obtained because the intensity and uniformity of the surface waves are dependent on the diameter and distance of the holes in the holey plate.

In this paper, we describe the development details of the holey-plate plasma source. Moreover, the key features of this source will be described, together with the plasma characteristics.

## 2. SOURCE CONSTRUCTION AND OPERATING PRINCIPLE

### 2.1. Coaxial-Type Open-Ended Dielectric Cavity

In order to produce uniform and high-density microwave plasma in a large area without using magnets, a plasma source employing surface wave radiation from a coaxial-type open-ended dielectric cavity has been developed.[2] The experimental apparatus of this plasma source is schematically shown in Fig.1. A shorted probe is used to couple the TE10 waveguide (54.5mm×109mm) to the transverse electromagnetic (TEM) coaxial waveguide. The coaxial waveguide (the outer conductor has an inner diameter of 49mm, and the inner conductor has an outer diameter of 21mm) is connected to a 45°

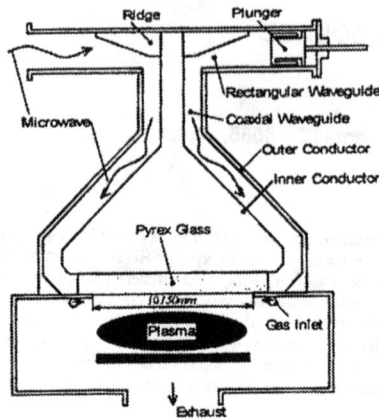

**Figure 1.** Schematic diagram of the coaxial-type open-ended dielectric cavity.

tapered pipe. The opposite side of the pipe is connected to the coaxial-type cavity (the outer conductor has an inner of diameter 254mm, and the inner conductor has an outer diameter of 199mm). The coaxial waveguide is made of nickel-plated brass.

A Pyrex glass disk of 20mm thickness and 175mm diameter is placed at the open end of the coaxial-type cavity, and the flat surface of the Pyrex glass disk is in contact with the inner conductor. Thus, the microwave transmitting space of the coaxial waveguide leads to the circumference of the Pyrex glass. The Pyrex glass is used as both a dielectric medium in which microwaves are propagated, and as a vacuum-sealed window.

Isotropic microwave power at a frequency of 2.45GHz is propagated through the 45° tapered coaxial waveguide, and into the circumference of the Pyrex glass. The microwaves are propagated through the Pyrex glass from the surrounding edge toward the center. Therefore, the surface wave radiates into the chamber from the glass surface and generates microwave plasma.

The electron density, electron temperature, and ion saturation current density of Ar plasma are measured using a single Langmuir probe. The electron density of $8 \times 10^{11} cm^{-3}$ is obtained at an Ar gas pressure of 6.7Pa and at a microwave power of 800W. The electron temperature is 4-6eV.

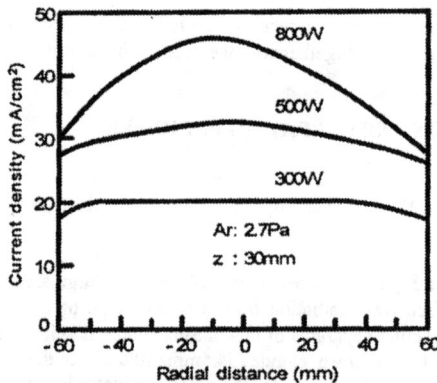

**Figure 2.** The ion saturation current density as a function of the distance from the center at an Ar gas pressure of 2.7Pa for various microwave powers.

Figure 2 shows the ion saturation current density as a function of the distance from the center at an Ar gas pressure of 2.7Pa for various microwave powers. Current density is maximized near the center. The dispersion of the current density in the radial direction increases with increasing microwave power because the power ratio between the surface wave radiating from the dielectric disk surface and the microwaves propagating within the dielectric disk can be determined by the dielectric thickness.

This source can generate high-density and uniform plasma at pressures of 1.3-40Pa. However, plasma is not maintained at low pressures because the electrons are not efficiently confined in the plasma.

## 2.2. Coaxial-Type Holey-Plate Surface-Wave Cavity

An electromagnetic surface wave can be defined as an electromagnetic wave that propagates along an interface between two media. A dielectric plate on a metal plane can obtain a surface wave structure having the necessary isotropy in the plane. This source is called a dielectric-plate surface-wave (DPSW) structure. The plasma source shown in Fig.1 is also of this structure. To investigate the relationship between the surface wave radiating from the dielectric disk surface and the microwaves propagating within the dielectric disk, we have developed a 77D coaxial-type open-ended DPSW cavity. [3]

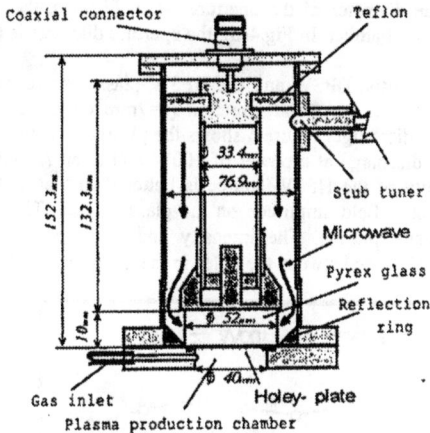

Figure 3. Schematic diagram of the coaxial-type holey-plate surface-wave cavity.

Another surface wave structure is a parallel plate structure with a holey top plate.[4] This source is called a holey-plate surface-wave (HPSW) structure.[5] A schematic of a 77D coaxial-type open-ended HPSW is shown in Fig.3. A microwave of 2.45GHz is supplied from a 77D-coaxial waveguide through a flexible coaxial cable and then transferred to the dielectric disk at the open end of the cavity. The outer conductor of the waveguide has an inner diameter of 77.6mm and a length of 152.3mm, and the inner conductor has an outer diameter of 33.4mm and a length of 132.3mm. A Pyrex glass disk of thickness 10mm and diameter 52mm is placed at the open end of the cavity. One side of the Pyrex glass disk is in close contact with the edge of the inner conductor. The Pyrex glass is used both as a dielectric disk in which microwaves are propagated and as a vacuum-sealed window. The glass is then covered with the holey-plate, which is made from a 0.5mm thick stainless steel sheet. The hole diameter and pitch are 1.7mm and 4.0mm, respectively.

The microwave enters through the circumference of the Pyrex glass toward the center, as illustrated in Fig.4. Figure 4(a) shows the operating principle for the DPSW. The microwave propagates through the glass, reflecting at the boundaries of both glass surfaces. When the microwave is perfectly reflected from

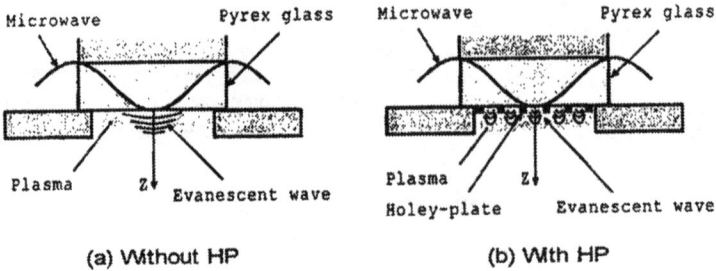

(a) Without HP                   (b) With HP

Figure 4. Schematic for producing an evanescent wave.

the glass-vacuum boundary, an evanescent electric field with a large amplitude gradient is produced from the glass into the plasma production chamber. In the case of the HPSW shown in Fig.4(b), the microwave propagates through the glass, reflecting at the top and bottom plates. The evanescent electric field is then produced in the plasma production chamber from the holey top plate. The penetration depth of the evanescent field depends on the diameter of the aperture.[4] In Fig.4(a), the aperture diameter is the diameter of the plasma production chamber. In Fig.4(b), the aperture diameter is the diameter of the holes in the holey-plate.

Discharge comparison between the DPSW and the HPSW plasma sources is made using a double Langmuir probe which is inserted in the radial direction (z=7mm downstream from the Pyrex glass surface) toward the center of the discharge. Figure 5 shows the plasma density as a function of the radial position from the center of the discharge at a power of 100W and at an Ar gas pressure of 13Pa. It is found that the plasma uniformity of the HPSW source is better than that of the DPSW source. This uniformity is related to the electric field amplitude on the glass surface. The HPSW structure is very useful for producing high-density plasma. The intensity and uniformity of the surface-waves are determined by the diameter and distance between the holes in the holey-plate.[6]

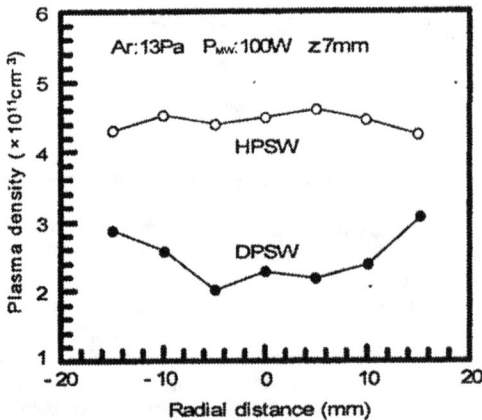

Figure 5. The plasma density as a function of the radial position from the center of the discharge at a power of 100W and at an Ar gas pressure of 13Pa.

## 2.3. Partial-Coaxial Cavity-Type Holey-Plate

The HPSW plasma source is not easily scalable to large diameters because the microwave enters through the circumference of a dielectric disk, which necessarily acts as a coupler. In order to scale this source to large diameter, we have designed a partial-coaxial cavity-type holey-plate (HP) plasma source without using a dielectric coupler.[7] A major difference between the HPSW and the HP sources is that the dielectric coupler used to generate surface waves is necessary for the HPSW source.

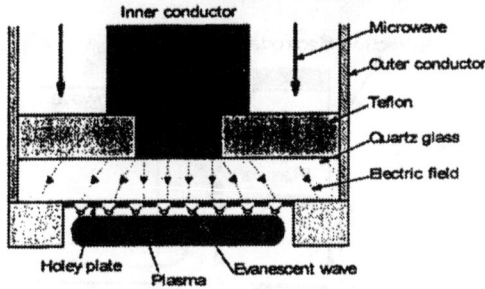

Figure 6. Schematic diagram of the partial-coaxial cavity-type holey-plate plasma source.

A schematic of the HP plasma source is shown in Fig.6. Microwave power at 2.45GHz is supplied from a 77D coaxial waveguide. The outer conductor has an inner diameter of 77.6mm, and the inner conductor has an outer diameter of 33.4mm. A quartz disk of thickness 10mm and diameter 76mm is placed at the open end of the cavity. One side of the quartz disk is flush with the edge of the inner conductor. The other side of the disk is covered with the holey-plate, which is made from a 0.5mm thick stainless steel sheet. The hole diameter and pitch are 1.7mm and 4.0mm, respectively. The total number of holes is 167. The diameter of the plasma production chamber is 54mm.

The microwaves enter the base of the quartz disk and propagate toward the center. The evanescent electric field is then produced in the plasma production chamber from the holey-plate. The effect of the microwaves on the holey-plate can be given to explain induced dipoles inside a hole.

Figure 7 shows the plasma density as a function of the radial position from the center of the discharge at a power of 100W and at an Ar gas pressure of 2Pa. A double Langmuir probe is inserted from right to left. The plasma density along the circumference of the plasma production chamber is lower than that of the center.

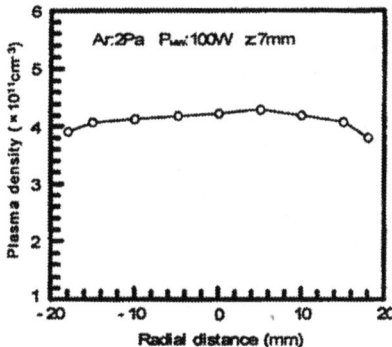

Figure 7. The plasma density as a function of the radial position from the center of the discharge at a power of 100W and at an Ar gas pressure of 2Pa.

## 2.4. Partial-Coaxial Cavity-Type Holey-Plate Using a Magnetic Field

The HP plasma source is designed to determine the most efficient method of coupling between the plasma and the surface wave. However, in this source, plasma is not maintained at low pressures and at low microwave powers because the electrons are not efficiently confined within plasma. For the stable confinement of the plasma at low gas pressures, magnetic multipoles are used to produce high-density plasma. We have developed an ion source that incorporates permanent magnet multicusp fields with the HP plasma source to provide the necessary plasma confinement.[8]

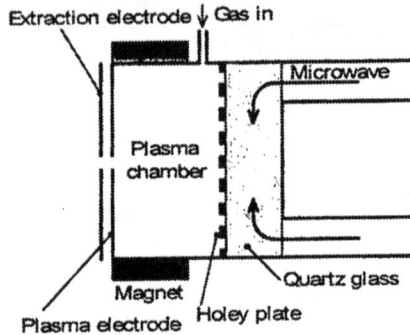

Figure 8. Schematic diagram of the partial-coaxial cavity-type holey-plate plasma source using a magnetic field.

A schematic of the HP ion source is shown in Fig.8. A microwave of 2.45GHz is supplied from a 77D coaxial waveguide and then converted into an evanescent mode through the use of a holey-plate located the end of the cavity. A quartz disk of thickness 10mm and diameter 76mm is placed at the open end of the cavity as a vacuum-sealed window. The disk is then covered with the holey-plate. The holey-plate is made from a 0.5mm thick stainless steel sheet. The hole diameter and pitch are 2.5mm and 3.6mm, respectively. Holes account for 36% of the area of the holey-plate. The diameter of the plasma production chamber is 54mm. Eight permanent magnets are mounted longitudinally to provide octopole fields for plasma confinement. In this arrangement, Sm-Co alloy magnets (2mm×4mm cross section and 20mm long) are employed. The magnetic field is 325G at the plasma chamber wall.

An ion-extraction system with 31 holes consists of a screen grid and an accelerator grid. Each grid is made from a 1-mm-thick stainless steel sheet. The screen grid holes are 2mm in diameter with 40% transparency. The distance between the holey-plate and the screen grid is 20mm. An 18-mm diameter ion beam can be extracted.

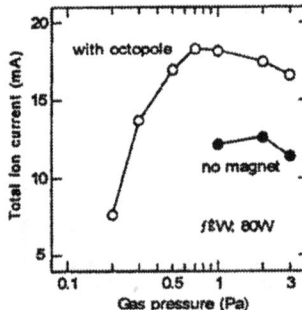

Figure 9. The total ion current versus the gas pressure for an extraction voltage of 1kV and an input microwave power of 80W.

Figure 9 shows the total ion current versus the gas pressure for an extraction voltage of 1kV and an input microwave power of 80W. The total ion current represents a drain current indicated in the extraction voltage source. Using the octopole magnetic field, the discharge becomes unstable under a gas pressure of about 0.1Pa. The ion current decreases at pressures higher than 0.7Pa because the mean free path in the process chamber decreases with increasing pressure. Without a magnetic field, the discharge becomes unstable under a gas pressure of about 0.9Pa. The effect of the magnets is most notable at lower gas pressure levels. Without magnets, the ion current of 12mA is obtained at an argon gas pressure of 1Pa. On the contrary, the same current is achieved at an argon gas pressure of 0.25Pa when magnets are used. The ionization efficiency increases about four times when magnets are used.

## 2.5. Rectangular-Waveguide Type Holey-Plate

Before this section, we have described in detail how the partial-coaxial cavity-type HP plasma source came about. Drawing from the research already conducted on this source, we have built a rectangular waveguide type HP plasma source (RHPS) for large wafers.[9,10]

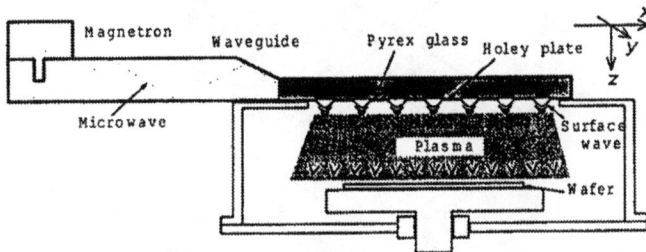

Figure 10. Schematic diagram of the rectangular-waveguide type holey-plate plasma source.

Figure 10 shows a schematic view of the RHPS. This plasma source is composed of a microwave power unit that supplies a 2.45GHz microwave to a Pyrex glass via a rectangular waveguide (109.2mm×54.6mm). The glass itself is located inside the waveguide. The end of the waveguide is tapered down to a height of 25mm, a width of 200mm, and a length of 300mm to accommodate the geometry of the glass sheet. A 160mm×160mm holey-plate is placed on the side of an H-face facing the process chamber. The holey-plate is made from a 1mm thick stainless-steel sheet. The hole diameter and pitch are 4mm and 6mm, respectively. Holes account for 45% of the area of the holey-plate. When evaluating the suitability for plasma processing, ashing rates of photoresist polymer in oxygen plasma are measured.

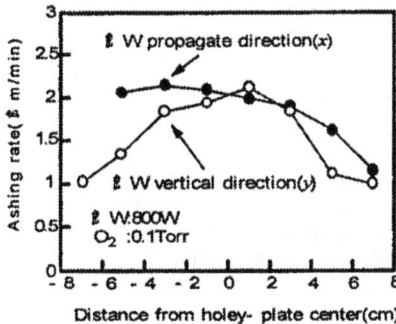

Distance from holey- plate center(cm)
Figure 11. The distribution of the ashing rate with respect to the microwave propagation direction (x), and the vertical direction (y). The microwave power is 800W and the gas pressure is 13Pa.

Figure 11 shows the distribution of the ashing rate with respect to the microwave propagation direction (x), and the vertical direction (y). The microwave power is 800W and the gas pressure is 13Pa. The distance between the wafer surface and holey-plate is 44mm. Also, gas is supplied at a flow rate of 50sccm. The uniformity of the ashing rate is ±23% in the direction of x. The microwave propagates from left to right. The uniformity is ±24% in the direction of y. This uniformity seems to be affected by the propagation modes of the microwave inside the Pyrex glass.

## 2.6. Parallel-Plate Type Holey-Plate

The RHPS is not easily scalable to larger plasma sources because the glass becomes larger and heavier, which becomes a significant loss material for microwaves. In order to scale this source to a large area, we have designed a parallel-plate type holey-plate plasma source (PHPS) without using glass as a dielectric coupler.

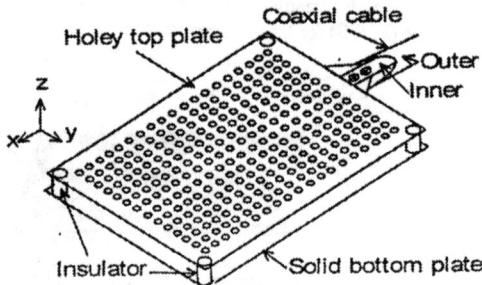

**Figure 12.** Schematic diagram of the parallel-plate type holey-plate plasma source.

Figure 12 shows a schematic view of the PHPS. This plasma source is composed of a microwave power unit that supplies a 2.45GHz microwave to two metal plates via a coaxial cable connected at the inlet of the plates. The two plates run parallel with a 10mm space between them. The plates are made from a 2mm thick aluminum sheet (100mm×140mm). One of the plates is a holey top plate; the other is a solid bottom plate. The hole diameter and pitch are 3mm and 6mm, respectively. Holes account for 22% of the area of the holey-plate. An inner conductor of the coaxial cable connects with the holey-plate, and an outer conductor connects with the solid plate.

Figure 13 shows the ion saturation current density with respect to the microwave propagation direction (x), and the vertical direction (y). The microwave power is 120W and the Ar gas pressure is 3Pa. A double Langmuir probe is inserted into the plasma near the holey-plate. The distance between the probe and the holey-plate is 10mm. In the direction of x, the profile of the current density is maximized at both ends. In the direction of y, the profile of the current density is maximized near the center.

## 3. CONCLUSION

The development of the holey-plate plasma source has been briefly reviewed. The holey-plate source is able to produce high-density sheet plasma. Moreover, the effect of the uniformity of the stationary waves in the waveguide on the plasma uniformity is affected by the holey-plate. Thus, this system naturally leads itself to achieving a large plane source delivering uniform plasma. The key features of this source in achieving large areas of uniform plasma with the parallel-plate structure assume that a strong and uniform electromagnetic field will be generated on the outside of the holey-plate. Our research shows promise for the development of the next generation of plasma sources that can operate under low pressures, allow independent control of ion flux and ion energy, and provide high density, uniform plasma over large wafers.

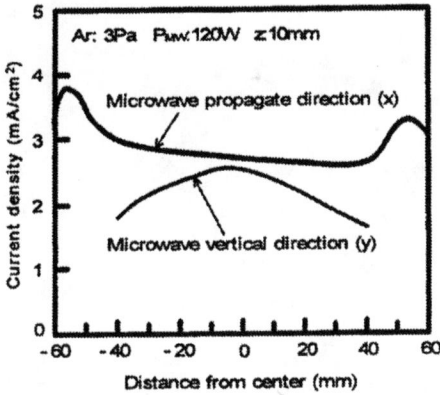

Figure 13. The ion saturation current density with respect to the microwave propagation direction (x), and the vertical direction (y). The microwave power is 120W and the Ar gas pressure is 3Pa.

# References

1. Yoshida Y., Plasma Source Sci. Technol., 1996, **5**, 275.
2. Kimura T., Yoshida Y., Mizuguchi S., Jpn. J. Appl. Phys., 1995, **34**, L1076.
3. Yoshida Y., Miyazawa T., Kazama A., Rev. Sci. Instrum., 1997, **68**, 79.
4. Walter C., IRE Trans. Antennas & Propag., 1960, **8**, 508.
5. Yoshida Y., Sakurai T., Jpn. J. Appl. Phys., 1998, **37**, 5746.
6. Yoshida Y., Rev. Sci. Instrum., 1998, **69**, 2032.
7. Yoshida Y., Rev. Sci. Instrum., 1999, **70**, 1710.
8. Yoshida Y., Takeiri Y., Rev. Sci. Instrum., 2000, **71**, 66.
9. Ogura H., Miyazawa K., Yoshida Y., Proc. 15th Symp. on Plasma Processing Ed. M. Kand: The Japan Society of Applied Physics, 1998, 558.
10. Yoshida Y., Ogura H., Vacuum, 2000, **59**, 459.

# MICROWAVE PLASMA APPLICATIONS

# MICROWAVE DISCHARGE FOR PLASMA CATALYSIS

## A.Babaritskyi, M.Deminsky, E.Gerasimov, S.Dyomkin, V.Jivotov, A.Knignik, B.Potapkin, E.Ryazantzev , V.Rusanov, R.Smirnov

Russian Research Center "Kurchatov Institute", Kurchatov sq., 123182, Moscow, Russia

**Abstract.** This paper is devoted to properties and plasma parameters in microwave pulse-periodic discharge in hydrocarbons at atmospheric pressure. This type of discharge is of fundamental interest because of its non-stationary and non-uniform nature and possible applications, for example, for ozone generation or for treatment of halogen containing impurities. The most interesting property of the discharge is its activity as a catalyst in the process of hydrocarbon decomposition into hydrogen and carbon [1, 2, 3]. The discharge structure development, energy absorption characteristics, spectroscopic diagnostic measurements as well as catalytic discharge properties are reported.

## 1. EXPERIMENTAL INSTALLATION

The scheme of the experimental plasma chemical reactor is shown in Fig. 1. The modulator generates high voltage pulses with pulse duration of $\tau_p = 0,1 \div 1\mu s$ and period of 1ms (1KHz). These pulses are the power supply for the magnetron, which generates microwave pulses with the wave type of $H_{01}$ in rectangular waveguide, frequency of 9,04GHz (wavelength 3cm) and pulse power value up to $W_p = 100kW$. Microwave radiation is directed into the discharge chamber by the waveguides. In the wave transformer the wave type is transformed into the type $H_{11}$ in cylinder wave- guide. The discharge chamber is a 10-cm section of cylinder waveguide with inner diameter of 20mm. A metal plate that serves as a shorting plunger and creates standing wave in the discharge chamber closing the end of the discharge chamber. The matched load has water-cooling; it is used for measurements of average microwave power. The shape of the microwave pulse is measured with the aid of the detector. The tungsten needle can be introduced into the discharge chamber in the maximum of the standing wave magnitude and serves to initiate the discharge. Maximum value of the electric field in the discharge chamber without the needle does not exceed 30kV/cm, so that discharge does not appear at atmospheric pressure without needle initiation. We determined discharge energy absorption characteristics as the difference of power values indicated by the matched load and the detector in the case without discharge and with discharge.

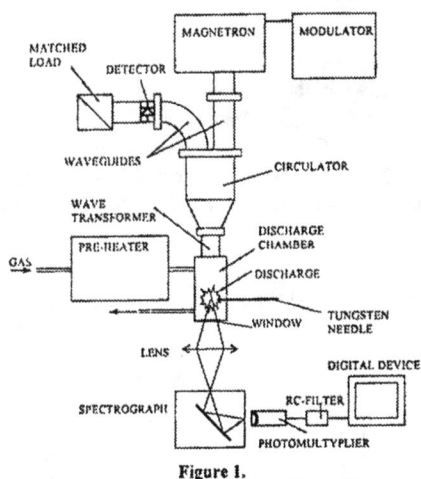

Figure 1.

## 2. DISCHARGE STRUCTURE

Visually the discharge has the same structure in the different hydrocarbons, air, hydrogen and carbon dioxide. It forms a bunch of several (about ten) thin plasma filaments that grow simultaneously along the electric field lines from the needlepoint. Fig. 2,3 show the discharge structure for different values of pulse duration and different pulse powers. At each new pulse plasma filaments do not repeat their previous tra-

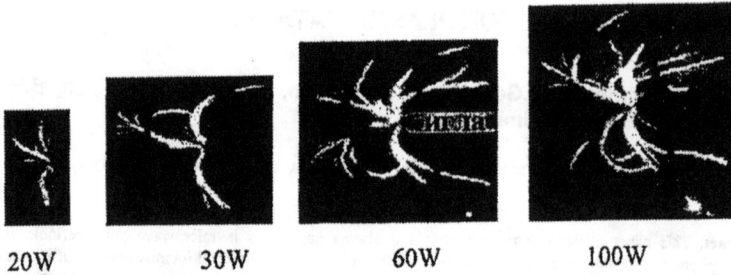

Figure 2. Discharge under different average power of magnetron and pulse duration $\tau=1\mu s$. Scale is shown on Fig. 3.

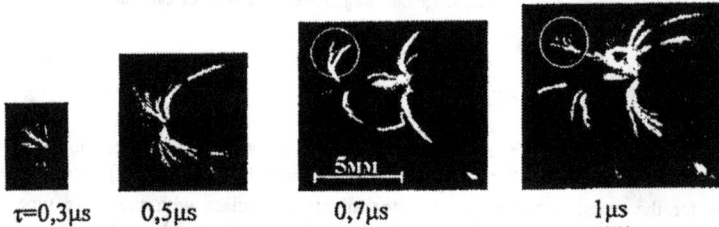

Figure 3. Discharge at different pulse duration and average power 60W.

jectories so that the discharge treatment of reactant gas is uniform on average. Total volume of all the filaments for a single pulse is roughly proportional to the energy E of the microwave pulse. For E = 0,03 J ($W_p$ = 30kW, $\tau_p$ = 1μs) total volume is 1,7mm$^3$, for 0,06 J (60kW, 1μs) - 3mm$^3$. Length of the plasma filaments is ~1sm, visual diameter is ~0,2mm. It should be mentioned that the plasma filaments have a property of ramification into several new filaments (in Fig. 3 this fact is marked by circles). So there is a certain resemblance between the filament structure at the needlepoint at early moments (0,1μs) and the structure on the tip of a single propagating filament.

## 3. PLASMA CATALYSIS OF HYDROCARBON REACTIONS IN PULSE MICROWAVE DISCHARGE

This division is dedicated to plasma catalysis effect of pulse microwave discharge described above in the case of endothermic reactions of hydrogen and hydrogen rich gas production such as methane, propane, butane decomposition and methane and ethanol steam reforming processes. Process energy requirements are covered in this case mainly by low potential heat while plasma is using for chemical reaction acceleration only via active species generation. Experiments clearly demonstrated an ability of microwave plasma to accelerate chemical reaction at the relatively low temperature level. Possible mechanisms of plasma catalytic were analyzed as well.

### 3.1. Introduction

Currently industrial demand for hydrogen and hydrogen rich gas is continuously growing. Conventional catalytic technology, essentially if we are speaking about small and moderate scale portable applications, has certain problems because of relatively low specific productivity, high metal capacity and equipment size. Even thermal plasma, being very high energy density media, is giving an attractive alternative for the hydrogen and syngas production. In this approach plasma replaces catalysis and accelerates chemical reactions mainly because of high temperature effect. The advantages of plasma chemical method are extremely high specific productivity of apparatus, low investment and operation costs. However relatively high electric energy consumption, related with gas heating applies certain restrictions on possible applications of thermal plasma approach. Non-thermal plasma can accelerate chemical reactions at the low temperature as well because of active species generation by fast electrons. If active species, generated by non-thermal plasma, are capable to promote many cycles of chemical transformation when high specific productivity of plasma can be combined with low energy consumption of traditional catalyst. That is why this so-called plasma catalysis approach is the subject of permanent, strong interest of several last decades.

Hydrocarbon conversion processes and in particular methane decomposition into hydrogen and carbon black, which are a subject of our current interest, are endothermic reactions and to cover process enthalpy and to shift chemical equilibrium to hydrogen and carbon, methane should be heated in any case. Nevertheless temperature level required to shift chemical equilibrium is relatively low (600K-1000K) and one can use for the process relatively low potential heat while plasma will be applied only as a catalytic agent for active species generation.

### 3.2. Plasma catalysis of methane decomposition

3.2.1. Experiment

Experimental setup consisted of two main bocks. First, one was a gas reagent pre-heater. In particular, methane gas with flow rate $30cm^3/s$-$250cm^3/s$ was heated at the atmospheric pressure in this block up to the temperature in the rage 400-800°C. Temperature during heating process was controlled by thermocouple measurements. Preliminary heated methane came to a plasma block to be treated by pulse microwave discharge. The section of axial waveguide was at the same time reactor volume. Gas temperature at the inlet of discharge zone was measured by movable thermocouple. For gas cooling and reaction products quenching a heat exchanger combined with carbon black separator was used. The concentration of gas components at the inlet and outlet of the system such as H2, CH4, C2H6, C2H4, C3H8, C3H6, C2H2 was determined by gas chromatography.

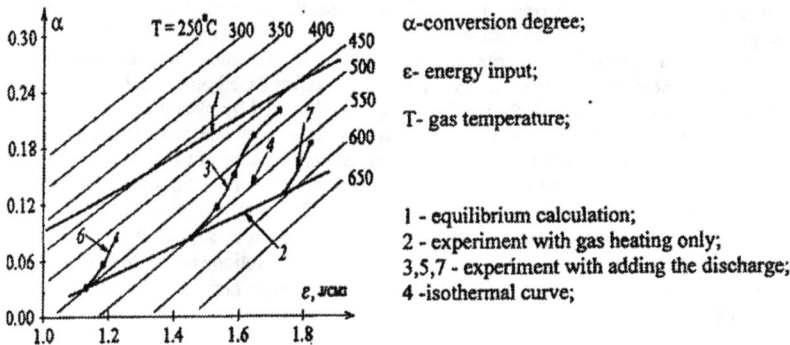

Figure 4. Plasma catalysis process of methane conversion

The measurements have been carried out by the following way. Firstly for the certain regime of methane heating gas temperature and gas composition at discharge's inlet have been determined. After that, outlet gas composition was measured as a function of discharge power. It was found that in all interesting regimes the only important products were hydrogen and carbon black. Experimental results are presented at Fig. 4 as dependence of methane conversion degree on specific energy input that is on ratio of total (both thermal and plasma ones) energy deposit into the gas to the initial methane flow rate. The specific energy input was calculated based on measured values as:

$$E = E_t + E_p = \left[ (H_h - H_0) + W \right] \Big/ Q \tag{1}$$

Where $H_h$ and $H_0$ gas enthalpy at the inlet and outlet of the heater, which can be calculated, based on temperature and chemical composition measurements. W -plasma energy deposition, Q - methane gas flow rate. Taking into account that the only important channel of $CH_4$ transformation was methane dissociation into hydrogen and carbon black, gas enthalpy at the heater outlet can be determined as a function of the only parameter characterized chemical composition that is methane conversion degree $\alpha$.

As one can see the experimental conversion degree under pure thermal effect (curve 2 in Fig. 4) is lower than equilibrium value – curve 1. This discrepancy concerns with kinetic limitation in methane decomposition under experimental conditions. System behavior at the same total energy input changed drastically when microwave discharge has been applied (see curves 3,6,7 in Fig. 4). These curves correspond to different initial methane temperature. In particular curve 3 corresponds preliminary heating of methane up to 550 C. In all cases conversion degree approaches to its equilibrium value. It should be noted that process energy consumption is covered mainly by gas internal thermal energy and hence process acceleration is accompanied by gas cooling (curve 3).

It is important also that plasma catalysis effect can lead to the significant decrease of process energy consumption in several times (from 2,1 to 0,95 eV/mol. of $H_2$).

### 3.2.2. Theoretical Analysis and Computer Simulation

To explain experimental results the following possible mechanisms of plasma effect on $CH_4$ decomposition were taken into account:

- Thermal acceleration related with local overheating onto microwave streamer channels;
- Radical acceleration provided by possible chain reactions of methane radicals such as $CH_3$, H, $CH_2$ generated by plasma from matrix gas;
- Auto catalytic methane decomposition on the surface of carbon black particles generated in plasma zone;
- Ion acceleration caused by ion–molecular chain reaction of methane decomposition.

An estimation of thermal acceleration mechanism was made supposing that complete thermal energy recuperation took place under experimental conditions. Dependence of methane conversion degree on plasma energy input for thermal mechanism is shown at Fig. 4 (curve 4, temperature 550 C). One can see that this mechanism can not explain experimental results obtained. The same conclusion was made about radical acceleration mechanism. Kinetic modeling of plasma effect based on radical mechanism of $CH_4$ decomposition has shown that radical chain, which might be initiated by plasma via radical production, would have too small chain length to explain measured value of methane decomposition energy cost. Auto catalytic methane decomposition has been rejected as well because of too low (a discrepancy was more than several orders of magnitude) value of surface methane decomposition rate constant to explain process characteristic time under experimental conditions. The most probable candidate to explain catalytic effect of pulse microwave plasma on methane dissociation is supposed to be ion-molecular Winchester mechanism. This mechanism is characterized by low activation energy, high reaction rate and is considering now as the most probable mechanism of new phase generation in $SiH_4$ plasma. This mechanism can be described as:

Initiation reaction

$$CH_4 + e \rightarrow CH_3^- + H \tag{2}$$

Cluster growth reactions

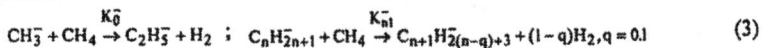

$$CH_3^- + CH_4 \xrightarrow{K_0^-} C_2H_5^- + H_2 \ ; \ C_nH_{2n+1}^- + CH_4 \xrightarrow{K_{n1}^-} C_{n+1}H_{2(n-q)+3}^- + (1-q)H_2, q = 0.1 \tag{3}$$

Chain termination:

$$CH_4^+ + C_nH_m^- \xrightarrow{K_r} C_{n+1}H_{4+m-x} + xH \tag{4}$$

Kinetics analysis has shown that this mechanism might explain experimental results both from the point of view of process energetic and process kinetics. The theoretical dependence of methane conversion on plasma energy input is presented at Fig. 4 (curve 3). The reasonable agreement between experimental and theoretical results is an argument in favor of ion-molecular mechanism.

### 3.3. Plasma catalysis of ethane and ethanol conversion

The results of analysis of experiments of ethane and ethanol conversion are present on Fig. 5,6. As in the case of methane, there are several curves corresponding to different conditions of experiment. One can see that in the case of ethane conversion the plasma energy input leads to increasing of conversion degree from 19% to 26% at the plasma energy fraction in total energy consumption only $E_P/E_T=0,05$ (point 3 in the Fig. 5). At the same time the temperature of the mixture is decreased from 470 C to 400 C. Thus, plasma catalysis decreases the energy cost of hydrogen production from 1.5 eV/mol. (point 2) to 1.2 eV/mol (point 3). It should be mentioned also that energy cost of additional hydrogen formed by means of plasma effect is only 0,26 eV/mol.

In the experiments of ethanol conversion (Fig. 6) the plasma energy fraction consists only 5% from overall energy input. Plasma catalytic effect leads to increasing of conversion degree from 23 to 62% and decreases temperature from 410 C to 370 C. Curve 1 characterizes only thermal decomposition of ethanol.

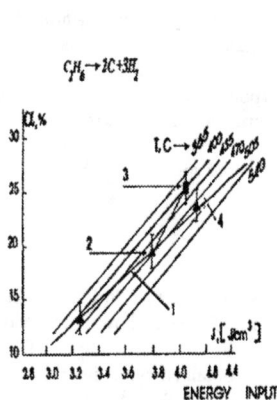

Figure 5. The dependence of ethane conversion degree on specific energy input. 1- experiment with pure thermal heating. 2,3 – experiments with MW discharge (heating temperature 470 C, microwave energy input is 5% of thermal heating). 4 – system evolution at the constant temperature

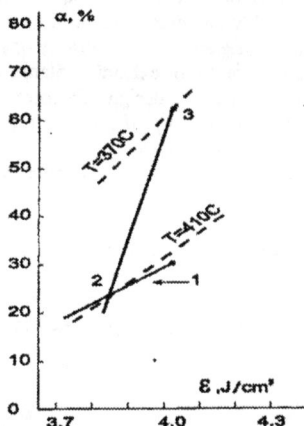

Figure 6. The dependence of ethanol conversion degree on specific energy input. 1- experiment with pure thermal heating. Point 3 – experiments with MW discharge

### 3.4. Summary

Thus, the effect of plasma catalysis is observed in process of hydrocarbon conversion. The same effect takes place in steam conversion of methane and ethanol. The wide spectrum of hydrocarbons, high efficiency and low range temperatures where effect of plasma catalysis is observed, permit regard this process as possible novel concept of hydrogen and hydrogen rich gas production.

## 4. DISCHARGE DIAGNOSTICS

### 4.1. Speed of the discharge propagation

Discharge propagation velocity along the plasma filament trajectories was measured in different gases with the aid of the photomultiplier by fixing the leading edge of plasma radiation pulse on the oscillograph. This radiation was observed from a small area (0,1×0,1mm) in the vicinity of the needlepoint. Then this area was shifted by a fixed space interval and the oscillograph measured time delay of the leading edge of plasma radiation pulse. The propagation velocity is the ratio of the interval and the time delay. Results of these measurements are presented in Tab.1.

| Gas | Velocity, ×10$^6$ sm/s |
|------|------|
| Air | 1,1 |
| $CO_2$ | 0,65 |
| $H_2$ | 2,8 |
| $CH_4$ | 1,3 |

Table 1.

The obtained value of speed was $1,1 \cdot 10^6$ cm/s for discharge in hydrocarbon gas and $2,8 \cdot 10^6$ cm/s for discharge in pure hydrogen. Let's remind that radius of chemical reactor equals 1.05 cm. Thus direct measurement of speed of development of discharge confirms the fact, established in section 1.2.1, that discharge has time to cross section chemical reactor during the established time of a pulse $= 1$ μs.

In frameworks of diffusion model of discharge propagation there is a ratio connecting speed of propagation of the ionization front (of the streamer head) with frequency of ionization and diffusion factor. As the frequency of ionization is quickly growing function of the reduced electrical field E/N (E - amplitude of an electrical field; N - density of gas), quasi-uniformity of growth of filaments in length tells us about a constancy of intensity of a field on tip of extending filaments. This conclusion is extremely important and proves the correctness of a choice of a design of a corona element (pointed needle), as an aggravation (the increase of an electrical field at an tip of a needle results in occurrence quickly extending microwave streamer, processing by the plasma reagent).

The results of account of this field value from the measured speed of discharge propagation are given below.

### 4.2. Microwave energy absorption

Microwave energy absorption characteristics of the discharge are shown in Fig. 7. Curve 1 displays the shape of microwave power pulse generated by the magnetron. Curve 2 shows microwave power absorbed

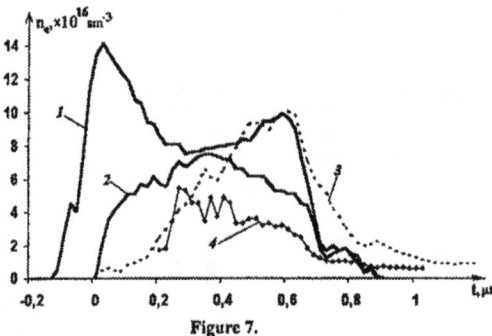

Figure 7.

by the discharge. These curves were obtained with the aid of the detector. Both curves 1 and 2 have the same scale on the ordinate axis, but units on this axis belong to the curve 4 (electron density). The origin on the time axis has been chosen at the moment when visible plasma radiation (curve 3) appears. It can be seen in Fig. 7 that energy absorption achieves almost 100% at t = 0,35μs, at this moment the discharge is a well matched impedance for the microwave circuit, then absorption decreases and becomes great again only on the trailing edge of the pulse. Average absorption can achieve 60%. It was measured as the difference of the energy absorbed in the matched load when the discharge does not appear and when the discharge is initiated. This value can also be determined as the ratio between areas under the curve 2 and under the curve 1.

193

## 4.3. Diagnostics of plasma parameters

### 4.3.1. Amplitude of a microwave field and electrons temperature in discharge plasma.

Actually value of an electrical field E and reduced electrical field E/N is not so evident, however, this value, recounted in value of electrons temperature $T_e$, evidently characterizes properties of plasma in dependence on the given pulse microwave power, which directly determines all these parameters. In case, which interests us now, when the plasma generates active particles with low energy expense, this electron temperature $T_e$ in plasma of our discharge should be not less than 2,5-3,5 $eV_,$.

The amplitude of a microwave field in discharge plasma in hydrogen and in hydrocarbon gas was measured by the form of contours of spectral lines $H_\alpha$ and $H_\beta$, which presence is typical for the categories in these gases. The method of modeling of a contour was used, in which the measured value of parameters corresponds to the best concurrence of experimental and modeling contours.

The value of an electrical field in the plasma filament in a range from 20 up to 30 kV/cm was result of these measurements. For a time interval 0÷50ns the intensity of radiation was not sufficient for reliable registration of contours of spectral lines. However maximal value of a field is realized in this time interval

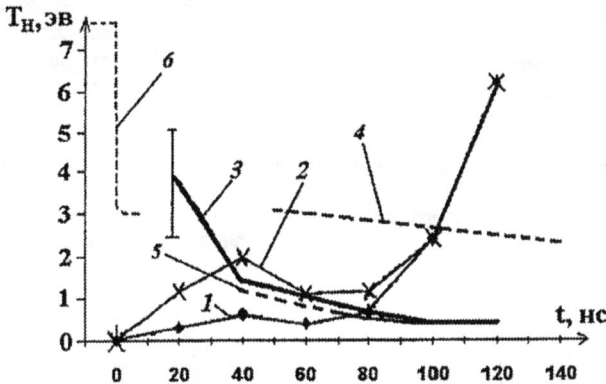

Figure 8. Electron temperature; 4 – average energy of electrons from measuring of value of microwave field; 6 – calculating average energy of electrons in microwave streamer head.
1 and 2- $H_\alpha$ and $H_\beta$ emission; 3 and 5 – excitation temperature.

(streamer head). Evolution of a field at the early moments of time was received by modeling front of ionization with use of the measured value of speed of streamer propagation (section 3.1.3). The maximal field in the streamer head achieves value 70-100kV/cm and quickly falls down (in time about one nanosecond) up to value 25÷30kV/cm.

In a Fig. 8 the temporary dependence of effective electrons temperature, determined through average energy in a case nonmaxwellian EEDF, is shown. Electronic temperature at an early stage (<10ns) was calculated taking into account the value of amplitude of an electrical field received from propagation speed of the streamer head. At a late stage (>50ns) electronic temperature was determined from amplitude of an electrical field measured by spectroscope methods.

A conclusion, which is possible to make from given on Fig. 8 dependencies of electrons temperature, is that the temperature is sufficient for effective process of plasma catalysis.

### 4.3.2. Measurement of concentration of the charged particles in discharge plasma and gas temperature in discharge.

These parameters (electrons concentration $n_e$ and gas temperature $T_0$ ) characterize a degree of development of the plasma filament and usually sharply grow in the end of some typical time from a beginning of the discharge pulse as a result of development so-called ionization-overheating instability.

Since the moment of time 250ns in hydrogen and 400ns in hydrocarbon sharp jump of half width of lines $H_\alpha$ and $H_\beta$ was observed, which is explained by sharp increase of concentration of the charged particles in plasma. Broadening by ions and electrons became the prevailing mechanism broadening of line contours in these conditions.

Processing of the experimental contours of lines made for definition of concentration of the charged particles, also consisted in modeling calculating contours and comparison them with experiment. The concentration of free electrons in plasma was estimated with help of broadening of the central components of a line $H_\alpha$ by electron impact. The received value of electronic concentration is close to the measured value of concentration of ions.

The measured temporary dependencies of concentration of charged particles are shown in a Fig. 9. The jump of concentration in hydrogen discharge occurs at t≈250ns, in hydrocarbon discharge the jump occurs a little bit later, at t≈400ns. The concentration of electrons in hydrocarbon discharge achieves value approximately twice-superior then value of concentration in hydrogen. As it is visible from a Fig. 9, the concentration falls down a little with time.

In first 0.4μs, when lines are broaden by a microwave electrical field, the electrons concentration in hydrocarbon discharge was estimated by the absorbed power P (Fig. 7, curve 4), measured value of propagation speed, amplitude of a field and spatial parameters. The received estimated value (curve 4 in a Fig. 9) approximately is constant in time and consists $1\div2\times10^{15} sm^{-3}$. Thus data, submitted in a Fig. 9, characterize dynamics of behavior of concentration of the charged particles in plasma practically during all time of a pulse (1 μs), including the moment of development of ionization-overheating instability (0,3-0,4 μs).

This conclusion proves to be true also by spectral measurements of temperature of gas in plasma of the pulse discharge.

In hydrocarbon discharge the measurements of rotational temperature of a molecule $C_2$ on Swan's strips (transition $V_{0-0}$, 516 nm) were carried out. Rotational temperature of molecules is usually close to

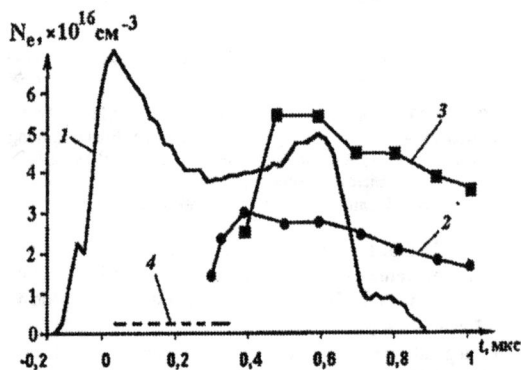

Figure 9. . Temporary dependence electron concentration.
1 – microwave pulse; 2,3,4- electron concentration;2 – discharge in pure hydrogen, 3 – discharge in hydrocarbon, 4 – electron concentration from measurement absorbed power and field.

transitional temperature, therefore these temperatures were identified. Temperature was measured by means of modeling spectra and comparison them with experimental.

The results of measurement of temporary dependence of rotational temperature are given in a Fig. 10 (curve 2). From fig. 10 it can be seen, that rotational temperature grows from room up to 4000÷5000K in time about 0.4 μs, practically coincident with the moment of growth of concentration of the charged particles. Further, behind cut of a microwave-pulse during time about 2μs, while the intensity of a spectrum is sufficient for reliable registration, decrease of temperature was not observed.

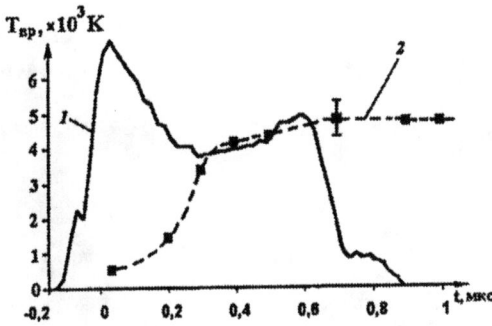

Figure 10. Temporary dependence of rotational temperature $T_{rot}$ :
1 – microwave pulse, 2 – $T_{rot}$.

## 4.3.3. Summary of the diagnostics results.

| N | Discharge parameter | DISCHARGE PARAMETERS | | | |
|---|---|---|---|---|---|
| | | Time interval | | | |
| | | $t \sim 1ns$ | $t < 50ns$ | $50 < t < 250 \div 400ns$ | $400 < t < 800ns$ |
| 1 | Electron density $N_e$ | - | $3 \times 10^{14} cm^{-3}$ (from the model of streamer body) | $\sim 10^{15} cm^{-3}$ (from the discharge power absorption) | $3 \div 5 \times 10^{16} cm^{-3}$ (by Stark spectroscopy of $H_\alpha$ and $H_\beta$ lines) |
| 2 | Magnitude of microwave field in plasma E | $\sim 100$ kV/cm (from the model of streamer head) | $\sim 30$ kV/cm (from the model of streamer body) | $30 \div 20$ kV/cm (by Stark spectroscopy of $H_\alpha$ and $H_\beta$ lines) | - |
| 3 | Electron energy from E/N parameter | $7 \div 8$ eV (from the model of streamer head) | 3 eV (calculated from electric field magnitude) | $3 \div 2$ eV (from measured electric field magnitude) | - |
| 4 | Gas temperature $T_0$ | 300K | $\sim 300$K | $600 \div 4000$K (form rotational spectrum of $C_2$) | 4500K |
| 5 | Discharge propagation velocity $V_t$ | $1 \div 2 \times 10^6$ cm/s (experiment) | | | |

## References

1. Rusanov V., Etivan K., Babaritskii A., Baranov I., Demkin S., Jivotov V., Potapkin B., Ryazantsev E. Dokl. Akad. Nauk, 1997, **354**, 213.
2. Babaritskii A., Deminskii M., Demkin S., Etivan K., Jivotov V., Potapkin B., Potekhin S., Rusanov V., Ryazantsev E. Khim. Vys. Energ., 1999, **33**, 59.
3. Babaritski A., Deminskiy M., Etievant K., Jivotov V., Potapkin B., Rusanov V., Ryazantsev E. Proc. of ISPC 14, Prague, August 2-6, 1999.

# PULSED HEATING OF THIN LAYERS OF POWDER MIXTURES BY HIGH-INTENSITY MICROWAVES

G. M. Batanov, N. K. Berezhetskaya, I. A. Kossyi, A. N. Magunov, and V. P. Silakov*

General Physics Institute, Russian Academy of Sciences, Moscow
*Institute of Applied Mathematics, Russian Academy of Sciences, Moscow

**Abstract.** Pulsed heating of thin (0.05 cm) layers of conductive and dielectric powders by microwaves with intensity about 10 kWcm$^{-2}$ has been studied. For intensities of 10 kWcm$^{-2}$ and up, and a pulse duration of 5—8 ms, in a sequence of 5 pulses at most, melting of the powder layer occurs and the characteristic absorption depth is equal to ~0.05 cm, whereas it is about 1 cm at a low microwave power. Measurements showed also that the sample absorption coefficients decreased during the microwave pulse. The possibility is considered for effective absorption of microwaves because of the surface breakdown and formation of a plasma in gaps between powder grains throughout the powder volume.

## 1. INTRODUCTION

In recent years, much attention have been given to technologies of microwave processing of various materials (see, e.g., [1]). In the case of ceramic powders of oxides of various metals, efficiency of microwave heating is low because of rather weak absorption and, in order to increase the efficiency, it is necessary to pass to mm-wavelength range [2]. More problems are encountered in achieving an efficient heating of thin powder layers.

It should be mentioned that rather high absorption coefficient over a wide range of frequencies was observed for materials produced by fusing together powders of metals and their oxides at high temperatures [3]. Unfortunately, for the present, the characteristic absorption depth for such materials and the conduction mechanism in the microwave frequency range have not been determined, it is known however that these materials conduct direct currents.

On the other hand, in [4], at high (10$^4$ Wcm$^{-2}$ and up) radiation intensity, surface discharges were observed on irradiated metal—dielectric targets. The target was a dielectric plate coated with a metal sawdust stuck with its surface. In this case, the energy dissipated locally near the surface because breakdown at the metal—dielectric interface was accompanied by the formation of a narrow plasma layer adjacent to the surface.

In view of the said above, two questions may be formulated. First, how strong is the absorption in mechanical heterogeneous mixtures of metals with their oxides? Second, whether or not nonlinear absorption processes can occur in such mixtures at high temperatures, such as occur during the surface breakdown on heterogeneous surfaces.

## 2. EXPERIMENTAL

As a subject of experimentation, we used powders of titanium, aluminum, tin oxide, cupric oxide, aluminum oxide and their mixtures. The powder-grain sizes were in the range 10—40 μm. Powder samples were packed between two glass plates 2 mm thick and the placket was squeezed to make a uniform thin layer 0.2—0.5 mm thick. The samples were mounted in a special holder and places inside the caustic of a gyrotron beam, transverse to the microwave-beam axis (Fig. 1). The wavelength of the gyrotron radiation was 4 mm, the power was in the range 120—180 kW (the intensity at the beam axis was 10—15 kW/cm², respectively), a Gaussian beam being used. To monitor the gyrotron power signal and the signal reflected from the sample, a quasioptical coupler was included in the scheme of measurement. The microwave-pulse duration varied from 1 to 8 ms. When a pulse sequence was used, the intervals between pulses lasted 2 min.

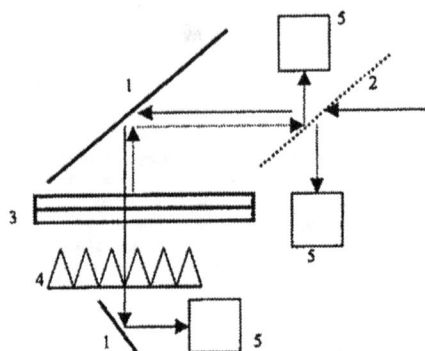

**Figure 1.** Schematic diagram of the experiment on heating thin layers of dielectric--conductive powder mixtures: (*1*) reflecting metal mirrors, (*2*) mica plate of the quasioptical coupler, (*3*) irradiated sample, (*4*) absorbent loads, and (*5*) microwave detectors. Arrows indicate directions of propagation of incident and reflected beams.

Packing the powder between two plates transparent to radio rays, we avoided the situation when the powder surface is directly exposed to the action of the beam in which case a discharge initiated by microwave breakdown inside the powder layer begins to propagate toward the gyrotron.

The characteristics of powder lays were measured at a low (mW) power level in the specially created testing device by the quasioptical scheme. The testing measurements with quartz plates of different thickness showed that the error in determining the sum of the reflection and transmission coefficients did not exceed ±0.5%.

The measurements of the characteristic of typical glass plates used as substrates for powder layers in the previous experiments at a high power level gave the values of the reflection coefficient R = 0.25-- 0.32 and the transmission coefficient T = 0.46--0.53, which corresponded to the absorption coefficient A ≈ 0.22--0.23. For the packet of two glass plates, the corresponding values are T = 0.4, R = 0.4--0.43, and A ≈ 0.23--0.26. With so high values of the reflection and absorption coefficients, it makes no sense to conduct measurements of the characteristics of powder layers, because they cannot be measured with a sufficient accuracy against the background of high values of reflection and absorption in the substrates. For this reason, the glass substrate was changed to a half-wave (1-mm-thick) fused-quartz plate (refractive index of silica equals to ≈1.94 at λ=4 mm). The measurement results for powder mixtures of different composition are listed in the table.

As is seen from the table, no-absorption condition take place for thin layers of the dielectric powders such as plumbic oxide and cupric oxide, and the observed difference between 1 and the sum R + T lies inside the error boundaries. The values of the reflection coefficient for these powders are also low.

In an aluminum powder, which is a highly conductive material, the absorption was also insignificant, which may be attributed to the presence of contacts between the grains; as a consequence its reflection properties are similar to those of a solid medium. However, it is reasonable also to point to a rather high level of microwave power passing through the layer of disperse material.

In semiconductor (silicon and tin oxide) powders, the absorption is substantial. Their characteristics turn out to be comparable with the characteristics of a thin (0.45 mm) phosphorus-doped silicon single crystal with a resistivity of 4.5 Ω·cm.

Such a behavior of semiconductor powders, which is similar to the behavior of the aluminum powder, is explained by the presence of contacts between grains.

As follows from Table 1, the mixtures of conductive and dielectric powders do not demonstrate an appreciable absorption, and the values of the reflection coefficient appear to be rather high (~0.5).

Table. Reflectance, transmittance and absorbance for powders.

| Composition | R | T | R + T | A |
|---|---|---|---|---|
| PbO | 0.143 | 0.907 | 1.043 | -0.043 |
| CuO | 0.112 | 0.928 | 1.04 | -0.04 |
| Al | 0.912 | 0.138 | 1.05 | -0.05 |
| Si | 0.313 | 0.295 | 0.608 | +0.392 |
| $SnO_2$ | 0.38 | 0.16 | 0.54 | +0.46 |
| 0.5 Ti + 0.5 CuO | 0.646 | 0.422 | 1.068 | -0.068 |
| 0.5 Si + 0.5 $Al_2O_3$ | 0.525 | 0.422 | 0.947 | +0.05 |
| 0.5 $SnO_2$ + 0.5 $Al_2O_3$ | 0.617 | 0.389 | 1.006 | -0.006 |
| 0.5 Al + 0.5 PbO | 0.759 | 0.257 | 1.016 | -0.016 |

## 3. DISCUSSION

The action of a single microwave pulse of an intensity above and duration of 5—7 ms on the silicon—plumbic oxide (1:1) mixture produces one or two fused points less than 1 mm in diameter in the mixture. During the series of subsequent pulses, the local melt zones enlarge and merge together after 5 pulses. A similar behavior is observed in the tin oxide—plumbic oxide (1:1) mixture. Figure 2 shows a photograph of the formed melt layer. Note that the layer is melt nonuniformly across the layer thickness: on the side of the incident beam, the melt material is firmly bound with the glass plate, whereas on the side of the second glass plate, the powder is melt nonuniformly and some regions are little affected.

If the pulse duration is shortened to 2 ms, then no melt zones were observed after 4—5 successive pulses. The melting effect was not also observed if the pulse duration remained 7—8 ms, but the intensity was reduced to 4—5 kWcm$^{-2}$.

A somewhat different picture was observed in the titanium—plumbic oxide (1:1) mixture. Even at intensities of 10 kW·cm$^{-2}$, a continuous melt zone arose after 2 pulses as short as 4 ms.

A vigorous exothermic reaction accompanied by the formation of metallic copper was observed in the titanium—cupric oxide (1:1) mixture. The reaction of this kind was initiated when the microwave intensity was lowered to 5—5.5 kW·cm$^{-2}$ while the pulse duration was equal to 1 ms. To initiate this reaction at lower levels of the intensity, the pulse duration should be increased.

A completely different situation occurred in the titanium—aluminum oxide (1:1) mixture. At an intensity of 8—9 kW·cm$^{-2}$ and pulse duration of 7 ms, only isolated melt points arose in the mixture. In this case, in each pulse, the powder between the glass plates was sputtered in the radial direction, which may be attributed to heating of the gas between grains. The sputtering of the powder between the glass plates in the direction from the beam center toward the periphery was observed also for the silicon powder.

a

b

**Figure 2.** Photographs of irradiated samples:

(a) 25%Si + 75%PbO powder mixture. The intensity is ~ 10 kWcm$^{-2}$. 5 pulses of duration 7 ms.

(b) (b) 50%Ti + 50%PbO powder mixture. The intensity is ~ 10 kWcm$^{-2}$. 1 pulse of duration about 8 ms.

As follows from the time behavior of the reflected signal from the targets, the reflection varies during the microwave pulse (Fig. 3). It should be noted that the reflection signal varies during the entire pulse

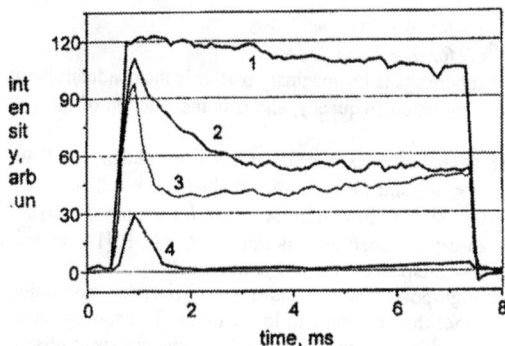

Figure 3. The reflected pulse for the 25%Si + 75%PbO mixture: (1) the first, (2) second, (3) third, and (4) forth pulses.

sequence. As seen from oscillograms illustrated in Fig. 3a, some slow decrease in the reflection is observed even in the first pulse. In the next pulse, the reflected signal reduces by one half for 2—3 ms. In the subsequent pulses, the signal falls sharply for 1 ms and the final level of the reflected signal decreases from a pulse to pulse.

To detect variations in the electric parameters of the powder mixtures affected by microwave pulses, we carried out special measurements with the help of a magnetron microwave source with the following parameters: the radiation wavelength was 2.5 cm, the intensity was about 10 $kW \cdot cm^{-2}$ and the pulse duration was 5 $\mu s$. Two concentric electrodes were inserted into the powder sample between the plates to measure a voltage pulse generated under the microwave action. A negative voltage pulse was detected at the central electrode only for the samples of silicon powder or its mixture with cupric oxide powder (Fig. 4). The average signal amplitudes in a 1-$M\Omega$-load were on the order of 0.2 V for the silicon powder and its mixture with cupric oxide powder in the volume ratio 1:3. For the silicon—cupric oxide mixtures with the ratios 1:1 and 3:1, the average pulse amplitude increased to 1—1.5 V. For 1:1 and 1:3 mixtures, bright scintillation points were observed over the powder surfaces. In the silicon powder and its mixture with cupric oxide, no scintillation points were observed.

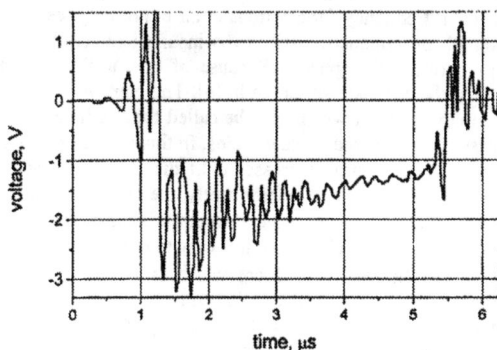

Figure 4. Voltage pulse between the external ring and the central electrode in the 1-$M\Omega$ load. 50%Si + 50%CuO mixture.

According to [5], the characteristic absorption depths in a plane layer consisting of both spherical particles that do not absorb incident radiation and conductive particles of radius $a_0 << \lambda$ with a density $n_0 = 3\alpha/4\pi a_0^3$ are defined by the expressions

$$h_0 \approx \lambda\,|\varepsilon|^2/18\pi\alpha\varepsilon_i \quad\quad for \;\; \delta = c/(4\pi\sigma\omega)^{1/2} > a_0$$
$$h_0 \approx (4a_0/9\alpha)(\lambda\sigma/c)^{1/2} \quad for \;\; a_0 > \delta \; and \; |\varepsilon| >> 1.$$

Here $\varepsilon$ is the dielectric constant of conductive spheres, $\varepsilon_i$ is its imaginary part, $\sigma$ is the conductivity of the material of conductive spheres, $\omega$ is the cyclic radiation frequency, and $\alpha$ is the fraction of the volume occupied by the conductive spheres.

For the silicon—aluminum oxide powder, we have $\delta > a_0$, and for $\alpha = 0.5$, we obtain $h_0 \approx 0.44$ cm; for titanium—cupric oxide mixture, we have $\delta < a_0$, and for $\alpha = 0.5$, we obtain $h_0 \approx 0.4$ cm when the diameter of conductive spheres is equal to $4 \cdot 10^{-3}$ cm. It follows from here that, for the powder layer with a characteristic thickness of $h = 0.05$ cm, the absorption coefficient is equal to $A = (1-R)[1-\exp(-h/h_0)] \approx$ 0.05-0.06 which coincides with the results of measurements at the low power level.

The powder melting in a sequence of high-power pulses shows the absorption coefficient at intensities about 10 kWcm$^{-2}$ is significantly higher than its value in lower fields. The above-mentioned fact of nonuniformity of melt across the powder thickness evidences that the characteristic absorption depth became smaller than the layer thickness (0.05 cm), i.e., it fell by one order of magnitude in comparison with the characteristic absorption depth at low intensities.

In evaluating the absorption coefficient of powder layers, it is necessary to take into account the coherence of waves scattered by individual grains, because the distance between grains is much less than the radiation wavelength. The value of the absorption coefficient must be proportional to the scattering cross section of a grain. For a conductive sphere, this cross section (see. [5]) is given by the expression $S_t = 8\pi(2\pi)^4 a_0^6/3\lambda^4$. For an absorbing dielectric sphere, this cross section is smaller by a factor of $(\varepsilon - 1)^2/(\varepsilon + 2)^2$. For quartz, in our case, this factor is equal to 0.25. Apparently, this is why the absorption coefficients obtained at low power levels for the powders of plumbic and cupric oxides differ from those for the powder mixtures consisting, by halves, of highly conductive grains. On the other hand, based on the difference in scattering of high-intensity microwaves by dielectric and conductive grains, we can make some inferences about the processes occurring in the powders irradiated by high-amplitude microwaves. As seen in Fig. 3, the reflected signal falls sharply during the high-power pulse. This rises the question as to the reason of a decrease in the scattering cross section of a highly conductive spherical grain. It may be suggested that we have the effect of surface microwave breakdown similar to that observed in [4]. Indeed, the wave electric field (~ 3 kWcm$^{-1}$) is much lower that the breakdown field in air (~ 30 kWcm$^{-1}$). The breakdown field between the powder grains is even higher than this level because, in gaps of size 10--40μm, the electron diffusion loss frequency $v_d \approx D_e/a_0^2 \approx 10^9$ s$^{-1}$ is significantly higher that the three-body attachment frequency for electrons in air at normal pressure $v_a \approx 10^8$ s$^{-1}$. A tree-fold increase in the field strength on conductive spheres is insufficient for the critical breakdown field to be exceeded in the gaps between the spheres. Because of this, a plasma can be produced only by breakdown on the dielectric surface, as was observed in [4]. The gaps between grains are then filled with the plasma to form a complex medium which may be called the "cellular" plasma, because it includes the cells filled with the plasma between the powder grains. In this case, as the charge-particle density in the cells grows, the microwave absorption increases until the skin depth becomes smaller that the cell diameter. Then, the absorption decreases and, consequently, the density cannot grow further. It may be assumed that there is an optimal condition when the absorption is most effective: $\delta \approx a_0$, since the cell dimensions are on the order of $a_0$. From this condition, it is easy to deduce the minimum value of the effective absorption depth: $h_{min} \approx 2^{1/2}\lambda/9\pi\alpha = 0.02$ cm which is consistent with the experiment. In this case, we have the three-component system similar to that considered in [6].

The above assumptions are supported by observations of the unipolar voltage pulse generated at the concentric electrodes in the powder (Fig. 4). Note, the signals of this kind were observed in pulsed microwave discharges in gases at high pressures [7].

To verify the assumption that the high absorption coefficients may be attributed to a plasma generated by surface breakdown, we calculated the value of energy absorbed in the plasma-containing medium and compared it with the energy needed to melt the components of the powder layer. Assuming the absorption coefficient of the plasma-containing medium to be equal to $A_{pl} \approx 1$, it is easy to calculate the energy density absorbed in the powder layer during a microwave pulse:

$$q_0 = s\tau/h \approx 10^4 \cdot 7 \cdot 10^{-3}/0.05 \approx 1.4\ kJcm^{-3},$$

where $s$ is the intensity and $\tau$ is the duration of the microwave pulse.

The melting energy can be estimated by the formula

$$q \approx \alpha(\Delta H_l + C_{pl}\Delta T_l)\rho_l/M_l + (1-\alpha)\ C_{p2}\Delta T\rho_2/M_2,$$

where $\Delta H_l$ is the heat of melting for the material with the lower melting temperature, $C_p$, $\rho$, $M$ are the specific heat, the mass density, and the molecular weight of the powder components.

Thus, for the silicon—plumbic oxide mixture, we obtain $q \approx 1.7\ kJcm^{-3}$. This estimate is consistent with the energy expenditure (the microwave pulse intensity and duration) in our experiments. Note that, the formation of point melt zones during the first pulses directly indicates that, for these regions, the total absorption cross section is larger than the geometric area. The fact that the melt area enlarges during the following pulses may be explained by the formation of a phase with semiconductor properties.

High values of the absorption cross sections in the local regions are convincingly demonstrated by the experiment in which a solid-phase oxidation-reduction reaction occurs in the 0.5Ti +0.5Cu mixture. Flaming of this mixture can occur only at high temperatures ($\sim 10^3$ K). As follows from measurements of the minimal irradiating energy, the average temperature in this mixture, even assuming $A \approx 1$, cannot exceed 200 K.

One more evidence of the local energy release is the formation of point melt zones in the 0.5Al + 0.5 $Al_2O_3$ mixture. To melt aluminum in this mixture, it is necessary to ensure the energy density about 3 $kJcm^{-3}$, whereas the average energy density in the experiment could not exceed 1.6 $kJcm^{-3}$.

## 4. CONCLUSION

Let us summarize the main results of our study. The measurements of the reflection and transmission coefficients of thin (0.05 cm) layers of dielectric powders and their mixtures with semiconductor and metal powders, which were performed by the quasioptical method at a low level of microwave power, have given low values of the absorption coefficient (no more than 0.05) which coincide with the results of calculations for the mixtures under study. The reflection coefficients for the mixtures containing conductive particles turns out to be several times higher than for the dielectric powders.

At intensities about 10 $kWcm^{-2}$, in the mixtures of conductive and dielectric grains, there existed local melt zones (1 mm in diameter) and merged regions after a sequence of 4—7 pulses of duration 5—8 ms. Experiments demonstrated a high local heating of the material and nonuniform melting across the powder-layer thickness, indicating that the characteristic absorption depth fell to the value $h_0 \approx 0.05$ cm.

The other effects observed in the experiment are a decrease in the reflection during the microwave pulse and the generation of a voltage pulse at the electrodes inserted into the powder layer. The results of the experiments performed at high irradiation intensity point out that, in the powders, surface breakdowns occur in the contact points of conductive and dielectric grains, thus producing a plasma that fills the gaps between them. In our opinion, it is precisely the process of plasma formation leads to the effective microwave absorption.

### Acknowledgments

The authors thank L. V. Kolik and V. A. Plotnikov for assistance in the gyrotron experiments and N. I. Malykh for help in the quasioptical measurements. This work was supported in part by the International Science & Technology Center (project. no. 908).

## References

1.  Agrawal D., J. Mater. Edu., 1999, **19**, 49.
2.  Bykov Yu.V. and Eremeev A.G. In: High-Frequency Discharge in Wave Fields, IPF, Gorkii, 1988, p. 265 (in Russian).
3.  Kovneristyi Yu.N., Lazareva I.N., and Ravaev A.A. Materials Absorbing Microwaves, Moscow: Nauka, 1982 (in Russian).
4.  Batanov G.M., Bol'shakov E.F., Dorofeyuk A.A. et al. J. Phys. D. Appl. Phys., 1996, **29**, 1641.
5.  Landau L.D. and Lifshits E.M. Electrodynamics of Solid Media, Moscow: Nauka, 1982 (in Russian).
6.  Galstyan E.A. and Ravaev A.A. Izv. Vyssh. Uchebn. Zaved. Radiofiz., 1987, **30**, 1243 (in Russian).
7.  Gritsynin S.I., Dorofeyuk A.A., Kossyi I.A., and Magunov A.A., Teplofiz.Vys.Temp., 1987, **25**, 1068 (in Russian).

# DEVELOPMENT OF MICROWAVE POWERED ELECTRODELESS LIGHT SOURCE IN MEPhI

A.Didenko, B.Zverev, A.Koljaskin, A.Prokopenko

Moscow state engineering physics institute, Kashirskoe shosse, 31, 115409 Moscow, Russia

Abstract. The paper is dedicated to application of microwave plasma in a visible light source. The phenomena of efficient conversion of microwave energy to broad spectrum visible light by molecular emission from sulfur is considered and the experimental apparatus used to generate this light is described. Processes in such sources are reviewed. Settlement of the problems, of such lamp operation are offered.

## 1. INTRODUCTION

At the end of the 20th century microwave power engineering was enriched by a new direction, connected with development of high-efficient, microwave plasma light source. The essence of this problem is the following. It is known, that radiation temperature of a solar crown is $T_k = 5800°C$, and spectral density $I_v$ has a view as shown in fig. 1.

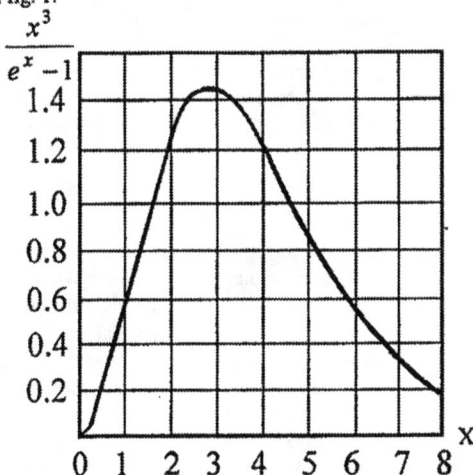

Figure 1. Planck function of $I_v = \dfrac{2(k_B \cdot T)^3}{c^2 \cdot h} \dfrac{x^3}{e^x - 1}$, where $x = h\nu/kT$.

As we can see from this figure, 31 % of radiated energy lies within the range of human eye responsivity. It means, that as a light source the sun has efficiency equal to almost 31%. In earth requirements it is impossible to receive light sources with a major efficiency on the basis of incandescent lamp because melting points of all elements at the periodic table have lower $T_k$, and wave length of maximum of radiation $\lambda \sim 1/T$. For this reason all incandescent lamps have a conversion efficiency of electrical energy to light about (2-3) % (fig. 2.).

Coloring Rendering Index (CRI) — vertical axis: 100, 75, 50, 25

Luminance
Watt — horizontal axis: 25, 50, 75, 100, 125, 150

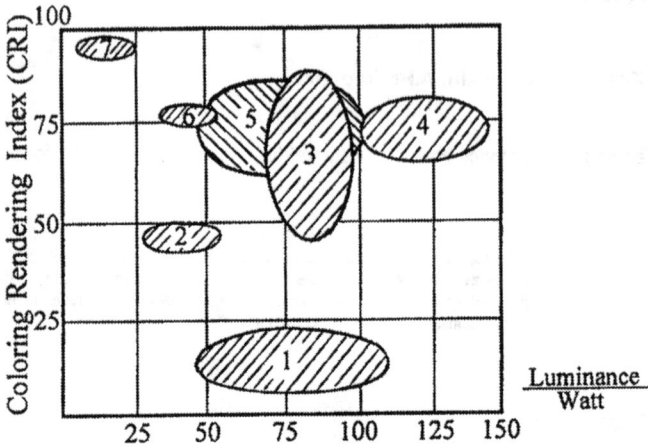

**Figure 2.** Lamp System Efficacies:
1. High Pressure Sodium; 2. Hg Vapor;
3. Full Size Fluorescent; 4. Sulfur; 5. Metal Halide;
6. White HPS; 7. Incandescent lamp.

Luminescent Hg lamps have higher efficiency ~ (10-15)%, but their spectral characteristics essentially differ from spectral characteristics of sun and therefore they have a low colour rendering index ~ 40-50%. If some restricted plasma volume heats up to $T = T_k = 5800°C$ it is possible to create high efficiency lamp, i.e. if the simulated sun is created under laboratory conditions, it would have the transformation efficiency of energy inserted in it at the level of 31%. Unfortunately, this prime problem has no solution interesting for practical application. Actually, the plasma will radiate as an ideal black body, if radius of plasma ball R fulfils the requirement

$$R > \frac{10^{23} \cdot T^{\frac{1}{2}}(K)}{Z^2 \cdot n_i \cdot n_e}, \qquad (1.1)$$

where T – temperature, $n_i$ and $n_e$ – electron and ion densities. At $T_e = 6000°C$ radius of a plasma ball is of the order of 1cm, if $n \geq 3 \cdot 10^{18} 1/cm^3$. The plasma radiates as an ideal black body, if it is dense enough. However this problem is not crucial. The fact is that the plasma blob of radius R = 1cm heated up to 6000°C would radiate power $W = \sigma_T \cdot T^4 \cdot S = 92.3$ kW. It is clear, that we can not speak about any light sources on this basis. All above-stated testifies that the development of light sources is possible only on the basis of nuclear and molecular spectrums of particular elements or their compounds. Unfortunately, the nuclear spectrums of all elements lay in a more short-wave range.

Nowadays, scientific community pays attention to discharges in mixed gases as the base of a possible high-performance light source. Microwave powered discharges have been used for nearly three decades as industrial sources of UV light, but there has been no actual application to the production of visible light. In 1992, the discovery of unique plasma system on the base of sulphur molecular spectrum has enabled the production of lighting systems with outstanding performance [1,2]. The experts of the American Company Fusion Lighting Corporation have created such microwave powered lamp systems. The plasma-forming medium of these lamps is high pressure sulfur or selenium vapor which produces light by molecular radiation. This result is achieved in a design, which provides a combination of high brightness, excellent stability and good efficacy (fig. 2.). First commerce samples of lamp system SOLAR 1000 were already utilized in USA. The researches of a microwave discharge in vapour of sulphur are carried out in Russia [3-6].

## 2. SULFUR LAMP

The conversion phenomena of microwave energy to broad spectrum visible light by molecular emission from sulphur was developed in Moscow State Engineering Physics Institute (MEPhI). The theoretical analysis of plasma processes in microwave fields testifies to high power efficiency of such lamps, and the tests confirm these conclusions.

Argon filled bulb at pressure 150 Pa (corresponding to the minimum of Pashen curve at frequency 2450 MHz (fig.3.)) and powder of sulphur is located in a microwave field.

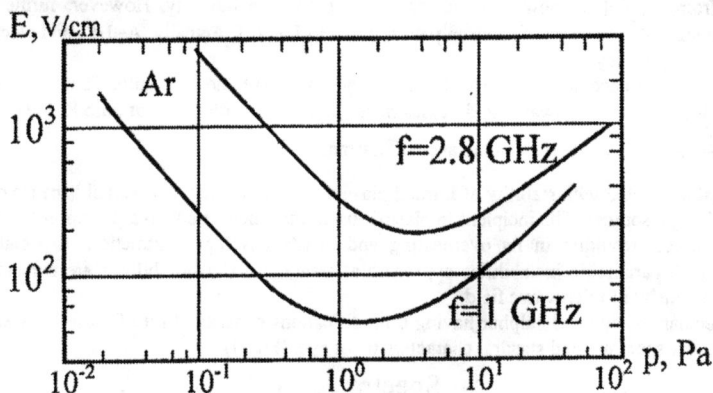

**Figure 3.** Thresholds of microwave disruption for argon.

The argon serves as medium of a microwave discharge and the electrons, torn off from its atoms, excite atoms and molecules of sulphur. The sulphur begins to sublimate at $T_s=444,6$ °C and its molecules transfer into an atomic state at temperature 1200 °C, however such temperature is not reached in the bulb. In its simplest form, the only components of the bulb are a quartz envelope, argon and sulphur. The bulb contains no mercury. There is no evidence of sulphur loss or reaction over many thousands of hours of life testing, resulting in no measurable shift in the output spectrum over the life of the bulb.

Microwave discharge with sulphur vapour allows to obtain not only highly effective light source (~ 20%), but also light source with a high light coefficient (70-85 %).

The choice of frequency of a high frequency field is a crucial question. It is known that high frequency discharges are possible at some kHz (induction discharge), MHz (HF discharge) and GHz (microwave discharge) frequency ranges. Let's compare these ranges for the purpose of definition of frequencies requiring minimum power. The induction discharge is induced by electric field E~$\omega$. Low power (hundreds of Watts) is not enough for the disruption in kHz range. In order to define working frequency of a HF or microwave oscillations it is necessary to compare 3 values: the diameter of a bulb D, free path l and amplitude of electrons oscillation A. The electrons do not collide inside envelope and can not provide ionisation at l>>D, then D~(1-2) cm. It takes place at pressure p of ~ $10^{-2}$ Torr. It means that for effective ionisation it is necessary to use pressure of buffer gas above this value. The Ar discharge, vapour of electronegative element sulphur being present, is possible as soon as electrons will make oscillations in a quartz bulb, i.e. when $A < \frac{D}{2}$. Capture of part of electrons by sulphur molecules can be compensated only in case of very intensive ionisation. The following relation takes place for electron oscillation amplitude in HF and microwave fields

$$A = \frac{e \cdot E_0}{m \cdot \omega \cdot \sqrt{\omega^2 + v_c^2}}, \tag{2.1}$$

where $v_c$ - collision frequency; $E_0$ - amplitude of an electric field; $\omega$ - oscillation frequency. If $v_c \ll \omega$ than $\frac{1}{2 \cdot \pi^2} \cdot \frac{e \cdot E_0 \cdot \lambda^2}{m \cdot c^2 \cdot D} < 1$ and at $E_0 \sim 200$V/cm these inequalities are valid at $\lambda < 2$m. If $v_c \gg \omega$ inequality $\frac{2 \cdot e \cdot E_0}{m \cdot \omega \cdot v_c \cdot D} \approx \frac{0.14 \cdot E_0 \left( V/cm \right)}{f(MHz) \cdot D(cm) \cdot p(Torr)} \leq 1$ should be fulfilled. It means, that the boundary frequency, hence the disruption intensity of electric field increases, depend on a diameter of quartz bulb, pressure at gas inside bulb and intensity of HF or microwave fields. It is necessary to mark that microwave range is preferable from a point of view of minimization of disruption intensity. However, further frequency increase is undesirable because of electron path is equal to $L = A \cdot f$. Actually, $A \sim 1/f^2$, and electron path is decreased with frequency.

Apparently, it is necessary to realize the discharge in all volume of a bulb. That is possible, if the penetration depth $\delta$ of a microwave field in plasma is comparable with radius of bulb R. Penetration depth is $\delta = \frac{c}{\sqrt{2 \cdot \pi \cdot \sigma \cdot \omega}} \geq R$, where $\sigma$ - conductance of plasma.

The question related with stability of formed plasma was considered. It is actual because of a durable work of such light sources. The incipient in plasma instabilities occurs due to a prompt motion of plasma to a bulb surface, resulting in its overheating and quartz envelope destruction. Especial danger is represented with parametric instabilities appearing at $\omega = \omega_{plasma}$. The instabilities are expanded even at rather small intensity of microwave field.

The molecular spectrum of sulphur having 6 modifications consists of set of close lines and is rather close to the solar spectrum and spectral characteristic of eye. (Fig. 4)

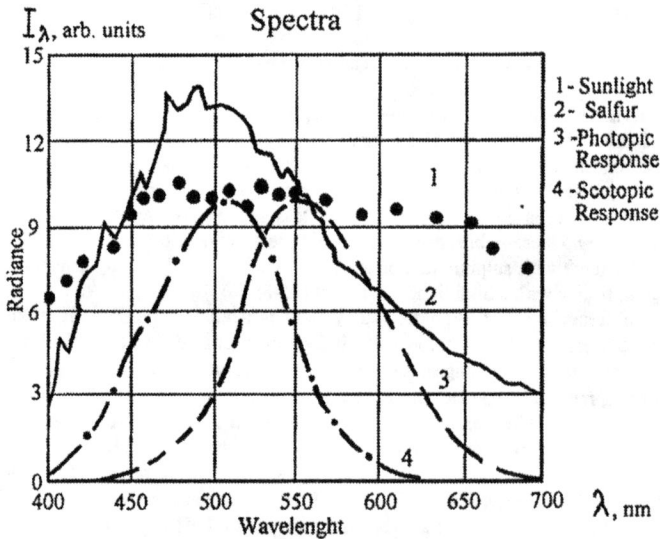

Figure 4. Spectral irradiance of sulphur lamp, solar spectra,
scotopic and photopic eye responses.

The power of a microwave generator required for discharge take place in buffer gas and for discharge sustenance in plasma after evaporation and ionisation of sulphur is an important question.

Minimum disruption voltage for buffer gas - Ar at f ~ 3.0 GHz and pressure 1-2 Torr is (200-300) V/cm. For instance, such intensities are achieved in a resonator with one light-transparent wall. Electric intensity $E_0$ is proportional to $\sqrt{100 \cdot P[Watt] \cdot Q}$, and microwave power at Q = 100 should be ~ 10 W. Effective intensity of a variable field $E_{eff}$ is less then $E_0$.

$$E_{eff} = \frac{E_o}{\sqrt{2}} \cdot \frac{v_c}{\sqrt{\omega^2 + v_c^2}} > E_{min}. \tag{2.2}$$

Therefore, it is necessary to use the generator with power 100 W for plasma excitation.

The discharge beginning in argon atmosphere leads to temperature rise of quartz envelope and subsequent formation of sulphur vapour and plasma. After that, the argon luminescence turns into a luminescence of polymorphic sulphur, which spectral characteristics are close to spectral characteristics of sun (fig. 4).

The pressure of sulphur vapour is defined by sulphur quantity in a quartz bulb. Usually, it is enough to have ~ (2-5)mg/cm$^3$ of sulphur on the bulb volume.

The discharge sustenance in plasma requires essentially smaller field gradient. This electric intensity $E_d$ and electronic temperature of plasma are linked by relations [7]:

$$\sqrt{\frac{k_B \cdot T_e}{I}} \cdot e^{\frac{I}{k_B T_e}} = const \cdot (pR)^2 = 1.27 \cdot 10^7 \cdot \tilde{c}_T^2 \cdot (pR)^2; \quad E_d = \frac{m}{e} \cdot \sqrt{\frac{3 \cdot k_B \cdot T_e}{M} \cdot \left(\omega^2 + v_c^2\right)}, \tag{2.3}$$

where I - potential of ionisation, R - radius of discharge channel, M - atomic mass of gas. The constant value $\tilde{c}_T$ depends on sort of gas being for example $\tilde{c}_T = 4 \cdot 10^{-2}$ for argon. The dependence between electronic temperature and $\tilde{c}_T$, p, R is presented in fig. 5.

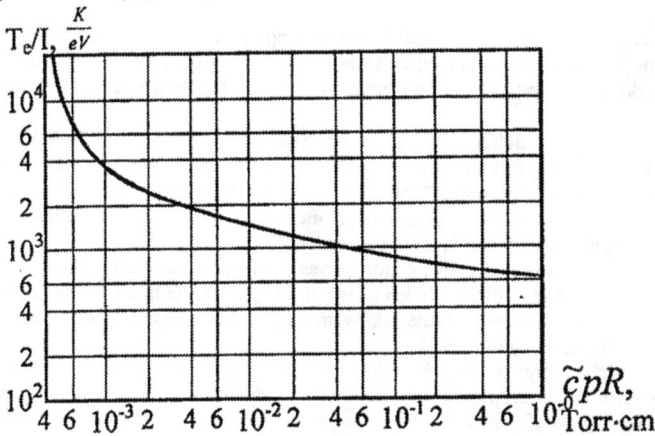

Figure 5. Universal curve for evaluation $T_e$ in a positive column.

As it is seen from fig. 5, electronic temperature $T_e$ is approximately equal to $T_e \approx 2 \cdot 10^3 \cdot I$ at p~(1-5)Torr and radius R~1cm, i.e. it is $3 \cdot 10^4$ K and decreases, then pressure increases. We substitute this value $T_e$ in expression $E_d$ (2.3). It is easy to define that on an incipient state (when $v_c \ll \omega$) $E_d$ equals

$$E_d = \frac{m \cdot \omega}{e} \cdot \sqrt{\frac{3 \cdot k_B \cdot T_e}{M}} = \frac{2\pi \cdot m \cdot c^2}{e \cdot \lambda} \cdot \sqrt{\frac{3 \cdot k_B \cdot T_e}{M \cdot c^2}}. \tag{2.4}$$

Molecular sulphur ($M \cdot c^2 = 50 GeV$) has $E_d$~0.4 V/cm. This quantity is incremented up to (1.0-2.0) V/cm with increase of $v_c$ (when $v_c \gg \omega$). However, this augmentation is insignificant, because $T_e$ is decreased then $v_c$ is increased. We should mark that formula (2.4) works for elastic collisions. If the inelastic collisions play essential role, it is necessary to increase value $E_d$ approximately by one order to provide discharge.

We made only inferior estimations of electric intensity, which are necessary for the discharge disruption and discharge support in the bulb with Ar-S mixture.

Our calculations and practical work display, that at frequency 2450 MHz for argon at pressure ≈ 1 Torr it is necessary to have electric intensity of the order 1 kV/cm, and the sufficient intensity is 150 V/cm for excitation of sulphur. We considered many systems which often use to produce such electric intensity, however, such intensity is easily achieved in microwave cavities at feeding power of hundreds watts. The block diagram of the microwave powered visible light source is presented in fig. 6.

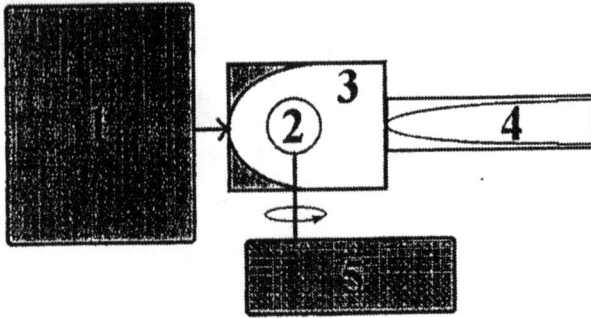

Figure 6. Main parts of microwave light source.

Its principal components are enumerated: 1. Power supply, microwave powered source generator and matching system with microwave cavity; 2. Quartz bulb with Ar-S mixture; 3. Microwave resonator together with light output system; 4. Light guide systems; 5. Rotation and forced cooling systems of quartz bulb.

Fusion Lighting sample of Sulfer Lighting System (SLS) is performed according to the similar plan.

The bulb is located inside a cavity that is opaque to microwaves but transparent to visible light. The microwave frequency used in this lamp, 2450±50 MHz, is designated for use in industrial, scientific, and medical applications. It is the band used around the world by microwave ovens, and inexpensive microwave oven magnetron is available for application in SLS.

Systems of forced cooling and bulb rotation serves to ensure a reasonable thermal mode of quartz envelope with Ar-S mixed gas filling at which it isn't destroyed. The bulb inside the microwave cavity in SLS rotates at 600 revolution per minute taking maximum advantage of cooling air. Such crockish cooling system is a shortcoming of SLS.

The extraction light energy system from building bag of installation should ensure maximal light output from the resonator at supporting of a microwave radiation safe level in an accommodation with installed SLS. The experts of Fusion Lighting company applied resonators manufactured from grid and additional mesh screens. The important parts of this system are the parabolic reflector, and prism light guide. Parabolic reflector serves for practical means to illuminate the prismatic hollow light guide. The parabolic geometric constructions are characterised by a vertical optical axes and focal distance. The bulb accommodated in the resonator is located in parabolic reflector focus. The system of light guides executed on special technology is used for high-performance light transfer to consumers. The technique involves the use of the emitting light-guide surface, which is partially longitudinally reflective and partially longitudinally transparent in order to generate the desired output intensity distribution. Parameters of a wave-guide are the following: material - acryl; walls thickness – 3mm; exterior diameter – 250mm; length 28m. Such wave-guide is connected to parabolic reflector and has posts of fastening for accommodation installation.

Such system set in Aeronautics museum of USA has shown magnificent results in economy of the electric power and illumination level improvement.

# 3. PROGRESS IN SULFUR LAMP DEVELOPMENTS.

When developing and constructing microwave powered visible light sources the authors solved the following problems:

a) to provide better reasonable thermal mode of quartz envelope with Ar-S mixed gas filling at which it isn't destroyed at absence of its forced cooling; b) to remove light energy from building bag of installation at ecological safety till a microwave radiation without light guides; c) to provide operational stabilities of a feeding magnetron on a resonator loading with a variable input resistance.

For the solution of the first problem it is offered to use working cavities with axial-symmetric electromagnetic fields in order to reduce an electron flow to walls. The thermal mode of a quartz bulb is essentially improved, then it is coaxially located in the resonator with oscillations $H_{01p}$ and the electric field $E_{\varphi}$ is tangential its surfaces. Especially good results are gained at use of a toroidal quartz bulb. Practically complete extract of a light is carried out through out-of-limit for microwave field holes in resonators.

The different types of resonators were surveyed at sulphur lamp development. As it was already noted earlier, the preference was given to resonators with axial-symmetric electric field of type $H_{01p}$. Among them it is possible to accentuate a circular cavity, light output from which is carried through out-of-limit for microwave field holes in resonators (fig. 7).

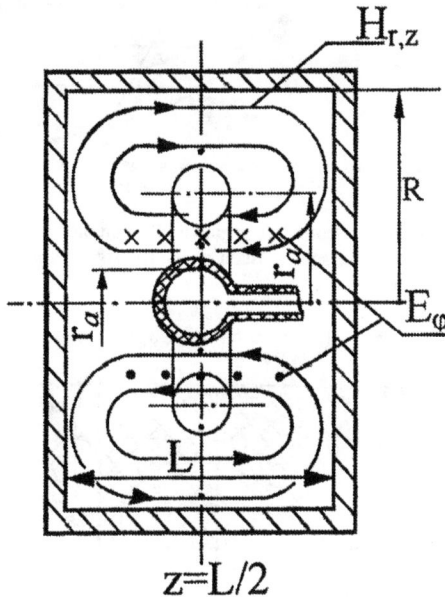

Figure 7. Microwave working chamber with spherical and toroidal bulbs on the basis of a cylindrical cavity with oscillations $H_{011}$.

Spherical or toroidal bulbs are used in the resonator. If bulbs are coaxially located in the cavity, electric field $E_{\varphi}$ is tangential to their surfaces. Parameter of electric intensity for cylindrical cavity is noted:

$$\xi_{\varphi} = \frac{E_{\varphi}}{\sqrt{PQ}} = \sqrt{\frac{2\lambda_0 \cdot Z_0}{L}} \frac{J_1((\mu_{01} \cdot r)/R)}{\pi \cdot R \cdot J_0(\mu_{01})} \sin\left(\frac{\pi}{L}z\right), \qquad (3.1)$$

where P – feeding power; Q – quality factor; $Z_0$=377 Ohm - free space resistance; R and L – radius and length of cavity; $J_0$, $J_1$ and $\mu_{01}$ - Bessel functions and their root. Toroidal bulb appears to be preferable in this case (3.1).

The conducting paraboloid of revolution, which mirror is overlapped by a fine conductive grid, can be the resonator with axial-symmetric field (fig.8). If quartz bulb is located in focus of paraboloid of revolution, the resonator frame also will ensure directional light output. Such system also maintains properties of thermic mode facilitation.

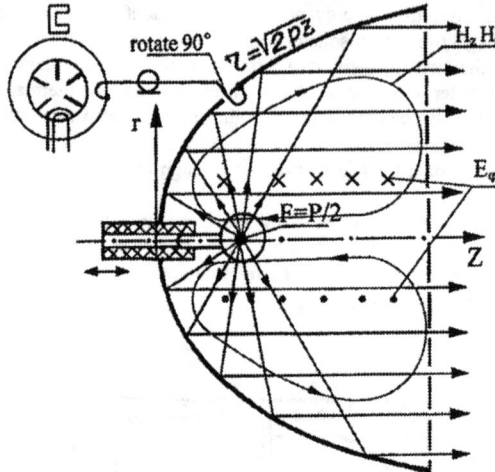

**Figure 8.** Microwave working chamber with spherical bulb on the basis of a parabolic cavity with oscillations $H_{011}$.

The dielectric resonator based on a half-wave segment of round dielectric wave guide, restricted metal plate, also can be used for a microwave source. Such resonator (fig. 9) has axial symmetry at oscillations of $H_{011}$.

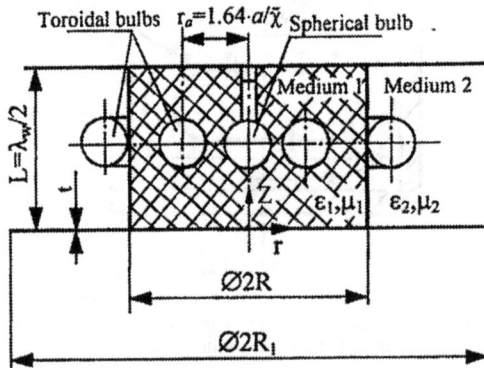

**Figure 9.** Microwave light-transparent working chamber with spherical end toroidal bulbs on the basis of a dielectric cylindrical cavity with oscillations $H_{011}$.

Advantage of such resonator is the light output opportunity through lateral surface of a transparent dielectric. The use of a toroidal quartz bulb gives excellent results. The ecological safety belong a microwave radiation level is ensured by the radius $R_1$ choice of conducting plates.

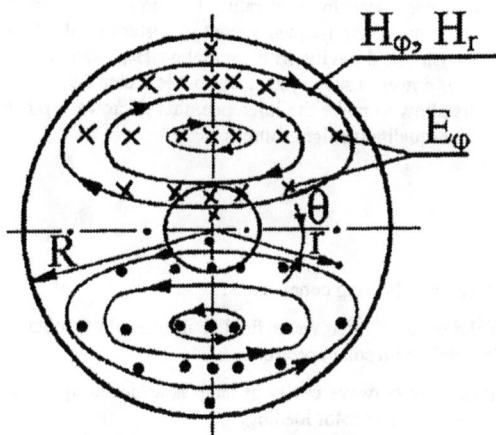

Figure 10. Microwave working chamber with spherical bulb on the basis of a spherical cavity with oscillations $H_{011}$.

The spherical resonator with oscillations $H_{101}$ (fig. 10) was explored as the building bag of microwave installation. Having all advantages of above mentioned resonators with axial-symmetric electric field, it has the greatest energy efficacy of disruption and provide discharges. Light output is carried out through out-of-limit for microwave field holes in resonator.

Experimental research of a microwave powered visible light source [3-6], resonators were excited on oscillations $H_{01P}$ with sole component of electric field $E_{\varphi}$, have shown, that power load of spherical and toroidal bulb envelopes is reduced 5 times in practical constructions as compared with an irradiation by a direct E-field.

The tested microwaves lamp operation at resonator feeding power 600 W and free convection air cooling of a bulb led to bulb envelope heating at the worse case to a dark red luminescence (at inexact concurrence of bulb and resonator symmetry axes). Relevant envelope temperature did not exceed 570°C and was insufficient for envelope destruction. The obtained results have allowed to set the patent on use of resonators with an axial - symmetric field as building bag and on use toroidal bulb in a microwave visible light sources.

The light output at ecological safety of a microwave radiation was also experimentally proved.
If we use cavity in microwave lamp, natural quality factor, resonant frequency of the cavity and resonators coupling coefficient are changed, then discharge occurs. It is important to ensure operational stability of a feeding magnetron on a load with different input resistance. The system of a frequency control of a feeding magnetron by a resonator loading (fig. 11) was suggested to ensure the operational stability of a microwave cavity of lamp.

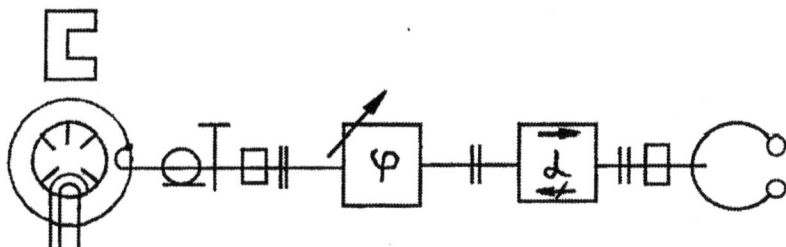

Figure 11. System of a frequency control of a feeding magnetron by a resonator load.

This system recommended itself well in accelerating technique. The authors experimentally proved opportunity of operational stability of the frequency control system on the base of magnetron M105-1 without a ferrite gate for a cylindrical cavity with oscillation $H_{011}$, spherical, dielectric and parabolic resonators. The transfer ratio of power in such system was not less then 0.9.

Thus, the obtained results allow to make the inference, that microwave powered sources of a visible light are high-efficient and high-qualitative light sources.

## 4. CONCLUSION

These results allow to deduce the following conclusions:

1. It was shown that on the base of microwave field it is possible to obtain efficient conversion of microwave energy to visible light with solar spectrum.

2. The operational stability of a microwave cavity of lamp is achieved by use of system of a frequency control of a feeding magnetron by a resonator loading.

3. Practically complete light output is carried out through out-of-limit for microwave field holes in resonators or by uses of light-transparent cavities.

4. The thermal regime of a quartz bulb is essentially improved, when the electric field is tangential to its surfaces. It was suggested to use working cavities with axial-symmetric electromagnetic fields. It allows to ensure an inconvertible thermal duty of the quartz bulb without its rotation and forced cooling.

These researches are fulfilled at RFBR support within the scope of project № 00-02-16170

References

1. MacLenan D.A., Doland J.T., Ury M.G. Soc. Inform. Display Intern. Symp. Digest Techn. Pap., 1992, 23, 460.
2. Doland J.T., Ury M.G., MacLennan D.A. Proc. VI Intern. simp. on science and technology of light sources. Budapest, 1992. 301-311.
3. Didenko A.N., Zverev B.V. Microwave Power Engineering. Moscow: *Nauka*, 2000, 264. (In Russian).
4. Didenko A.N., Zverev B.V. et.al. Izvestia RAS. Energetics. 1997, 6, 134 (In Russian).
5. Didenko A.N., Zverev B.V. et.al. Izvestia RAS. Energetics. 1998, 1, 147 (In Russian).
6. Didenko A.N., Zverev B.V. et.al. Izvestia RAS. Energetics. 1997, 6, 129 (In Russian).
7. Raizer Yu.P. Osnovu sovremennoi phisiki gasorazriadnyh processov. Moscow: Nauka, 1980, 416. (in Russian)

# NEGATIVE ION-ASSISTED SILICON OXIDATION IN DOWNSTREAM MICROWAVE OXYGEN PLASMA

**Takashi Fujii, Hitoshi Aoyagi, Takuya Urayama,
Yasuhiro Horiike\* and Haruo Shindo**

Department of Applied Physics, Tokai University, Hiratsuka, 259-1292, Japan
\*Department of Materials Science, Tokyo University, Tokyo,113-8656, Japan

**Abstract.** For low-damage silicon oxidation at low temperature, negative-ion-assisted oxidation was studied by employing negative ions in oxygen plasma. In the downstream region of microwave oxygen plasma, the silicon oxidation was examined under various radio frequencies of the substrate bias. The oxidation depth showed a strong frequency dependence and had a maximum around the ion plasma frequency and thus, it was concluded that the oxidation was due to the negative ions. An X-ray photoelectron spectroscopy analysis revealed less suboxide in the negative-ion irradiation compared with the positive-ion oxidation. Further, the directional oxidation was studied in a silicon trench for a cell isolation layer in memory.

## 1. INTRODUCTION

A low-temperature and low-damage silicon oxidation technique[1] is highly required in various ultralarge-scale-integrated circuit (ULSI) processes. In particular, for trench isolation of a memory cell to realize further integration, the oxidation should be ion-assisted for directionality but with low damage. For this purpose, a new method of negative-ion-assisted silicon oxidation has been proposed employing microwave oxygen plasma, and high-rate and low-temperature silicon oxidation has been demonstrated.[2,3]

The objective of this work is to study silicon oxidation by negative ions under the radio-frequency (RF) bias and to show a direct evidence of negative-ion oxidation. For practical application of this method, the oxidation should be proceeded under the RF bias, because DC bias becomes ineffective as the oxide film is grown. Therefore, it is also important to find out an optimal condition in the frequency of the substrate bias. In this work, silicon oxidation was investigated under a transformer-coupled RF bias. This device enables us to irradiate the negative and positive ions separately onto the silicon surface. In particular, the major role of negative ions in oxidation was mainly focused on under various bias frequency conditions.

## 2. EXPERIMENTAL

Microwave oxygen plasma was produced in a 6-inch stainless-steel chamber, as shown in Fig.1, and the downstream plasma was mainly considered because in this region the negative ion was highly populated [2]. Silicon oxidation was made on a substrate stage, placed in downstream of the plasma, and the stage was biased by RF voltage as well as DC bias to irradiate both negative and positive ions. The RF bias voltage was applied to the stage with a cored-transformer, and its secondary was also biased by DC voltage at the same time. Thus the substrate voltage could be varied from the negative to the positive during the cycle, centered on an artificially settled DC potential. The oxidation was made at the substrate temperature of 300 to 400 C, and no temperature change was observed during ion irradiations. The substrate temperature was monitored at the back of Si wafer employing a thermocouple.

The oxide film quality thus produced was analyzed by X ray photoelectron spectroscopy (XPS). The oxide film thickness was measured by the XPS and ellipsometry methods.[2] The sample used for oxidation was a P type crystal Si (100) and its surface was treated by 5% B-HF before oxidation. The oxygen plasma condition used for oxidation was 150 W of microwave power and 30 to 50 mTorr of pressure. Hereafter the experimental results will be shown as a function of the axial distance Z from the microwave entrance window.

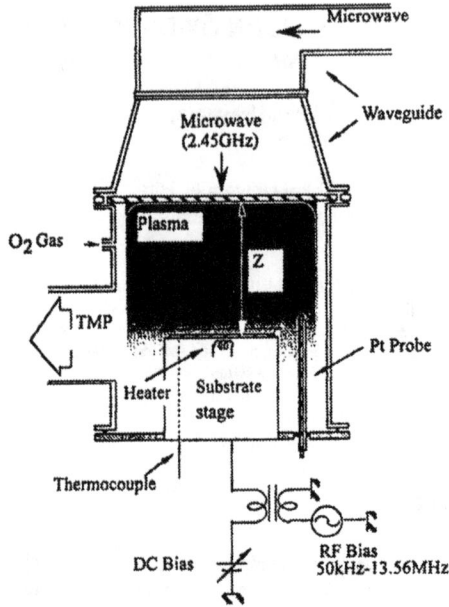

**Figure 1.** Schematic illustration of the experimental apparatus employed in this work.

## 3. RESULT AND DISCUSSION

A typical example of the oxygen plasma parameters are shown in Fig.2, where the axial variations of the electron density and temperature are plotted for two pressures of 30 and 50 mTorr. The electron density becomes as high as $10^{11}$ cm-3 in both the pressures and it is reduced with the axial distance Z. However, the electron temperature does not simply behave for the distance, and it has an tendency of increase at the downstream below Z=15 cm, especially at 50 mTorr. A temperature inorease at the downstream may imply that the low energy electron attachment to the excited oxygen molecule occurs, producing a negative oxygen ion O- and atom O by dissociation.[4] Thus the temperature increase is due to low energy electron loss above stated, and correspondingly the negative ion can be much produced in thsese region.

In order to confirm the above statement, the silicon oxidation is actually made, and the results are plotted in Fig.3. In the figure, the silicon oxidation depth obtained at 30 and 50 mTorr is plotted for the axial distance Z. A remarkable increase of the oxidation depth at the downstream is seen at 50 mTorr, and this behavior is very much consistent with that of the electron temperature shown in Fig. 2. A clear difference between the pressures is also seen in the figure. This difference between 30 and 50 mTorr of pressure strongly suggests that an optimum condition for the negative ion production mechanism above cited [4].

For a further detailed study of the negative ion behavior, a frequency dependence of silicon oxidation under the RF bias was examined, and a typical result is shown in Fig.4, where the oxidation depth obtained at Z=20 cm is plotted against the frequency. The frequency of the bias was varied with keeping the RF voltage of 65 V peak-to-peak and DC voltage of +30 V. This value of +30 V is just the plasma potential observed2) at the distance Z=20 cm. The oxidation depth in this case was determined from the step observed on the sample surface etched by B-HF. It is clearly demonstrated, in Fig.4, that the oxidation strongly depends on the frequency and there is a limitation of oxidation at both sides of frequency. In particular, a high frequency limitation of oxidation should be primarily concerned. Since the limitation begins at about 1 MHz which is close to the negative ion plasma frequency,[2] it makes clear that the oxidation is negative ion-assisted. Although an exact value of the negative ion density has not been determined yet, the plasma frequency of the negative ion O- yields about 2 MHz at Z=20 cm if the density is equal to that of the positive ion. It can be stressed, therefore, that the data in Fig.4 is one of direct evidences for negative ion oxidation as well as the DC voltage dependence shown previously.[2]

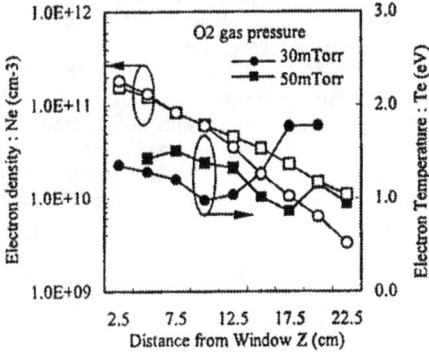

**Figure 2.** Electron temperature and density

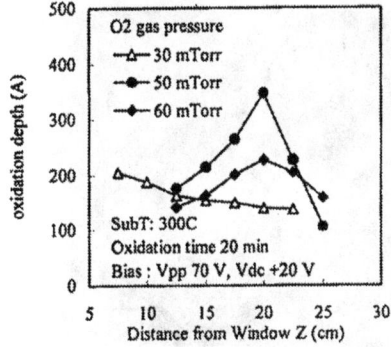

**Figure 3.** Silicon oxidation depth

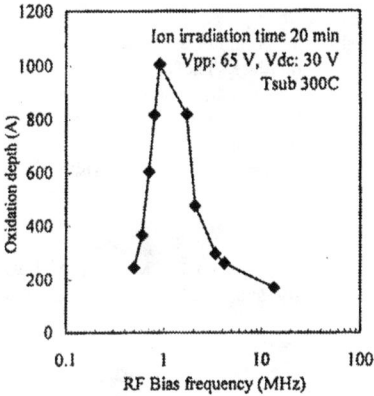

**Figure 4.** Frequency dependence of oxidation

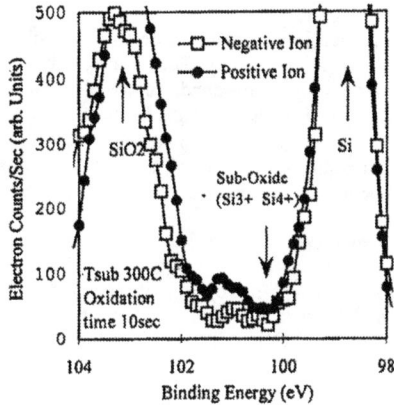

**Figure 5.** XPS spectra of oxidation films

In Fig.5, a typical example of X-ray photoelectron spectroscopy (XPS) spectra is shown, where the spectrum obtained for negative ion oxidation is compared to that for the positive ion one. The intensity of the spectra in the figure is normalized by the Si(2P3/2) peak. Since the SiO2 peak observed with the positive ion oxidation film is fairly shifted and widened, the figure clearly demonstrates that the film quality is improved by the negative ion irradiations with less suboxide. The fact that the film quality is improved by the negative ion irradiation may be explained as the following. The mass analysis[5] of the ion species using a quadrupole mass analyzer showed that the dominant negative ion was O- and its density was almost one order of magnitude higher than that of O2-, while the dominant positive ion was

O2+. The dominant negative ion O- is very atomically similar to the oxygen radical with a high chemical reactivity and this may be the origin of the high quality oxidation film shown in Fig.5, as well as the high oxidation rate shown in Figs.2 and 3. Further, one more advantage in negative ion assisted oxidation would be expected in conjunction with damage. The ionization energy of negative ions is relatively low and their neutralization on the substrate surface is an energy absorption process, in contrast to that of the positive ions. This kind of low damage silicon oxidation can also be expected to be involved in the spectrum shown in Fig.5.

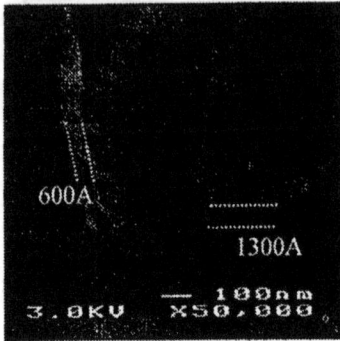

In Fig.6, a SEM photograph of cross section of trench oxidation is shown. The conditions of oxidation are as follows; Z=20 cm, DC bias +80V, RF bias 70 V peak-to-peak, substrate temperature 400C. The trench sample was realized by chroline RIE etching of silicon with SiO2 mask. The directional oxidation can be seen in a narrow trench of 300 nm, and the aspect ratio of this trench oxidation is about 2 and the oxidation thickness is 130 nm.

Figure 6. A SEM photograph of trench oxidation

## 3. CONCLUSION

The negative ion irradiation in the downstream of microwave oxygen plasma provides a high rate silicon oxidation with a high quality. The oxidation depth has shown a maximum at the downstream and this oxidation maximum corresponds to the electron temperature increase. The frequency dependence of oxidation depth was also examined, and the oxidation showed a maximum at around the ion plasma frequency, indicating that the oxidation was due to the negative ion. Since the negative ion is very atomically like the oxygen radical, the high rate oxidation with a good quality can be made by employing the negative ion. A directional oxidation was attained with the silicon trench of 300 nm width.

### Acknowledgements

This work was supported by a Grant-in-Aid for Scientific Research from the Ministry of Education, Science and Culture in Japan. This work has also been partly supported by THE RAINBOW 21 PROJECT of Tokai University.

### References

1. Taylor S., Zhang J. F. and Eccleston W. Semicond. Sci. Technol., 1993, 8, 1426.
2. Koromogawa T., Fujii T., Horiike Y.and. Shindo H. Jpn. J. Appl. Phys., 1998, 37, 5028.
3. Fujii T., Aoyagi H., Kusaba K., Horiike Y. and Shindo H. Jpn. J. Appl. Phys., 1999, 38,L1466.
4. Ishikawa T., Hayashi D., Sasaki K.and Kadota K. Appl. Phys. Lett., 1998, 72, 2391.
5. Shindo H., Koromogawa T.and Horiike Y. Abstr. 44th Nat. Symp. of American Vacuum Society, (San Jose, CA, 1997), p.208.

# INTERACTION OF MICROWAVE DISCHARGES WITH DIELECTRIC LiF CRYSTALS

**V. A. Ivanov and M. E. Konyzhev**

Institute of General Physics, Russian Academy of Sciences,
Vavilov street 38, Box 117942, Moscow, Russia
Phone: +7(095)132-8344, Fax: +7(095)1358011, E-mail: ivanov@fpl.gpi.ru

Abstract. In studying the interaction of dense microwave–discharge plasmas with dielectric crystals of lithium fluoride (LiF), a new type of breakdown of dielectrics has been discovered, specifically, electrodeless electric breakdown in the bulk of LiF crystals. This breakdown is initiated by a surface microwave discharge taking the form of a contracted discharge. It is very important that new type of breakdown occurs in the absence of metal electrodes contacting with the LiF crystals. The breakdown is observed at a weak threshold electric field of the incident microwave radiation (about 3 kV/cm) which is three orders of magnitude weaker than the threshold electrostatic field (about 3 MV/cm) needed for the high–voltage electric breakdown in the bulk of LiF crystals contacting with metal electrodes. A physical model is developed that explains electrodeless breakdown in the bulk of dielectrics in terms of a double plasma layer that appears at the dielectric surface and in which the quasi–electrostatic field is as strong as about 10 MV/cm.

## 1. INTRODUCTION

Historically the interest to studing electrodeless microwave discharges developing at the surface of solid–state dielectrics was appeared due to applied researches, oriented to increase the electric strength of the output windows of high–frequency vacuum devices and waveguides during the propagation of powerful pulsed microwave radiation along them [1, 2].

The empirically found methods of preventing the electrodeless breakdowns at the inner surfaces of vacuum systems, as a rule, were limited to reducing the level of the energy flux density of the incident microwave radiation at these surfaces. Special coatings and/or external electric and magnetic fields were applied also for suppression the microwave discharges at the surface of dielectrics, so in a number of cases it was possible to prevent the eventual development of the secondary–electron–emission microwave discharge to plasma stages of microwave discharges at the surface of dielectrics.

These procedures fruited some positive results. However, despite of these successes, the problem of increasing of an electric strength of dielectrics, semiconductors and composite metal–dielectric materials in strong microwave fields becomes more and more important due to the considerable advance in the development of powerful nonrelativistic and relativistic superhigh–frequency electronics engineering [3].

The necessity of solution of more common problems of a powerful superhigh–frequency electronics engineering, concerning the generation, transporting and transformation of microwave radiation, initiated a development of basic and applied studies of electrodeless microwave discharges developing at the surface of solid–state dielectrics in the vacuum. Investigations of nonlinear physical phenomena occuring in the plasma of the microwave discharges, interacting with dielectrics, were performed [4–6].

Of special interest are those physical processes in the near–surface layer of dielectric crystals which bring about the accumulation and relaxation of high concentrations of radiation–induced point defects, the arising of luminescence of induced color centers, the appearance of induced electrical conductivity, the stimulation of strong absorbtion of microwave power, the occuring of both a contracted discharges and an electrodeless breakdowns of crystals in a field of microwave radiation [7–10]. A classification of different types of electrodeless pulsed microwave discharges is under development.

Here, we report the new phenomena – the electrodeless electric breakdown and local destructions in the bulk of dielectric crystals of lithium fluoride (LiF) excited by contracted microwave discharges.

## 2. EXPERIMENTAL SETUP AND MEASUREMENT TECHNIQUE

Fig. 1 shows the experimental layout and Fig. 2 shows the evolution of main processes during the interaction of microwave discharges with LiF crystals.

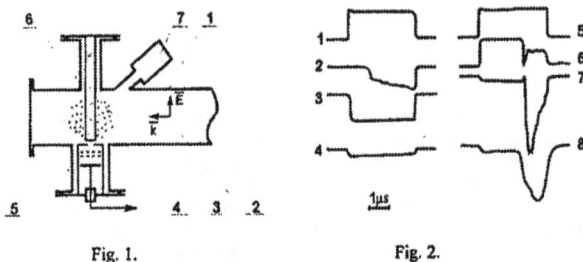

Fig. 1.                    Fig. 2.

Figure 1. Experimental layout: 1 – waveguide; 2 – incident microwave radiation; 3 – LiF crystal; 4 – microwave discharge; 5 – multigrid electrostatic analyzer; 6 – cutoff section with a diameter of 24 mm; 7 – photomultiplier FEU–79 and electron-optical image converter KADR–4.

Figure 2. Typical oscillograms of some signals during an evolution of microwave discharges. SEEMD: 1 – incident microwave power; 2 – electron current from the discharge when a field is near to the threshold one for the onset of the SEEMD ($P_0 = 45$ kW, $E_0 = 2.3$ kV/cm); 3, 4 – electron current and luminescence from the discharge when the threshold fiel for a SEEMD is exceeded considerably ($P_0 = 280$ kW, $E_0 = 5.7$ kV/cm). The discharge transformations (SEEMD → SMB → PFMD): 5, 6 – power of the incident and reflected microwave radiation; 7, 8 – electron current and luminescence from the discharge for the transformation from SEEMD to PFMD due to SMB ($P_0 = 300$ kW, $E_0 = 5.9$ kV/cm). The gain levels for traces 2, 3, and 7 form the rations 260:13:1, respectively. Here SEEMD is a secondary-electron-emission microwave discharge, SMB is a surface microwave breakdown (in a form of a contracted discharge), PFMD is a plasma–flare microwave discharge.

Microwave discharges at the surface and in the bulk of dielectrics were investigated experimentally with pure LiF single crystals 70×10×5 mm in size (either cleaved along the cleavage planes or polished). The crystals have small absorption coefficient for microwave radiation at the frequency $v = 2$ GHz, the tangent of the dielectric loss angle was about $10^{-4}$. In our experiments, LiF crystals were placed at the maximum of the electric field of a standing $TE_{10}$ wave excited in a 120×57 mm rectangular metal waveguide by pulsed microwave radiation from a magnetron. The wavelength of microwave radiation in the waveguide was $\lambda_g = 20$ cm. Pulsed microwave radiation with an oscillation frequency of microwaves $v = 2$ GHz, a pulse length $\tau$ from 1 to 4 μs, a pulse power $P_0$ up to 0.3 MW, and pulse repetition rate $f = 0.1$ Hz was delivered from the magnetron to the vacuum section along a microwave line. The waveguide was previously evacuated by oil–free titanium pumps to a pressure of air of $10^{-4}$ Pa. High–speed photographs of microwave discharges were taken with a help of an electron–optical image converter (KADR–4) operating in the frame–by–frame mode (four frames with a frame exposure time of 50 ns, the time interval between the frames was 50 ns). The electron current from microwave discharges was measured by a multigrid electrostatic analyzer. X–ray emission with quantum energies from 3 to 20 keV was measured by scintillation detectors and a pinhole camera, with using the calibrated metal filters. The plasma density in microwave discharges was measured by a microwave interferometer, the wavelength of the probing radiation was 8 mm. The incident and reflected microwaves were recorded by directional waveguide couplers and semiconductor detectors. Destructions produced on the surface and in the bulk of LiF crystals due to electrodeless electric breakdowns were investigated with the help of a JEOLS scanning electron microscope and a BIOLAM optical microscope. The experiments were performed at room temperature.

# 3. CLASSIFICATION OF DIFFERENT TYPES OF ELECTRODELESS MICROWAVE DISCHARGES DEVELOPING AT THE SURFACE OF SOLID – STATE DIELECTRICS

As well known pulsed microwave radiation excites electrodeless microwave discharges of different types developing at the surface of solid–state dielectrics in vacuum [6–10]. There are three types of electrodeless microwave discharges developing at a single surface of the dielectrics in vacuum:
– secondary–electron–emission microwave discharge (SEEMD),
– surface microwave breakdown (SMB) taking a form of a contracted discharge,
– plasma–flare microwave discharge (PFMD).
The SEEMD transforms to the PFMD passing through the stage of the microwave breakdown on the surface of the crystals in vacuum:

$$SEEMD \rightarrow SMB \text{ (contracted discharge)} \rightarrow PFMD.$$

The mechanisms of these transformations have a great importance for the understanding of the exitation and evolution of different types of microwave discharges at the surface of dielectrics. We study these mechanisms experimentally.

## 3.1. Secondary–electron–emission microwave discharge (SEEMD)

The SEEMD arises when $\varepsilon = e^2 E_0^2 / 2m\omega^2 > W_1$ and exists during the breakdown delay time $\Delta\tau < \tau$. Corresponding threshold electric field $(E_0)_{thr}$ is determined from the condition: $e^2 [(E_0)_{thr}]^2 / 2m\omega^2 = W_1$, where $\varepsilon$ is the oscillation energy of electron, $E_0$ is the electric field amplitude of microwaves, $e$ and $m$ are the specific charge and mass of an electron, $\omega$ is the angular frequency of microwaves ($\omega = 1.23 \times 10^{10}$ s$^{-1}$ for the oscillation frequency of microwaves $\nu = 2$ GHz ), $W_1$ is the first critical potential for the surface of solids ($W_1 \approx 15$ eV for the LiF crystal), $\tau$ is the time duration of the microwave pulse, $\Delta\tau$ is the time duration of SEEMD, $(E_0)_{thr}$ is the threshold electric field for the onset both of the SEEMD and of the contracted discharge ($(E_0)_{thr} \approx 2$ kV/cm for the LiF crystal). Electrons of SEEMD have energies on the order of the oscillation energy. High pulsed electron current densities reach the value $j_e \sim 1$ A/cm$^2$. Microwave absorption coefficient (per unit area of the crystal) in the stage of the SEEMD is low:

$$\eta \sim eE_0 / m\omega c \sim 10^{-2}.$$

The high intensity luminescence spectra of short–lived $F_2$ and $F_3^+$ color centers, temporarily induced in the surface layer of uncolored LiF crystals by pulsed microwave discharges at the stage of SEEMD, were measured at room temperature. This luminescence was uniform over the crystal surface. It was found that the characteristic time of the luminescence growth/decay of LiF crystals, excited by pulsed microwave discharges at room temperature, was about 1 μs. These luminescence spectra were very close to the photo–luminescence spectra measured for LiF crystals previously colored by microwave discharges and containing stable $F_2$ и $F_3^+$ centers. It was shown that a density of induced aggregate $F_2$ and $F_3^+$ color centers is about a density of induced F centers. This fact testifies a high–intensive excitation of LiF crystals by microwave discharges [7, 9, 10].

A few thouthands of SEEMD pulses create an optically dense layer (on the surface of LiF crystals) containing induced color centers. Optical–absorption spectra of LiF crystals, previously colored by microwave discharges, were measured. The induced F, $F_2$, $F_3^-$, $F_3$, $F_3^+$, N color centers, stable at room temperature, were observed. The density of F, $F_2$ and $F_3^+$ centers (stable at room temperature for years) can reach the values of $10^{19}$–$10^{20}$ cm$^{-3}$, and the thickness of the layer is not larger than $10^{-4}$ cm. The density of defects that are induced by the SEEMD electrons is much larger than the density of defects that are induced by X–rays, γ–rays, and relativistic–electron beams [7, 9, 10].

## 3.2. Microwave breakdown on the surface of dielectrics (SMB)

Surface microwave breakdown is initiated by the SEEMD. When the electrons of the SEEMD bombard the dielectric crystals, high density of electronic excitations (electron–hole pairs, excitons, plasmons) arises in the surface layer. Decay of the excitations leads to creation of "metastable" color centers with lifetime ~ 1 μs at room temperature. During the breakdown delay time $\Delta\tau$ an accumulation of these centers takes place. An increase in the length $\tau$ or the intensity S of the microwave pulse results in increase the concentrations of the centers up to $10^{19}$ cm$^{-3}$ in the thin near–surface layer of the crystals. Recombination of the centers leads to the formation of an electron–hole plasma with a such high carrier density, that a local electric field intensification arises near inhomogeneities in the plasma density. Absorption of microwave energy in these plasma regions produces phonons, so that a local heating of the crystal lattice stimulates the explosive recombination of short–lived centers and plasma production again. As a result of these processes the plasma channels intergrow in a near–surface layer of the crystals and stretch out along the electric field vector of the wave. The seed electron density in the conduction band of the crystals required to initiate this process is $n_e$ ~ $10^{15}$ cm$^{-3}$. The concentrations of "metastable" centers on the order of $10^{17}$ cm$^{-3}$ are reached in our experiments, so that these values are sufficient here [8].

The breakdown delay time $\Delta\tau < \tau$ obeys the empirical law: $(S - S_{thr}) \times \Delta\tau \approx Const$, where $S_{thr}$ is the threshold intensity for the onset of the SEEMD. The value of the Const is determined by the nature of the crystals: Const $\approx$ (0.04 – 0.8) J/cm$^2$, Const ~ ω. The breakdown is accompanied by the absorption of microwave energy up to 100 % with a time scale $\delta\tau$ ~ (0.05–0.1) μs, an intense flash of light, and a burst of electron and ion currents from the contracted discharge [6, 8].

## 3.3. Plasma–flare microwave discharge (PFMD)

PFMD is produced as a result of an evolution of the contracted microwave discharge and is accompanied by the formation of dense plasma with an electron concentration which is much more than the critical one. PFMD is characterized by a strong absorbtion of microwave power ($\approx$ 50 %). The following phenomena take place in PFMD [4, 6]:

- excitation of strong Langmuir waves in a plasma resonance region (where an electron plasma concentration is the same that the critical one),
- acseleration of electrons due to a self–breaking of strong Langmuir waves,
- acseleration of ions due to a positive potential jump in a plasma resonance region,
- transformation of microwave power to the power of quasistationary electric current in the plasma.

## 4. EXPERIMENTAL RESULTS AND DISCUSSION

In the field of pulsed microwave radiation from a magnetron with the parameters ν = 2 GHz, $P_0$ = 0.3 MW, τ = 4 μs, and f = 0.1 Hz, the following types of electrodeless microwave discharges were observed to develop in consecutive order at the surface of LiF crystals [8]: a secondary–electron–emission microwave discharge (over the time interval $\Delta t_1$ from 0 to 2 μs), a contracted plasma discharge (over the time interval $\Delta t_2$ from 2 to 2.5 μs), and a plasma–flare discharge or a plasma flare (over the time interval $\Delta t_3$ from 2.5 to 4 μs).

During a secondary–electron–emission discharge, the power of microwave radiation was absorbed by electrons, which bombarded the surface of a LiF crystal. The relative power deposited in the electrons was no higher than 1 %, the characteristic electron energy was about 0.2 keV, and the maximum electron current density in the discharge was estimated to be lower than 1 A/cm$^2$. After the time interval $\Delta t_1 \approx 2$ μs starting from the onset of the secondary–electron–emission discharge, highly luminous contracted plasma discharge was observed to form at the surface of a LiF crystal. High–speed photography made it possible to reveal several very bright–lighting pointlike plasma microdischarges in the plasma filaments of the contracted discharge at the crystal surface, the characteristic sizes of lighting regions of microdischarges were about 1 mm (Fig. 3). In different microwave pulses, individual lighting plasma microdischarges were observed to occur in the same local regions of the plasma filaments of contracted discharges (in other words, microdischarges were "topographically" located with reference to some fixed local sites on the crystal surface).

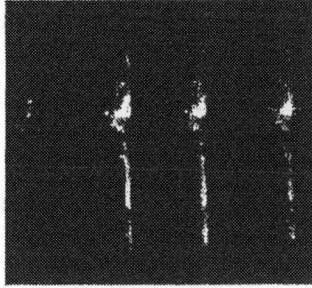

Figure 3. Four high–speed photographs of evolution of single–pulse microwave discharge (both a contracted discharge and pointlike microdischarges) at the surface of a LiF crystal. The photos were taken with a help of electron–optical image converter operating in a frame–by–frame mode (four frames with a frame exposure time of 50 ns, the time interval between the frames is 50 ns).

During the contracted discharge, a considerable fraction (about 30–100 %) of the incident microwave power was deposited in thin long plasma filaments (1–10 μm in diameter and 3–5 cm in length) at the crystal surface. The electron density inside the filaments was estimated to be $n_e > 10^{16}$ cm$^{-3}$ (the electron–ion and electron–neutral collision rates are $\nu_{ei} \approx \nu_{en} \approx 10^{11}$ s$^{-1}$). The expansion of a dense filamentary plasma was accompanied by the onset of a plasma flare, which began to interact with a crystal. Interferometric measurements showed that, after a time interval of about 0.5 μs starting from the onset of the plasma flare, the average electron density of plasma near the crystal surface exceeded the value $2 \times 10^{13}$ cm$^{-3}$, which was almost three orders of magnitude higher than the critical electron density $n_e = \pi m_e \nu^2/e^2 \approx 4 \times 10^{10}$ cm$^{-3}$ of the plasma irradiated by microwaves with the frequency $\nu = 2$ GHz. The electron density in the plasma flare was high enough to shield the region, where the plasma filaments interacted with the LiF crystal, from the incident microwave radiation, thereby forcing the contracted plasma discharge to decay.

A comparison of photographs of microwave discharges taken at different stages with the help of high–speed photography (with time resolution of 50 ns) and time–averaged photography (with time resolution of 4 μs) made it possible to establish that both plasma filaments and microdischarges were most luminous and most highly localized on the crystal surface precisely at the contracted–discharge stage (during the time interval $\Delta t_2 \approx 0.5$ μs, Fig. 3) rather that at the plasma–flare stage (during the time interval $\Delta t_3 \approx 1.5$ μs). Hence, we can conclude that plasma microdischarges occur exclusively at the contracted–discharge stage, whereas, at the plasma–flare stage no new microdischarges appear all over the crystal surface.

Investigating the time evolution of plasma parameters at the stage of contracted discharge, we also revealed that the dynamics of luminescence of plasma filaments and microdischarges (Fig. 3) correlated with the dynamics of absorption of the incident microwave power by the plasma filaments on the crystal surface: (1) at the very beginning of the contracted plasma discharge (i. e., after the time interval $\Delta t_1 \approx 2$ µs starting from the onset of the secondary–electron–emission discharge), the plasma filaments and microdischarges started to develop and the incident microwave power started to be absorbed; (2) approximately 100–150 ns after the discharge evolved into the contracted stage, the plasma filaments and microdischarges were the brightest and the microwave power absorption was the most intense; and (3) approximately 0.5 µs after the onset of the contracted discharge, plasma filaments and microdischarges decayed and the incident power ceased to be absorbed by plasma filaments. Consequently, strong absorption of the incident microwave power by the plasma of the contracted discharge gave rise to plasma filaments and microdischarges; the subsequent decay of the plasma filaments and microdischarges was governed by such factors as the ejection of charged and neutral particles from the filaments, the onset of a plasma flare, and the shielding of the region of an interaction of plasma filaments with crystal from the incident microwave radiation.

At the stage of the contracted discharge and during the expansion of the plasma flare, we recorded a jump in the positive electric potential near the crystal surface and the generation of hard bremsstrahlung, which are both attributed to the movement of high–energy electrons from the plasma. The highest electric potential (up to 10–15 kV) and the hardest bremsstrahlung (the energy of X–ray photons being 15–20 keV) were observed only during a time interval of about 0.2 µs at the contracted discharge stage. X–ray pinhole photography of microwave discharges showed that the most luminous discharge regions corresponded to plasma filaments of the contracted discharge. Hence, the local interaction of a plasma with a crystal bombarded by high–energy electrons was most intense in the regions where long thin plasma filaments were formed on the crystal surface at the contracted discharge stage.

In each experiment, contracted plasma discharges and microdischarges were initiated by $10^3$ microwave pulses. Strong local interactions of periodically produced dense plasmas of these discharges with LiF crystal resulted in the partial destruction of the crystal in the form of craters on its surface (Fig. 4) and the related inward dendrites in the bulk of crystal (Fig. 5). A comparison between the photographs of microwave discharges and examination of crystals with electron and optical microscopy showed that erosive craters with the characterize diameter $\phi_1 \approx 50$–200 µm appeared at the local sites where the luminous plasma microdischarges were in contact with the crystal surface. The inward dendrites, which originated from every crater, were observed as branched breakdown channels, each having a diameter $\phi_2 \approx 10$–200 µm and a length of up to 10 mm. We emphasize that the dendrites were predominantly confined to the cleavage planes of LiF crystals. In our experiments, LiF crystals were destroyed essentially in the same manner as dielectrics that experienced high–voltage electric breakdowns in their bulk at electric field with a strength of 1–10 MV/cm between the metal electrodes [11, 12] and as dielectrics that were previously charged in their bulk (by irradiating them with relativistic electron beams) and then experienced electric breakdowns between charged regions and external metal electrodes [13, 14].

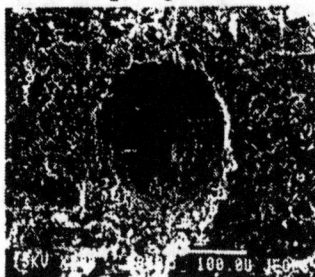

**Figure 4.** Microphotograph of a crater on the surface of a LiF crystal after $10^3$ pulses of microwave discharges (both contracted discharges and pointlike microdischarges). The diameter of the crater is 150 µm.

**Figure 5.** Photograph of the fragment of a group of dendrites in the bulk of a LiF crystal after multiple electrodeless electric breakdowns initiated by $10^3$ pulses of both contracted microwave discharges and pointlike microdischarges at the surface of the LiF crystal. The left edge of the photograph corresponds to the crystal face from which the dendrites originate. The frame in the figure is about 3×4 mm in size.

## 5. DISCUSSION OF THE MECHANISMS FOR ELECTRODELESS BREAKDOWN IN THE BULK OF DIELECTRICS

It is well known that electric discharges in the bulk of solid–state dielectric single crystals can only be initiated in sufficiently strong electric fields, the threshold for electric breakdown is 1–10 MV/cm [11–14]. The two main types of such discharges are (a) high–voltage electric breakdown in the bulk of dielectrics contacting with metal electrodes [11, 12] and (b) breakdown in the bulk of dielectrics that are first charged in the bulk of material by irradiating them with relativistic electron beams and then are spontaneously discharged into external metal electrodes [13, 14]. An important point is that, in both cases, high–voltage electric discharges (breakdowns) are initiated in the presence of metal electrodes in contact with dielectrics; the breakdowns irreversibly destroy solid–state dielectric crystals, giving rise to branched breakdown channels (inward dendrites) originating at the points at which metal electrodes are in contact with the dielectric surfaces [11–14].

We consider two physical processes that can initiate electrodeless breakdown in dielectric LiF crystals, specifically, the electric and local thermal mechanisms for the breakdown.

The electric mechanism for breakdown of dielectric LiF crystals implies that, during the interaction of a microwave discharge plasma with a crystal, the electric field generated in a dielectric surface layer exceeds the threshold electrostatic field for breakdown in the bulk of LiF crystals (3.1 MV/cm) [11]. For this reason, we analyze two possible ways in which a strong (above 3 MV/cm) quasi–electrostatic field can be generated in the surface layer of a dielectric crystal influenced by microwave–discharge plasma.

One way of generating a strong field is associated with the spatial separation of electric charges (electrons and ions) in a double plasma layer that arises at the dielectric surface under the interaction of a microwave discharge plasma with a crystal. In fact, the electrons moving from the plasma toward the dielectric surface outrun much less mobile ions, thereby giving rise to a double plasma layer in which the electric field (produced due to charge separation) is aligned with the normal to the surface. In a steady state, the total electric current $J$ from the double plasma layer toward the dielectric surface equals zero, $J = J_e + J_i = 0$, because the negative electron current $J_e$ is neutralized by the positive ion saturation current $J_i$. In this case, the ion saturation current density $j_i$, the quasi–electrostatic field $E_{plasma}$ in the plasma, and the characteristic time scale $\Delta t_E$ on which this field grows during the formation of a double plasma layer near the dielectric surface have the form

$$j_i \approx (eZ_i n_i /4)(8T_e /\pi m_i)^{1/2}, \quad E_{plasma} \approx (8\pi n_e T_e)^{1/2}(m_i/m_e)^{1/4}, \quad \Delta t_E \approx 10\,(R_D/V_{is}),$$

where $n_e$ and $T_e$ are the electron plasma density and temperature, $m_e$ is the mass of an electron, $n_i$ is the ion plasma density, $Z_i$ is the average ion charge number in the plasma, $m_i$ is the average mass of ions, $R_D \approx (T_e/4\pi n_e e^2)^{1/2}$ is the electron Debye radius, and $V_{is} \approx (ZT_e/m_i)^{1/2}$ is the ion acoustic velocity in the plasma. Since the plasma filaments in a contracted discharge at the surface of a LiF crystal are characterized by the parameters $n_e > 10^{16}$ cm$^{-3}$, $T_e \approx 5$–$10$ eV, and $(m_i/m_e)^{1/4} \approx 12$ [8], a double plasma layer is formed on the time scale $\Delta t_E \approx 10^{-14}$ s, which is many orders shorter than both the time scales corresponding to the inverse electron–ion and electron–neutral collision rates $(\nu_{ei})^{-1} \approx (\nu_{en})^{-1} \approx 10^{-11}$ s and the lifetime $\Delta t_2 \approx 0.5$ µs of the contracted discharge. In this case, the electric field in the double layer is as strong as $E_{plasma} > 4$–$6$ MV/cm, which exceeds the threshold electrostatic field $E_{static} = 3.1$ MV/cm, resulting in high–voltage breakdown and partial destruction of LiF crystals placed between the metal electrodes [11].

Another way of generating a strong field in the surface layer of a dielectric interacting with a microwave–discharge plasma is associated with charging the dielectric by high–energy electrons. We know that in microwave discharges the electrons are accelerated to characteristic energies of $10$–$20$ keV in the plasma resonance region as a result of the self-breaking of strong Langmuir waves excited by the incident microwave radiation [6]. Our investigations of the bremsstrahlung generated by high–energy electrons have shown that the electron bombardment of dielectric crystals is most intense at the sites of the plasma filaments of the contracted microwave discharge. Since the craters appear on the surface of LiF crystals also at the sites of plasma filaments, the charging and discharging of crystals by charged plasma particles should be analyzed just near the filaments. If we estimate the electric field $E_{charge}$ in the surface layer of a LiF crystal charged by a high–energy electron current with a density $(j_e)_{fast} \leq 10$ A/cm$^2$, taking into account the neutralizing effect of an ion saturation current with a density $j_i > 100$ A/cm$^2$, then we can see that, since $(j_e)_{fast} \ll j_i$, the electric field $E_{charge}$ will be several orders of magnitude weaker than both the threshold electrostatic field $E_{static}$ for breakdown in the bulk of LiF crystals and the quasi–electrostatic field $E_{plasma}$ in a double plasma layer:

$$E_{charge} \ll E_{static} < E_{plasma}.$$

Hence, in the interaction of a dense microwave–discharge plasma with a dielectric, the generation of a strong ($1$–$10$ MV/cm) quasi–electrostatic field aligned with the normal to the dielectric surface is attributed primarily to the formation of a double plasma layer with separated electric charges. The quasi–electrostatic field is excited on a very short time scale $\Delta t_E \approx 10^{-14}$ s and is characterized by full lifetime $\Delta t_2 \approx 0.5$ µs of the plasma filaments of the contracted discharge.

The local thermal mechanism for breakdown in dielectrics implies a strong local heating of the dielectric substance near the surface in the interaction of a dense plasma of a contracted microwave discharge with a dielectric. As a result, the ion conductivity of the dielectric increases and the threshold quasi–electrostatic field at the plasma–dielectric interface becomes substantially lower. The energy deposited by the electric–field–driven ion current in the locally heated dielectric regions is so high that the crystals are partially destroyed. Estimates of the highest local temperatures to which LiF crystals can be heated by the plasmas of contracted microwave discharges show that, at the very beginning of plasma microdischarges (i. e., during the time interval $\delta t \approx 0.05$ µs, Fig. 3), the local temperature of the crystals can reach the value of about 800 K, so that the threshold field for breakdown in dielectric crystals could be lower than 3 MV/cm. However, LiF crystals are locally heated within the characteristic depth $(\chi \delta t)^{1/2} \approx 0.3$ µm (where $\chi \approx 0.02$ cm$^2$/s is the temperature conductivity of LiF crystals), which is more than an order of magnitude smaller than the amount (about 10 µm) by which the average specific depth of the branched breakdown channels in the bulk of the crystal increases from pulse to pulse. Hence, we can conclude that the discharges in LiF crystals are initiated in local regions where the temperature remains close to the initial temperature (about 300 K). This indicates that the local thermal mechanism for breakdown in the bulk of LiF crystals contradicts our experimental data and thus fails to explain the high rate at which the depth of the dendrites increases from pulse to pulse during the electrodeless breakdown in the LiF crystal.

Our analysis of the possible mechanisms for electrodeless breakdown of dielectrics implies that the newly discovered breakdowns in the bulk of dielectric LiF crystals are of an electric (rather than a thermal) nature. As a result, the surface and internal electric fields generated in the interaction of dense microwave–discharge plasmas with LiF crystals are stronger than the threshold electrostatic field (3 MV/cm) for breakdown in the bulk of LiF crystals contacting with metal electrodes.

## 6. CONCLUSION

Our experiments revealed a new phenomenon – electrodeless electric breakdown in the bulk of dielectrics affected by microwave discharge plasmas – which can be briefly described as follows. A pulsed microwave radiation with a high–frequency electric field of about 3 kV/cm initiates microwave discharges at the surface of dielectrics in vacuum. The quasi–electrostatic fields generated within the dielectrics by dense plasmas of surface microwave discharges are stronger than 3 MV/cm. The electrodeless electric breakdown initiated by these fields in the bulk of dielectrics is accompanied by the partial destruction of the dielectrics in the form of surface craters and inward dendrites. Thus, we have shown that the dense plasma of a contracted microwave discharge strongly affect the local electric, thermal, and mechanical properties of dielectrics.

## ACKNOWLEDGMENTS

We are grateful to Professor G.M. Batanov for fruitful discussions.
This work was supported by Russian Fund for Basic Research (Project No. 99–02–16424; Project No. 99–02–04030), Deutsche Forchungsgemeinschaft (DFG) (Project No. 436 RUS 113544).

## References

1. Brown S.C. Basic Data of Plasma Physics. Cambridge, Mass. – London (England): M. I. T. press, 1966, 312 p.
2. Sazonov V.P. Electronic Technics, Series of Microwave Electronics[in Russian], 1967, 11, 47.
3. Strong Microwaves in Plasmas. Vols. 1, 2. Ed. A.G. Litvak. Nizhny Novgorod: Institute of Applied Physics, 2000.
4. Batanov G.M., Ivanov V.A., Kossyi I.A., and Sergeichev K.F. Sov. Phys. – J. Plasma Physics, 1986, 12, 317.
5. Batanov G.M., Ivanov V.A., Konyzhev M.E. et al. In the book : Strong Microwaves in Plasmas. Vol. 2. Ed. A.G. Litvak. Nizhny Novgorod: Institute of Applied Physics, 1991, 553.
6. Batanov G.M. and Ivanov V.A. In the book: Generation of Nonlinear Waves and Quasistationary Currents in Plasma. Ed. L.M. Kovrizhnykh. New–York: Nova Science Publishers, Inc., 1992, 227 p. (Proc. of the Institute of General Physics, Academy of Sciences of the USSR. Moscow: Nauka, 1988, 16, 46 [in Russian]).
7. Batanov G.M., Ivanov V.A., Konyzhev M.E. et al. Sov. Phys. – J. Tech. Phys. Lett., 1993, 19, 653.
8. Batanov G.M., Ivanov V.A., Konyzhev M.E. Sov. Phys. – J. Exper. Theor. Phys. Lett. (JETP Lett.), 1994, 59, 690.
9. Ter-Mikirtychev V.V., Tsuboi T., Konyzhev M.E. et al. Phys. Stat. Solidi B, 1996, 196, 269.
10. Batanov G.M., Ivanov V.A., Konyzhev M.E., and Letunov A.A. Sov. Phys. – J. Exper. Theor. Phys. Lett. (JETP Lett.), 1997, 66, 170.

11. Skanavi G.I. Physics of Dielectrics in Strong Fields [in Russian]. Moscow: Fizmatgiz, 1958. 908 p.

12. Vorob'ev A.A. and Vorob'ev G.A. Electric Breakdown and Destruction of Solid–State Dielectrics [in Russian]. Moscow: Vysshaya Shkola, 1966. 224 p.

13. Gromov V.V. Electric Charge in Irradiated Materials [in Russian]. Moscow: Energoizdat, 1982. 112 p.

14. Boiko V.I. and Evstigneev V.V. Introduction to the Physics of the Interaction of High–Current Beams of Charged Particles with Condensed Matter [in Russian]. Moscow: Energoizdat, 1988. 137 p.

# SPECTRAL CHARACTERISTICS OF THE CAPACITIVELY COUPLED MICROWAVE PLASMA OF ATMOSPHERIC PRESSURE

V.A.Kouchumov, V.V.Druzhenkov, Yu.I.Korovin, A.S.Antropov

State Scientific Center of Russian Federation, A.A.Bochvar All-Russia Research Institute of Inorganic Materials (VNIINM ), Moscow, Russia.

**Abstract.** Spectral and analytical characteristics have been investigated for the atmospheric pressure capacitively coupled microwave plasmas (CMP-discharge, frequency - 2450 MHz; power - 1.5 kW). The space toroidal discharge is initiated in the plasmatron designed as dual coaxial tube, with tangential injection of molecular plasma-generating gas (nitrogen or air at a gas flow rate of ~5 l/min) and the axial injection of another gas (argon, nitrogen or air at a gas flow rate of ~1 l/min), which transports the aerosol solution sample. The space distribution emission intensities of the analyt molecules, atoms and ions reveal the toroidal shape. The excitation of analyt atoms and ions occurs in the narrow central channel at the height of 10-15 mm above the electrode end (an analytical area). The injecvtion region is separated from the energy release area (skin-layer). The range of temperature excitation in the analytical area, measured by the relative atomic line intensity of various elements, is 4400 - $4700^0$K, whereas the gas temperature estimated by relative CN-molecular band line intensity lays in the range of 5100 -$5400^0$K. Relying on such investigations the plasmatron design has been developed as the dual coaxial tube. It has been used in the model of microwave plasma optical atomic emission spectrometer (MPIOAESp) which consists of "Chromatron-06" microwave generator, MFS-8 polychromator and the multielement simultaneous detection system with 9 line CCD "Toshiba" detectors. MPIOAESp - spectrometer is used for precision and rapid analysis of any complex materials and alloys, as well as technological solutions to define a content of more than 70 elements. Detection limits ($3\sigma$) of this method for several elements in solution are 2 –100 µg/ L.

## 1. INTRODUCTION

The microwave spectral sources of spectrum excitation have not found wide application on the background of development and application inductively-coupled plasma in spectral analysis. Two types of discharge are mostly known among microwave excitation sources. The first one is a microwave inductively coupled discharge excited usually in a quartz tube, with the input power conducted through a rectangular waveguide or resonator. This work deals with the second type of sources, so-called capacitatively coupled microwave discharges, being excited in a coaxial tube and formed on the tip of the central electrode. They are also called single-electrode torch discharges. A simplified design of a microwave plasmatron, which is used since 1970, is given in Fig.1. This plasmatron was designed by "Hitachi Ltd." (Japan). The discharge is formed on the tip of the central electrode; plasma-forming gas and sample solution are introduced at the outer side of the plasma.

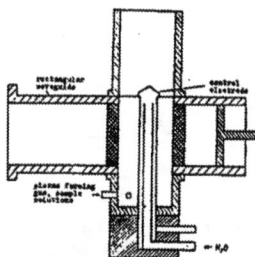

Figure 1. Microwave plasmatron

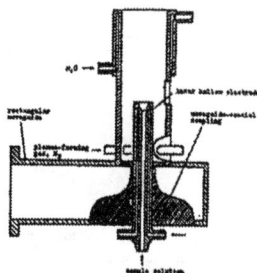

Figure 2. Design of microwave plasmatron for toroidal discharge

After Boumans work [1] published in 1975 in which he compared spectroanalytic characteristics of this plasmatron and an inductively-coupled torch, the interest in microwave plasmatrons has dropped considerably. Boumans showed that they have much worse characteristics than the inductively-coupled plasma. The influence of third components in them is stronger and the detection limits are substantially higher. However, according to the literature, the interest in these plasmatrons has grown considerably since 1995.

This is mainly due to the fact that the design of microwave plasmatron has been improved. This made it possible to use sample aerosols more efficiently and to improve their analytic characteristics. A series of papers has been published in Japan, USA, Canada, France and Germany, in which the plasmotron analytic characteristics and its application have been described [2,3,4,5]. In particular, steel componental analysis and sewage analysis for microelements content have been described

## 2. EXPERIMENTAL AND RESALTS

Using the experience of working with a microwave induced discharge and with ICP, having in mind its disadvantages, we tried to improve the plasmatron design in order to carry out the sample aerosol introduction into the discharge high-temperature region.

The design of the developed plasmatron is given schematically in Fig.2. This is a coaxial tube which is coupled with a rectangular waveguide. Microwave power is conducted to the discharge chamber by the rectangular waveguide. The aerosols of sample solution are introduced through the central channel of a hollow water-cooled electrode into the analytic region. Molecular gases, mainly nitrogen, are used as plasma-forming gases. It was impossible to ignite the discharge in argon because of a break-down in the interelectrode space. The discharge can be sustained in argon only with the addition of aqueous aerosols or at substantially less power values which are not suitable for the plasmatron operation. The consumption rate of the plasma-forming gas is 6-7 l/min; the flow rate of carrier gas (argon) is 1-2 l/min.

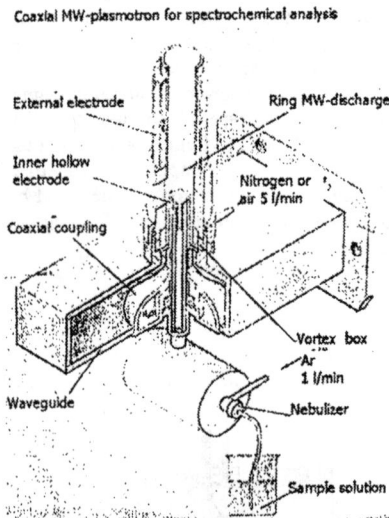

**Figure 3.** Coaxial MW-plasmatron for spectrochemical analysis of solutions

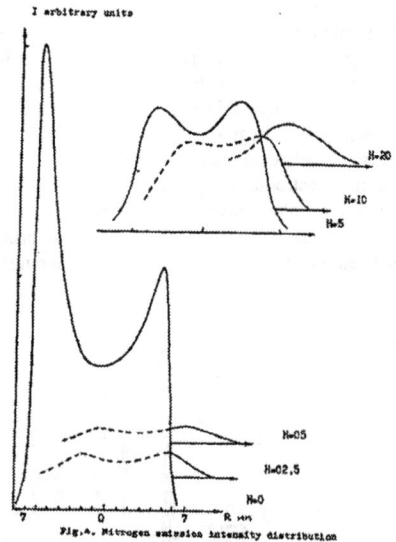

**Figure 4.** Nitrogen emission intensity distributions ($N_2$ 337,1 nm) along the discharge height and radius

The MW-discharge spectrum is very complicated. Molecular bands of nitrogen, nitrogen ions, CN and

-CH are excited in it. Our investigations made it possible to separate the region with molecular band excitation from the analytic region where the elements are excited. We have studied the spatial structure of the discharge, formed by means of the plasmatron. Fig.4 gives the longitudinal distribution of the $N_2$ band emission intensity. Radial distribution has pronounced the out-of-axis maximum and a fall in the center. At a height of 2.5 mm from the electrode tip the spectrum intensity drops substantially (~20 times), i.e. the whole molecular spectrum is formed in a narrow near-electrode region of the discharge. The calculations performed on the basis of the data obtained in the literature showed that the skin-layer formation is a characteristic feature for this type of discharge. The ring discharge occurs as a result of a skin- effect in vortex and axial gas flows specific for such plasmatron. This is, in particular, typical for the inductively-coupled plasma as well. At a larger height from the central electrode tip (10-15 mm) the distribution has a bell-like form, the molecular spectrum intensity substantially decreases.

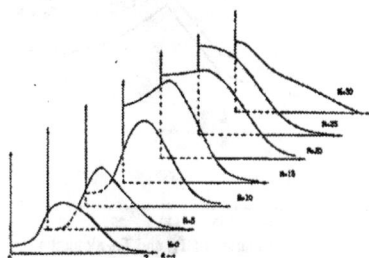

Figure 5. Spatial distribution of FeI emission intensity (373,487 nm)

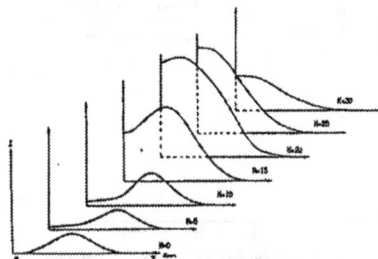

Figure 6. Spatial distribution of MgII emission intensity (279,553 nm)

Radial distributions of emission intensity of Mg and Fe introduced into the plasma are given in Fig. 5, 6. As it can be seen from the figures, the distribution is similar to that of nitrogen.

It is necessary to emphasize the asymmetry of the out-of-axis maximum that is due to the one-sided supply of the microwave power. The nitrogen intensity maximum is higher on that side of the discharge, which is closer to the input power, than that on the opposite side. Saddle-shaped intensity distribution is inherent to emission of all elements introduced into the plasma, but their maxima are placed closer to the axis. The maximum of molecular spectrum intensity is observed at a distance of 7 mm from the axis, whereas the intensity maxima of elements introduced axially are at distance of 4 mm. The intensity distribution acquires a bell-like character as the height from the electrode tip increases. This is inherent to all elements introduced into the discharge plasma.

Summing up the investigation of the spatial distribution of emission intensity of various plasma components and elements introduced into the plasma, one should ascertain that spatially non-uniform ring or toroidal plasma is being formed. Excitation of elements introduced into the plasma takes place in a narrow central region 4 mm in diameter. Thus we have carried out the task we set to achieve the efficient introduction of elements into the discharge plasma. The study of spatial distributions shows that a choice of an optimal region of atom and ion excitation for each element is possible, that is of essential importance for the analytical practice.

The excitation temperatures were measured for atomic lines of iron, copper, ionic lines of vanadium, and other elements (Fig.7). The effective temperature was measured using Lvov's method [6] that characterizes the distribution of atoms in lower lying levels. These investigations showed that excitation temperatures are sufficiently high and equal 5000°K. The maximum excitation temperature is observed in the near-electrode region where the microwave energy absorption takes place; the temperature drops as the height of observation increases. Besides, the effective temperature is substantially less than the excitation temperatures measured by the relative intensities of lines of the elements introduced into the plasma. This difference is especially substantial in the near-electrode region; as the height of observation increases, the temperatures are coming closer to each other. Note some more features. Rather high temperatures are obtained when CN are used as thermometric systems. The temperatures are of the order of 5500°K that is higher than the temperatures obtained by relative intensities of iron and copper lines. This is apparently due to the fact that CN excitation takes place in the regions other than the region of Fe and Cu atoms excitation and apparently in the regions adjacent to the regions of microwave energy release. The excitation temperature of 5000°K provides the effective excitation of most of the elements of the periodic system.

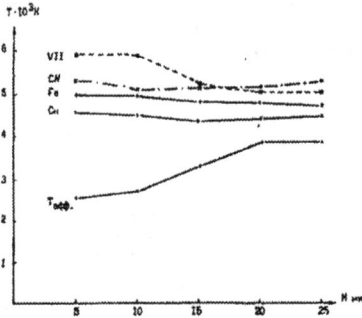

Figure 7.1. Distribution of $T_{ex}$ and $T_{eff}$ along the discharge height with K addition to the plasma ( K concentration in the solution is 2.5 g/l)

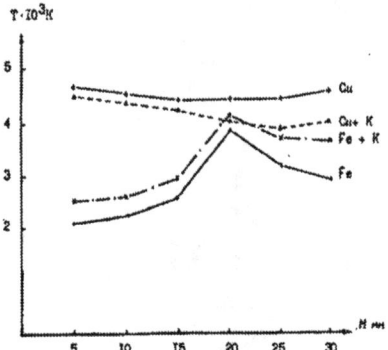

Figure 7.2. Distribution of $T_{ex}$ and $T_{eff}$ along the discharge height

Along with the temperature measurements we have evaluated electron concentration in the analytic region of the discharge. The traditional method of electron concentration measurements by Stark broadening of hydrogen lines could not be applied in this case as hydrogen was not practically excited. Addition of hydrogen or water lead to substantial increase of the NH line intensity though broadening of the H line was not observed. Therefore, the electron concentration has been evaluated by the relative intensity of atomic and ionic lines of Mg. This is a classical method, provided the local equilibrium in plasma is maintained. Electron concentration and its distribution along the discharge height are given in Tab. 1.

Table 1 Distribution of electron concentration along the discharge height

| H mm | $n_e$ cm$^{-3}$ |
|---|---|
| 5 | $2,2 \cdot 10^{13}$ |
| 10 | $1,5 \cdot 10^{13}$ |
| 15 | $7,4 \cdot 10^{12}$ |
| 20 | $7,1 \cdot 10^{12}$ |
| 25 | $6,9 \cdot 10^{12}$ |

Temperature measurements and their spacial distribution show that the plasma is non-uniform and it differs considerably from the local and thermodynamic equilibrium.

It was noted by Boumans [1] that the influence of third elements is a considerable disadvantage of microwave plasmatrons, in particular, he mentioned a noticeable effect of addition of easily-ionized elements. We have tested these effects for our plasmatron and they turned substantial. The temperatures measured by relative intensities of copper lines and by using Lvov method both in the presence of K additions and without them are shown in fig.7. As it is seen from this figure and other data we have obtained, K additions cause the excitation temperature decrease and the effective temperature increase; these temperatures move closer to each other in the analytic region of our plasmatron at a height of 15-20 mm above the electrode tip. Evaluations of the electron concentration obtained when additions of K into the plasma show that these additions in concentrations of 0.2 to 25 g/l should result in electron density increase in the discharge at 5000°K by a value, comparable with the values obtained for a discharge without K additions. If for the lower value of 0,2 g/l the addition of $n_e$ accounts 20%, then at K concentration in solution of 10-25 g/l one obtain the additions comparable with the values of electron concentration.

The distribution of the europium spectral line intensity along the discharge height for different K concentration in solution (from 0 to 25 g/l) are given in Fig.8. The distribution for K atomic line is shown in the upper figure. A similar distribution for the europium ionic line is presented in the bottom figure. Comparison of these two figures shows that K addition causes redistribution of the emission along the discharge height. The emission intensity decreases toward the electrode tip as K additions increase. The characteristic behavior of intensity of ionic and atomic lines is reflected in the figure. The intensity of ionic lines drops and that of atomic ones rises.

This mechanism of a simple shift of the ionization equilibrium is inherent for the flames too. One should note that such a behavior of the curve is given for optimal analytic conditions of the source operation both by the power introduced into the plasma and by gas consumption. More interesting effects are observed at low power values. Both the increase of atomic and ionic lines intensity and their suppression is being exhibited. This appearance is very complicated depending on the power introduced into the plasma, the height of observation and K concentration. Therefore in analytical practice it is necessary to use calibration specimens close in composition to those being analyzed or use the method of additions.

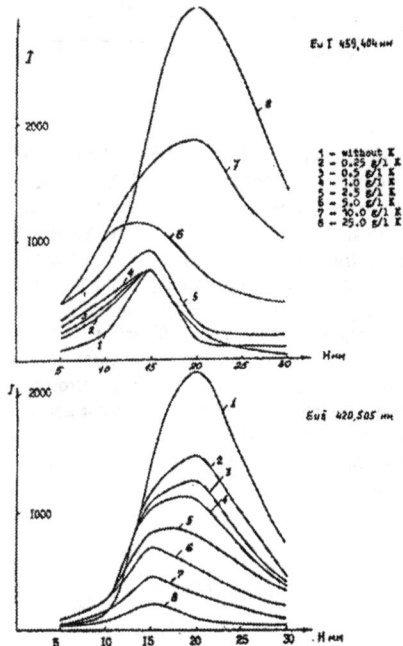

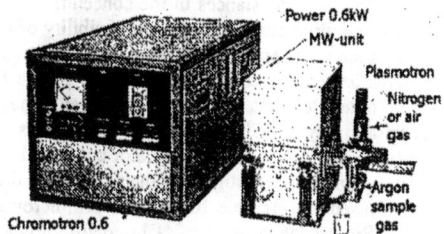

Figure 9. The microwave plasma generator for spectral analysis "Chromotron-06"

Figure 8. Effect of K addition on the intensity of europium spectral lines

The detection limits obtained by using this plasmatron as a source of spectra excitation are given in Table 2. Detection limits obtained by using our source are less than the limits for Hitachi Spectroscan-spectrometer, whose MW-plasmatron is on Fig.1, but higher than the values obtained by using inductively-coupled plasma.

Table 2. Detection limits of elements in solution with microwave CMP-discharge
atomic emission spectral method, mg/l.

| Element | Detection limits, µg/l. | Element | Detection limits, µg/l. | Element | Detection limits, µg/l. |
|---------|---------|---------|---------|---------|---------|
| Al | 20 | Ca | 2 | Sn | 100(200) |
| B | 30 (2000) | Co | 10 (50) | Pb | 30(100) |
| Be | 20 | La | 10 | Sc | 5 |
| V | 10 | Lu | 30 | Sm | 20 |
| Bi | 100 | Mg | 10 | Sr | 10 |
| Gd | 30 | Cu | 2(5) | Ta | 800 |
| Ho | 20 | Ni | 10(20) | Ti | 100 |
| Ge | 800 | Nd | 20 | Tb | 20 |
| Hf | 800 | Nb | 200 | Tm | 20 |
| Dy | 2.5 | Pr | 30 | U | 20 |
| Eu | 2.5 | Pd | 10(15) | Cr | 2(10) |
| Fe | 10(20) | Hg | 100 | Ce | 30 |
| Y | 5 | Rh | 12(20) | Zn | 500(10000) |
| Yt | 5 | Re | 100 | Zr | 10 |
| Na | 3 | Ru | 15(30) | Ce | 5 |
| Mn | 2(5) | Ga | 5 | Sb | 200 |
| Cd | 100(500) | Mo | 10 | | |

The detection limit values of elements are given in brackets for air-argon plasma.

## 3. CONCLUSION

For our experience, this plasmatron is a universal source of spectrum excitation and can be used for analysis of various substances in the concentration range of 0.01% to n·10%. Stability of the source of spectrum excitation ensures high reproducibility of the results in the optimal region that does not exceed 1%. This source has proved especially useful for high precision analysis of alloy components. Analytical procedures have been developed using this source for such matrices as zirconium, uranium, iron, rare earth and other elements. The total error usually does not exceed 2-5%. The microwave plasma generator for spectral analysis "Chromotron-06" (Fig.9) has 'been developed on the basis of our investigations. Commercial production of this generator began in 1994 at Production Union "Contact" (Saratov, RF).

Relying on such investigations the plasmatron design has been developed as the dual coaxial tube. It has been used to produce the model of microwave plasma optical atomic emission spectrometer (MPlOAESp), which consists of "Chromatron-06" microwave generator, MFS-8 polychromator and the multielement simultaneous detection system with 9 line CCD "Toshiba" detectors. MPlOAESp - spectrometer is used for precision and rapid analysis of any complex materials and alloys, as well as technological solutions to define the content of more than 70 elements.

## Acknowledgements

The work has been performed in the frame of ISTC Project 1052-98 and Project 1222-99.

## References
1. Boumans P.W.J.M., DeBoer F.J., Dahmen F.J., Hoelzel H., Meier A. Spectrochim. Acta, 1975, 30B, 449.
2. Patel B.M., Deavor J.P., Winefordner J.D. Talanta 1988, N 8, 641.
3. Atsuya I., Akatsuka K., Anal. Chim. Acta 1976, 81, 61.
4. Wunsch G., Hegenberg G., Czech N. Spectrocbim. Acta, 1983, 38, 1135.
5. Hanamura S., Smith B.W., Winefordner J.D. Anal. Chem, 1983, 55, 2026.
6. Lvov B.V., Orlov N.A., JAS, 1975, XXX, i 9, 1653.

# ELECTRODLESS MICROWAVE DISCHARGES AS SOURCES OF LIGHT AND UV-EMISSION FOR THE ILLUMINATION AND BIOMEDICAL APPLICATIONS

A.Kozlov[1], V.Perevodchikov[2], R.Umarhodzaev[3], E.Shlifer[2]

1 – IZMIRAN, Troitsk (Moscow region), 2 – ALEI (VEI), Moscow, 12, Krasnokazarmennaya, 3 – SRINP MSU

Abstract. The report contains some results of the researches and designing of Microwave-light modules for the Illuminating and Irradiating plants, which use the electrodless Microwave discharges in the different gases. The main problems and their solutions are considered for use in the plants of quasi-Sunlight and ultraviolet (UV) irradiation. The phenomenological picture of discharge excitation and stabilization is presented. The irradiation spectrums are shown. Some results and features of the creation the bactericidal plant for the complex Microwave, UV- and ozone treatment are presented.

## 1. INTRODUCTION

In 1992 experts of Fusion System Corporation (Maryland, USA) designed high efficient source of bright quasi sunlight radiation – electrodless microwave gas discharge sulfur tube. The spherical electrodless Argon-Mercury UV lamps with excitation by microwave energy in the frequency band 915-2450 MHz were prototypes of such device.

Main difference of the sulfur lamp from prototype is use of working medium (sulfur instead of mercury) which the power electrodless microwave discharge takes place.High-power light radiation of the sulfur lamp is characterized with continuous (molecular) spectrum close to sunlight spectrum, but with reduced UV and IR radiation levels.

In October 1994 Fusion Lighting Inc (FL) Company demonstrated in Washington two 'effective illumination systems – outdoor illumination of the DOE Headquarter (US Department of Energy) and indoor illumination system in the large hall of the National museum of aeronautics and space (NASM). These "presentations" attracted attention of experts-designers and perspective consumers to a new light source which introduction was accepted as the most considerable technological breakthrough in development illumination technology on the border of XXI century.

The first illumination systems on the base of sulfur tube used microwave excitation of power 3-4 kW (from two magnetrons of 1.7 kW), the illuminated light flux more than 400 klm and were characterized with the total output light efficiency not more than 75 lm/W. The highest results achieved in 1996 with the best power light device were: illumination flux of 480 klm, output light efficiency 95 lm/W.

However, at that time there was no mass production of such power light systems so we have not informed on mass production of such systems up to nowadays.

In 1996 FL company manufactured an analog of Solar 1000™ system (with light flux 138-140 klm at microwave power 800-1000 W), and by 1998 this company in cooperation with Sweden company IKL (Celsius Group) created prototypes of Light Drive 1000™ with the similar output parameters. The last light device has monoblock design that provides its use as autonomous light source so as light source for a hollow light guide.

The first prototypes of working microwave light devices in Russia (in Headlight projector design) were created in Russia in 1996 Open Stock Company "Pluton" in cooperation with VEI and some other enterprises. Some modifications were manufactured as for headlight projector applications so for light guide systems (designed by "Svetoch-PRO" and "Svetoch-SV" enterprises). Universal Russian demonstration monoblock model which is similar to devices Solar 1000™ and Light Drive 1000™ at first was demonstrated at the International exhibition "Interlight-98" in Moscow (December 1998) and won the first place in the nomination "Energosaving". At present time our works on development and modification of microwave light sources are being run ("Sveton" from "VEI" and "Pluton"). There are data on microwave light sources on the base of sulfur lamps presented in Table 1.

Table 1. Distinguishing features of Light devices.

| Specifications | Base example | NASM example 1994 | Example of 1995/96 | Solar-1000™ 1996 | Light-Drive 1000™ 1997-98 | Svetoch-PRO 1997 | Svetoch-SV 1998 | Sveton *) 1999-2000 |
|---|---|---|---|---|---|---|---|---|
| 1 | 2 | 3 | 4 | 5 | 6 | 7 | 8 | 9 |
| Total luminous flux, klm | 410 | 445 | 480 | 135 | 135-140 | 135 | 135 | 135 |
| Microwave power, kW | 3,4 | 3,1 | 3,1 | 1,0 | 1,0 | 0,9 | 0,9 | 0,9 |
| Input power, kVA | 6,3 | 5,9 | 5,1 | 1,375-1,425 | 1,4 | <1,325 | 1,325 | 1,325 |
| Total lighting efficiency, lm/VA | ~65 | ~75 | 94 | 91.5-98 | 98-100 | ~102 | 102 | 102 |
| Correlated color temperature | 6500 | 6300 | ~6000 | ~6000 | ~6000 | ~6000 | ~6000 | ~6000 |
| Color rendering index | 86 | 86 | 86 | ~79 | 78-79 | 78 | 78 | 78 |
| Starting time, s | 5 | 5 | 5-15 | 15-25 | 15-25 | 10-15 | 10-15 | 10-15 |
| Cooling condition | Forced air with rotating lamp | | | Conventional with rotating lamp | | | | |
| Reflector aperture | Round for light-guide "3M" Ø254 mm | | | | a) round for light-guide Ø254mm b) round Ø400mm c) rectangular | a) round for light-guide Ø254mm b) round Ø350mm for projector | round for light-guide Ø254 mm | a) round for light-guide Ø254mm b) round for asymmetrical and symmetrical re-reflectors |

**Continuous of Table 1.**

| 1 | 2 | 3 | 4 | 5 | 6 | 7 | 8 | 9 |
|---|---|---|---|---|---|---|---|---|
| Advantages | • spectrum close to sunlight <br> • very low level of UV and IR <br> • Small lighting body and uniform of illumination <br> • No environment hazards | | | | | | | |
| | • high level of luminous flux <br> • short starting time | | | • high lighting efficiency <br> • application of high quality magnetrons | | | | |
| | | | | • possibility of lighting drive | | • high stability of magnetron work <br> • low level of the fifth harmonic | | |
| | | | | | • light mono-block | | | • mono-block move-able |
| Disadvantages | • high level of acoustic noise <br> • difficulties with microwave leakage | | | • long time (about 5 min) for repeat starting | | | | |

<sup>*)</sup> "VEI" and "PLUTON"

If in illumination installations the energy of microwave oscillation is used only for pumping of an electrodless lamp, the combined irradiation installation developed in Russia (Microwave-Ultraviolet-Ozone) provide not only microwave pumping but also direct effect on irradiated object.

The results of experimental researches have shown considerable amplification of bactericidal effect of such installation in comparison with only microwave or ultra-violet radiation or ozone treatment effect. It is possible to use the some modifications of bactericidal installation as domestic polyfunctional Microwave Oven.

## 2. MECHANISM AND FEATURES OF WORK

Physical mechanism of work of foreign made and domestic made sulfur lamps consists of photons radiation with multiple transitions of energy states of evaporated sulfur molecules excited or ionized by microwave discharge in a small volume limited with a spherical quartz envelope.

Each transition has substructure due to rotational and vibrational substructures superimposed on each electronic state. Most probable optical radiation is in the UV-region, but because the optical depth of plasma "body" is large enough, the almost all radiation from bulb is in visible diapason and has continuum light emission due to the broadening of vibrational and rotating lines by high pressure and temperature of the working mixture of gas and vapor. Only during the "start up" phase of microwave discharge, when the sulfur vapor pressure is small, yet the UV-emission from the lamp is large enough, but this phenomena has a very short time, which depends on the microwave power input into lamp (or, more correct, on the intensity of electrical microwave field and its distribution). If the UV-emission is desired, for example, as bactericidal irradiation factor ($\lambda = 253,7$ nm), the envelope fill material may be argon (starting gas) and mercury. In this case the visible light is the "by-product", and irradiation spectrum is of continuum, but has a typical lines. The emission on the line $\lambda = 180$ nm make up the ozone in the surrounding air or oxygen.

The process of transformation of microwave energy (energy of pumping) in optical radiation (OR) is characterized by phenomenon sequence. After magnetron switching-on, as soon as amplitude of microwave field electrical component in a resonator (in a zone of a Sulfur lamp arrangement) reaches an appropriate potential of ignition, a microwave discharge arises in a buffer gas mixture (argon) and saturated vapor of sulfur which is while in a solid condition. At this stage rather lamp radiates a separate lines spectrum corresponding to typical energy transitions in atoms of argon and sulfur. With it there are rather noticeable levels of ultraviolet (UV) and infrared (IR) components presented in an optical radiation spectrum.

In a process of microwave energy absorption by low pressure discharge and growth of number of ionization acts, the concentration of charges in plasma grows, bombardment activity of envelope surface (coated with precipitated sulfur due to previous switching-on -off and cooling of a lamp) increases mainly for the account of the most mobile carriers of a charge – electrons. As a result of bombardment (mainly in a direction of microwave electric field force lines) envelope temperature grows fast, and sulfur vapor partial pressure grows also. This process passes a melting stage of different polymorphic sulfur modes (112,8°C; 119,3°C), and then full evaporation stage (boiling temperature: $T_{boil}$ = 444,6°C), when concentration of sulfur molecules in an envelope becomes rather high. In a stable plasma mode (in the high pressure discharge) the resulting optical radiation (OR) spectrum has a "molecular" character stipulated, as is marked above, plurality of energy condition, including rotating and oscillating conditions of molecules. Spectrum becomes continuous, and typical lines (including lines in UV and IR areas) are merged in this spectrum, though due to heating of an envelope in a continuous optical radiation IR-"wing" appears more elevated in comparison with the UV band. This feature of a spectrum is saved at a different levels of microwave energy pumping and at various initial amounts of sulfur in a lamp of the given size, though with other equal conditions common "displacement" of a continuous spectrum in "far blue" or "red" area can be controlled at the expense of initial amount of sulfur, and also at the expense of introduction of those or other "additives" in a working substance. If selenium is added as such "additive", a feature of stationary plasma mode process is that selenium has higher melting temperature (221°C) and boiling temperature (685,3°C) and, accordingly condensation temperature after deenergizing a microwave pumping. Therefore at different stages of plasma mode and envelope temperature the partial pressure of a vapor-gas mixture of argon, sulfur and selenium vary unequally, that finds its reflection in a character of optical radiation spectrum in transient and stationary modes.

In argon-mercury UV-lamps a phenomenon picture of stationary microwave discharge is more simple, as far already at room temperature a partial pressure of saturated mercury vapor in a mixture with argon is rather large.

If low pressure discharge is used, the mercury is not completely evaporated (working temperature ~ 60 –70°C). Radiation spectrum remains with typical resonance lines of argon and mercury, and in an UV-range ($\lambda$= 283,7 nm, $\lambda$ = 185 nm) optical radiation is the most high.

## 3. MAIN DIRECTIONS OF WORK ON USE ELECTRODELESS MICROWAVE-DISCHARGE SOURCES OF OPTICAL RADIATION FOR LIGHTING AND IRRADIATING SYSTEMS (DEVICES).

Depending on purpose of a system, such parts can be included into its structure as the main functional elements: a light-transparent microwave resonator or microwave screen, a former of an optical radiation flux (reflector) or an irradiation working chamber, which in its turn can execute functions of a microwave resonator, including non-transparent one. During work of such system microwave energy is partially absorbed by plasma and is transformed into optical radiation (useful effect), is partially reflected to a pumping microwave generator, rendering influence on the output power $P_{out}$, on the frequency of generated oscillations $f_{gen}$ ("pumping frequency"), and on operational mode. Besides, some part of microwave energy is radiated in ambient space. If as useful "product" it is necessary to receive just optical radiation, for example – for illumination or UV irradiation, microwave energy radiation should be considered as harmful effect requiring suppression up to biologically safe level.

Recently we have created irradiation installations, which both such "instruments": optical radiation in UV band and microwave radiation are used for deliberate effect on alive and lifeless objects completely are used. In this combination microwave energy alongside with a realization of pumping function is also independent " useful product ".

It is possible to speak today about forming three rather independent directions in construction lighting (illuminating) and irradiating devices, installations and systems.

First from these directions covers construction of autonomous lighting fixtures and UV-irradiators, for example, of projector type, with the specific form of optical radiation flux and its spectral distribution.

The second - construction of light sources, combined (integrated) with one or multi-channel light-distribution system on the base of light-guides.

The third - construction of complex systems of UV + microwave radiation for various objects and mediums.

There are general problems in all these directions of a basic essence character, and some specific features. At present it is possible to make their integrated grouping.

So, let us group main problems in appropriate blocks, that more or less visually problems, facing to the developers "were highlighted".

a) Problems relating to a lamp (torch).
There are the following problems in this block of problems:
- Choice and optimization of structure, quantitative parameters and frequency of filling media, responsible for start forming and stationary maintaining of microwave discharge, for character of optical radiation spectrum and it preservation during operation time.
- Choice of the envelope form, its sizes, material and processing technology, providing of required symmetry and accuracy of manufacturing of each copy and recurrence from a copy to a copy in production line.
- Determination and providing necessary temperature and its distribution on envelope surface in stationary operation mode and in cooling mode after deenergizing.

b) Problems relating to the microwave pumping system.
There are the following problems in this block:
- Choice of power level and form of microwave signal (continuous, amplitude-modulated).
- Creation of systems for "transportation" of microwave energy from a source (magnetron) to a load (electrodless lamp) and devices assigning (depending on oscillations mode) a topography of microwave field in space of its interaction with the lamp working substance (in its initial condition and in plasma mode).
- Providing (maintenance) stability of microwave generator (magnetron) work on a load essentially varying during development of microwave discharge (during time interval from start up to reaching stationary plasma mode).
- The prevention of an inadvertent microwave radiation in an environment at pumping frequency $f_w$ (in case if optical radiation is the only "useful product") and harmonic frequencies ($2f_w \ldots 5f_w \ldots nf_w$) etc. collateral oscillations, accordingly, in maintenance of ecological safety and electromagnetic compatibility (EMI), with it an accompanying complex problem is determination of the compromise between light-transparency and microwave resonator walls microwave-opasity, while interaction of microwave fields with plasma torch takes place in the resonator.

c) Problems relating to forming of an optical radiation stream (beam).
There are the following problems in this block:
- Choice form, sizes and reflector (reflectors) material with consideration separations or combining functions of the optical former and microwave resonator.
- Providing design compatibility of reflector with microwave resonator, light-distribution system, with cooling system and lamp rotation device , and in some cases - with irradiated objects (for combined UV-microwave radiators)

d) Problems relating to the power supply system, control and protection of optical radiation source as complex system (device).
There are the following problems in this block:

- Choice and optimization of magnetron power supply voltage form.
- Definition of objects which are being a subject of control and protection, kinds and character of the physical factors, which change the control (protection) system should operate.
- Choice of a kind of the carrier of transmitted information, appropriate sensors and receivers of control signal, converters of this signal in control command and actuators realizing a control operation.

## e) Problems relating to the operation process.

There are the following problems in this block:

- Providing resistance and stability to external influencing factors (climatic, mechanical) while in service lamps and systems as a whole.
- Providing operation with selected and given combination of different kinds of radiation and additional effects (for example, ozone effect) on irradiated object.

## f) Problems relating to the light-distribution system.

The are in this block, in particular, problems of choice and realization optical radiation delivery means to illuminated or irradiated objects (beam, hollow or fiber light-guides).

## g) Problems relating to the philosophy and measurements engineering:

There are the following problems in this block:

- Microwave measurement in a static operation mode ("cold measurements") including measurements parameters of microwave resonator, microwave adapter, microwave exitor and microwave tract in assembly with these elements.
- Microwave measurement in dynamic (transient) operation mode (with working pumping microwave generator and light torch), with evaluation of microwave power level, penetrating through transparent walls of microwave resonator at working frequency, measurement of relative levels of collateral oscillations and harmonics (EMI problem).
- Light and spectral measurements in dynamic (transient) operation modes in conditions complicated by microwave resonator or microwave screen and former of a light beam (flux);
- Thermal measurements including determination of temperature on different points of the light torch envelope surface during starting-up on stable duty of stationary burning, cooling repeated energizing.

This list does not include all complex of problems rising at creation of optical radiation systems, using the electrodless discharge lamps with microwave pumping.

In the field of creation lighting fixtures (illuminators) and illumination systems (within the framework of the first two directions) the achievements of foreign colleagues represented (illustrated) by systems Solar 1000 ™ and Light-Drive 1000 ™ allow to evaluate positively the ways, used by them, some from which confirm and supplement domestic works experience on quasi-sunlight sources (Tabl.1).

Speaking on creation irradiation bactericidal installations using electrodless microwave discharge lamps of UV-range (band) and useful in this case microwave effect on completely antiseptical treated objects, the desktop installation developed by us and manufactured at the "Pluton" Company demonstrates achievements and engineering solutions in this field.

The original devices and operation modes are realized in this system which provide positive results practically in all the listed blocks of problems.

## 4. SOME EXPERIMENTAL RESULTS. EXAMPLES OF DESIGNS.

The electrodless lamps, light torches with various filling composites, with various dosing and with various technological features of manufacturing were subjected to experimental studies. The interactions of microwave electromagnetic fields with vapor-gas mixtures in lamps of various forms and sizes in resonators working at various kinds of oscillations (including multi-mode resonators) and, accordingly,

241

having space distributions intrinsic to working operation mode ("topography"), electrical and magnetic components of microwave fields were investigated.

In particular, the resonators and oscillation modes with azimuth-symmetrical field (TEM, $TE_{01p}$), with azimuth-inhomogeneous field ($TE_{11p}$) and with superposition fields ($TE_{mnp} + TM_{mnp}$) were used. With it the features of work (behavior) and "output" parameters of optical radiation of lamps depending on amplitude and form of pumping "signal" were investigated.

There are some examples of radiation spectrum for sulfur and sulfur-selenium lamps in a stationary operation mode are presented in Fig.1. It is seen from Fig.1, that the adding selenium in composition of working vapor-gas mixture displaces a spectrum in the "red" band.

The spectrum of sulfur lamp at various microwave pumping power are shown in Fig.2. It is seen from Fig.2, that the change of pumping power almost in 3 times does not result in displacement of extremum on wavelength coordinate and changes practically proportionally the radiation flux.

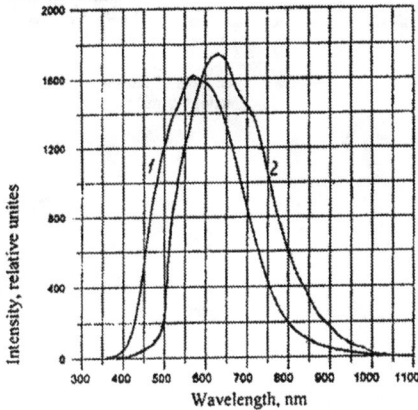

Figure 1. Spectrums of the lamps:
1 – sulfur,
2 – sulfur-selenium.
Stationary working conditions.

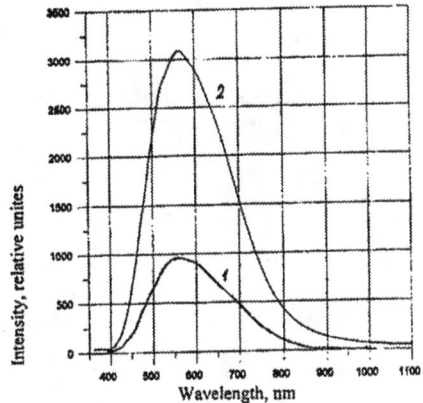

Figure 2. Spectrums of the sulfur lamp at microwave power:
1 – 565 W,
2 – 1400 W.

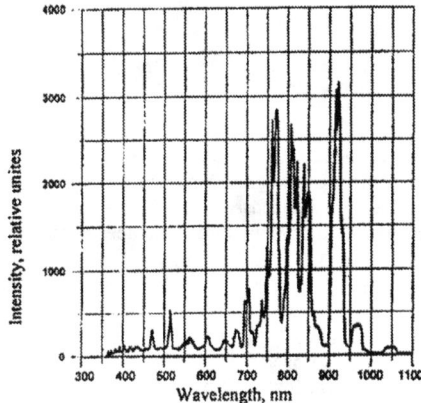

Figure 3. Spectrums of the sulfur lamp at the envelope temperature, $T_e=35°C$.

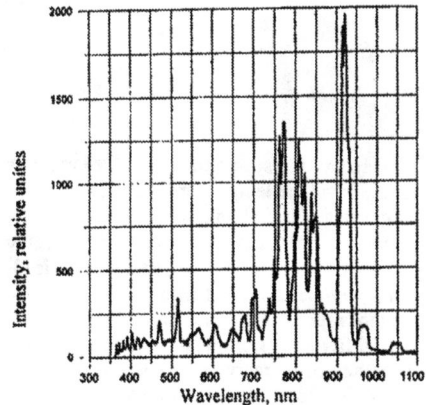

Figure 4. Spectrums of the sulfur lamp at the envelope temperature, $T_e=48°C$.

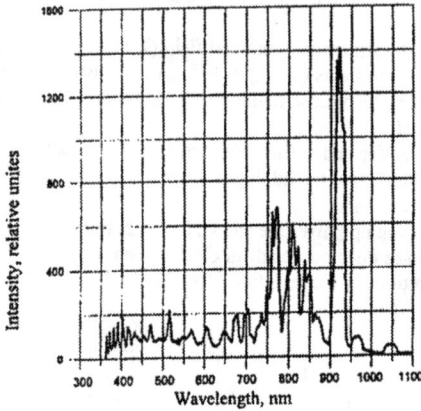

**Figure 5.** Spectrums of the sulfur lamp at the envelope temperature, $T_e = 56°C$.

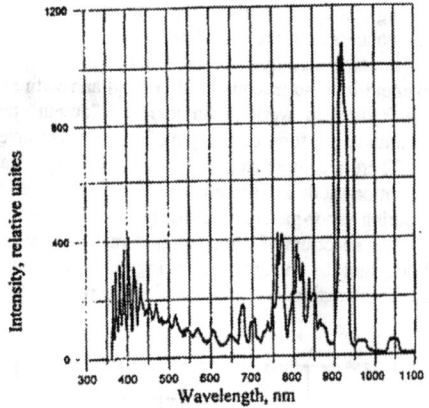

**Figure 6.** Spectrums of the sulfur lamp at the envelope temperature, $T_e = 75°C$.

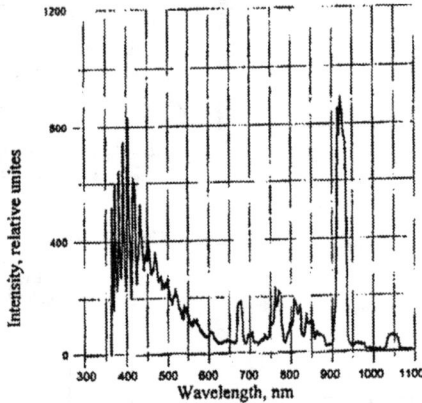

**Figure 7.** Spectrums of the sulfur lamp at the envelope temperature, $T_e = 112°C$.

There are the spectrums of sulfur lamp in transient regime are presented in Fig.3-7. It is seen, that with the envelope temperature growth and, therefore, with the pressure growth pressure, the reducing of start gas (argon and atomic sulfur) intensity lines takes place. Simultaneously in the "blue-violet" area the "molecular" spectrum is formed. Passing the intermediate conditions, the radiation spectrum is completely "smoothed out" in stationary mode ($T_1 = 700°C$) (Fig.1).

It should be noted that with use an off-standard crystal there are typical lines of impurities arise on a continuous spectrum (K, Li, Ca).

In difference from sulfur lamps, intended for the lighting (illumination) purposes, working temperature of mercury electrodless lamps intended for combined bactericidal effect (irradiation) (UV-Microwave- Ozone) is in limit up to 70°C (at least with using of the low pressure discharge), that ensures large freedom in choice the form of lamp, cooling method and quality of crystal.

The first modification of the bactericidal installation was designed as the scientific research instrument for the specialists in sphere of biomedical and others new technologies. There are some interchangeable accessories for the hanging surgical instruments, medical materials etc. It is possible to use different levels of Microwave power, UV-irradiation and treatment duration. The power supplied (~1200 W, 220 V, 50 Hz) is the same as the usual domestic Microwave Oven. Only 100 W expenditure of Microwave power is required for the excitation of electrodless discharges in two lamps, and UV-irradiated flux is ~48 Bact ($\lambda$=253,7 nm). The ozone is produced by UV-irradiation ($\lambda$=185 nm).

The prototype (model) of bactericidal installation assembled on the basis of household microwave oven SP-18 ("Pluton" Company) was comprehensively tested. The sketch diagram of this installation is presented in Fig.8.

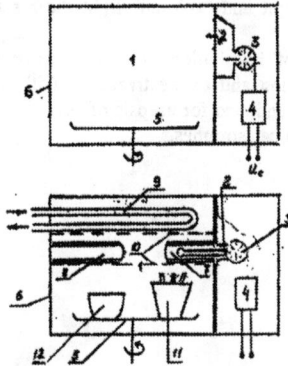

**Figure 8.** Microwave Oven (a), Bactericidal installation (b).
1 – Working chamber (multymode resonator), 2 – Microwave adapter, 3 – Magnetron M-155, 4 – Power supply block, 5 – Turning table, 6 – housing, 7 – UV-electrodless lamp (leading), 8 – UV-electrodless lamp (driven), 9 – quartz tube (coil), 10 – Light- and air-transparent RF-screen, 11 – UV-irradiated objects, 12 – Microwave heated object or liquid.

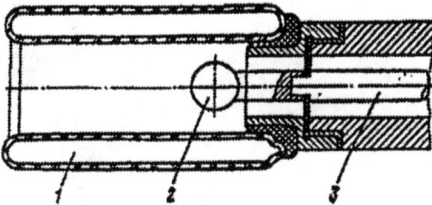

Figure 9. Ar-Hg electrodless UV-lamp
1 – lamp, 2 – Microwave exciter, 3 – coaxial guide (input).

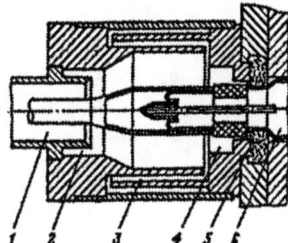

Figure 10. Coaxial microwave adapter
1 – Coaxial guide (output), 2 – 5th harmonic filter, 3 – Microwave drossel, connection, 4 – impedance transformer, 5 – contact gasket, 6 – magnetron output.

The module of irradiation unit consists of continuous operation mode magnetron (2450 MHz, 850 W), a coaxial microwave adapter, interchangeable electrodless Ar-Hg lamps of original design placed in the working chamber which is the multi-mode microwave resonator. The UV- and microwave transparent quartz serpentine is placed in this chamber for treatment medical solutions and other flowing liquids. A rotating table for installation irradiated objects (medical instruments, materials, vessels with different agents) is also placed in the working chamber. Electrodynamics principle of the coaxial microwave adapter provides stable operation of magnetron with variable non-stable microwave load and depression undesirable mode oscillations (including fifth harmonic) in the microwave spectrum.

The lamp and adapter constructions are shown at Fig.9, 10.

This adapter is used as in bactericidal plant, so in Microwave-lighting module.

## 5. CONCLUSION

The use of electrodless microwave discharge as a source of optical radiation in visible and UV-band of the spectrum is a perspective area of experts activity and promises in nearest and long-term future arising new updated products and systems for illumination and irradiation. There are, however, a lot of problems which are being a subject to solution, and today there is a lot of looked through possibilities. In the field of lighting systems of quasi-sun light and "colored" light the research and design works are already conducted.

The perspectives of creation of installation with microwave radiation power up to 20 kW in combination with strong UV-radiation and ozone treatment will be used for asepsis (purification) of drinking water or other substances, and also for asepsis of soil and reservoirs, having been infected, for example, in a result of the ecological catastrophes.

# RESIST ASHING EMPLOYING MICROWAVE OXYGEN PLASMA WITH HIGH PERMITTIVITY MATERIAL WINDOW

Kouta KUSABA, Keisuke Shinagawa*, Masakazu FURUKAWA*,
Katsufumi KAWAMURA* and Haruo SHINDO

Department of Applied Physics, Tokai University, Kitakaname 1117,
Hiratsuka 259-1292
*Canon Sales Corporation, Process Equipment Division, Kounan 2-13-29, Minatoku, Tokyo 108-0075

Abstract Oxygen microwave plasma production was studied by focusing an attention to high-permittivity material window. The plasma parameters were examined in conjunction with the permittivity of the dielectric window material. The results revealed that the higher permittivity windows yielded higher electron density in the regime above the cutoff of the 2.45 GHz microwave, insisting that the plasma production is due to the surface wave. The resist ashing experiments, performed in 8 inch wafer, showed that the rate was 2 times higher with the alumina of higher permittivity than the other, and a wafer damage analysis by DLTS measurement concluded that the high permittivity window material provided a large-diameter wafer ashing process with a high rate and low damage.

## 1. INTRODUCTION

Plasma processes, such as etching, deposition and resist ashing, are one of the key technologies for large-scale integrated (LSI) circuit fabrication. In these processes, a large-diameter plasma source is required to meet the forthcoming demand of 300-$\varphi$ wafer fabrication. Microwave plasma is a potential candidate as a large-diameter plasma sources. One issue in large diameter microwave plasma sources is related to the dielectric window material required for introduction of microwaves. As the plasma diameter is increased, the mechanical stress on the window material is enormously enhanced. Therefore, the thickness of the window material enlarges, making plasma production ineffective. To avoid this problem, the introduction of microwaves by slot antenna has been proposed.[1,2] This device enables suppression of the window area, but the plasma uniformity remains questionable. In this method, microwaves have been introduced in the surface wave.[1-3] In order to attain a good plasma uniformity, the microwaves should be introduced through a large-area window, and the window material should be as thin as possible.

In this work, microwave plasma production, as a large-diameter plasma generation method, was studied by employing a high-permittivity window material. The plasma properties in $O_2$ gases were closely examined in a viewpoint of the permittivity of the window material. If the microwave power is transferred into the plasma in a surface-wave mode, the plasma behaves differently depending on the permittivity of the window material. Experimental consideration on this point, therefore, is extremely important for future development of large-diameter microwave plasma sources. As one of typical applications of this kind of high density surface-wave oxygen plasmas, furthermore, resist ashing characteristics were also examined.

## 2. EXPERIMENTAL

The plasma was produced in an aluminum chamber of 240 mm in diameter as depicted in Fig. 1. At the one end of the chamber, a microwave of 2.45 GHz was introduced through a dielectric window of disc plate of 240 mm in diameter. Before introducing into the dielectric window, the microwave mode was converted from the $TE_{10}$ at the rectangular waveguide to the $TM_{01}$ at the circular one.

Three kinds of dielectric materials, quartz, alumina and two aliminitrides, were employed in this experiment; their permittivities were, respectively, 3.86 (14.9 GHz), 9.7 (10 GHz), 8.5 (14.9 GHz) and 10.9(14.9 GHz), where the frequencies used for the permittivity measurements are given in parentheses. The two aluminirides employed were both commercially available, but showed different values of the permittivity. The measured dielectric loss constants, tanδ, of these materials ranged from $1.31 \times 10^{-4}$ to $1.0 \times 10^{-4}$ and the values seem insignificant in plasma production.

The thickness of windows were 7 mm except of the quartz of 13 mm; one had to have the same thickness to compare the effect of permittivity, but the thickness of 13 mm was required for the quartz to safely perform the measurements.[4] The plasma parameters were measured using a Pt plane probe of 1 or 0.5 mm in diameter in $O_2$ plasmas.

The ashing experiment was also made in the same chamber, where a wafer stage of 8 inch size was newly equipped. The stage can be heated and also movable in the axial direction. The photo-resist of PFI-58, commercially available one for i line-resist was employed with the P(100) type of silicon substrate. The ashing rate of the photo-resist was measured at the substrate temperature of 200K. Ashing damage was analyzed for the silicon by a deep level transient spectroscopy(DLTS) measurement. Hereafter the experimental data will be shown as a function of the distance Z from the surface of microwave window.

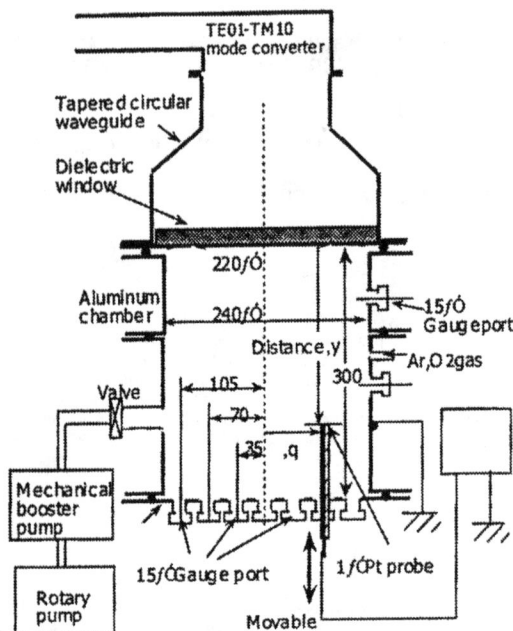

Figure 1. Schematic illustration of experimental apparatus

## 3. RESULTS AND DISCUSSIONS

In Fig. 2, the measured electron density is plotted against the value of the permittivity of the window materials, where the axial distance Z for measurement is taken as a parameter. It should be noted that the electron density values plotted in Fig. 2 are averaged one in each radial plane. It is seen that the electron density thus obtained has a great dependence on the window material, especially at lower Z, near the window, demonstrating a clear permittivity dependence. In the figure, the cutoff density for the 2.45 GHz microwave is also depicted by the broken line, and it is seen that the data point of density at the lowest permittivity (quartz) is below the cutoff even at Z=1 cm, suggesting not a pure surface-wave mode in this case. Since the dielectric flux density of the normal has to have the same value at the interface between plasma and window, the higher permittivity material window requires the higher density in plasma, and thus the surface wave mode is easily established in these conditions.

Another data which support the permittivity dependence shown in Fig.2 is obtained by a magnetic probe measurement. Figure 3 shows a typical example of the axial behaviors of the microwave magnetic field, where four window materials are compared. It is seen that these behaviors are very much consistent with the data in Fig.2; the microwave magnetic field in the case of quartz does not decay so much with the distance Z, and this is because the microwave is volume-mode in this case, being consistent with the low density as seen in Fig.2. In other window conditions, in the higher permittivity of Fig.3, the wave field decays rapidly with the distance Z, implying the surface wave-mode.

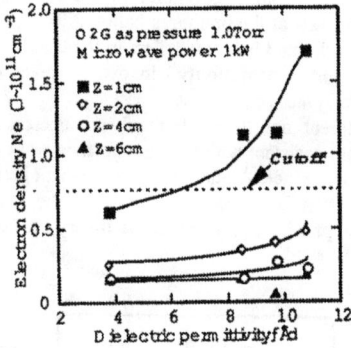

**Figure 2.** Permittivity dependence.

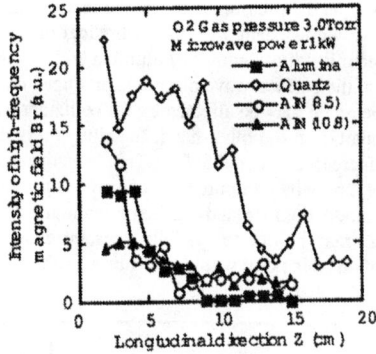

**Figure 3.** Microwave magnetic field.

In Fig.4 (a) and (b), the radial variations of the measured electron density are plotted and compared for the quartz and alumina. A radial variations of the density are fairly affected by the microwave mode, and therefore they become a little complicated. However, the major difference between the quartz and alumina is clear; in all radial positions the density becomes above the cutoff in the alumina, while in the quartz the density is almost below the cutoff.

As stated so far, the higher permittivity provides us a fairly overdense plasma, and this strongly suggests the microwave becomes a surface-wave mode. Then, we observed the optical emission mode of the plasma from the end of the plasma chamber at various $O_2$ pressures. The results of the optical emission mode number are plotted in Fig.5 as a function of the pressure, where the four window materials are compared. In the figure, the observed total mode numbers are plotted and each specific mode number (m,n,l) is also labeled. From the theoretical dispersion for the surface-wave eigen mode[2], the total mode number is just inversely proportional to the electron density at the constant permittivity,

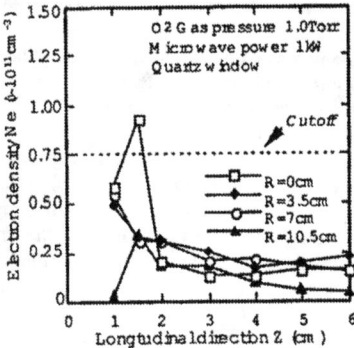

**Figure 4(a).** Quartz window

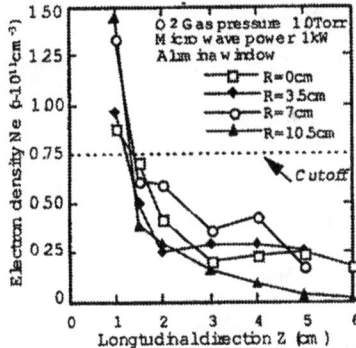

**Figure 4(b).** Alumina window.

while at the constant electron density the mode number decreases with the permittivity of the window material. The experimentally observed gas pressure dependence of the electron density decreased monotonically as the pressure is increased in all four window conditions. Therefore, the observed modes are qualitatively consistent with the theoretical dispersion [2] at the pressures below 2 Torr pressure in the higher permittivity windows of alumina and aluminitrides and below 0.3 Torr in the quartz. Thus we conclude that the surface-wave plasma is produced in the higher permittivity window. However, in higher pressures the mode may be affected by the collisions even in these conditions.

In the quartz, on the other hand, the situation is different, and the mode number is decreased as the pressure is increased above 0.5 Torr. This behavior is not consistent with the pure surface-wave mode[2] for TM, but it may be non-pure surface-wave with a volume mode. A quantitative agreement between the theoretical mode number and the experimental observation is unacceptable. One of the reasons for disagreement may be due to spatially constant density approximation cited in the theory[2], but further detailed study of this point would be a future problem

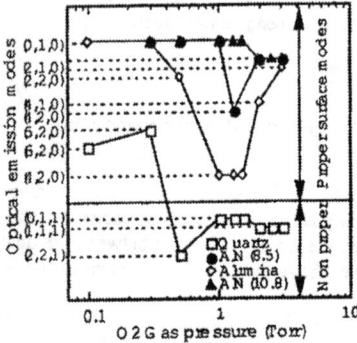

| Figure 5. Optical emission mode. | Figure 6. Resist ashing rates. |

As described above, the higher permittivity window materials produce higher electron density plasma due to the surface-wave, thus leaving a low level microwave filed in plasma volume. The plasma thus produced becomes so-called a downstream plasma, and this provides us a powerful tool for silicon wafer processes and other material processing. In this work, a resist ashing process are examined as one of applications of surface-wave plasma.

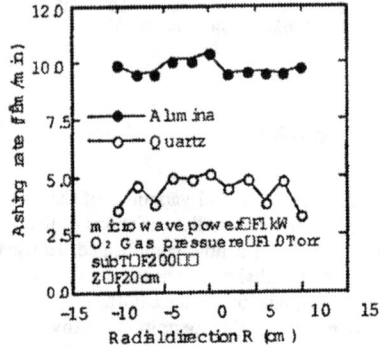

In Fig.6, the resist ashing rates are compared for the two cases of the quartz and alumina, where the radial variations of the rates obtained at 1 Torr $O_2$ pressure are plotted. A difference is clear, and the rate obtained with the alumina is almost two times higher than with the quartz, and it is obvious that this high ashing rate is originated from the high density plasma described above.

A wafer damage induced by ashing process is also important, and therefore a damage analysis of the silicon wafer after ashing was also made by a deep level transient spectroscopy (DLTS) in this work. In Fig.7, a typical example of the DLTS measurement is shown, where the depth profile of trap level density is compared for the alumina and quartz windows. In the figure, the data obtained at the several axial distances are also compared. The trap level density Nt observed in the quartz window becomes high in all axial positions, insisting a relatively higher wafer damage. On the other hand, in the alumina the trap level density is fairly reduced and its value is almost two orders lower than in the quartz, in spite of the higher ashing rate as shown in Fig. 6. This level of the trap density obtained in the alumina window can be seen tolerably low. In Fig.8, the trap level density normalized by the carrier density is plotted for the axial distance. In the figure, the energy of the trap level from the conduction band is also shown. The normalized trap level density is lowered as the distance is increased, and it becomes as small as 1% or less in the downstream below Z=15 cm in the case of alumina, while in the quartz it is still high level even in the downstream. Thus we conclude that the high rate and low damage resist ashing is possible by employing the higher permittivity window materials like alumina and aluminitride, and it should be emphasized that this excellent ashing performance is brought about by the surface-wave plasma.

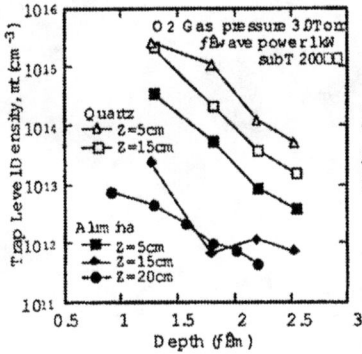

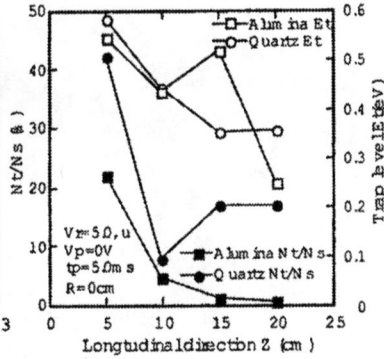

Figure 7. Depth profile of trap level density.　　　　Figure 8. Normalized trap level density.

## 4. CONCLUSION

Microwave oxygen plasma was studied by focusing an attention of the permittivity of microwave entrance window materials. The plasma produced by employing higher permittivity window material like alumina showed a fairly high density, and this is found to be due to surface-wave mode. As one typical application of this surface-wave plasma produced with high permittivity window, a resist ashing was examined, and results showed that the plasma could provide an excellent downstream and a high rate resist ashing was possible with a low damage.

References

1. Nagatsu M., Xu G., Yamage M., Kanoh M. and Sugai H. Jpn. J. Appl. Phys., 1996,35, L341.
2. Ghanashev I., Nagatsu M. and Sugai H. Jpn. J. Appl. Phys., 1997, 36, 337.
3. Zakrezewski Z.and Moisan M.: Microwave Discharges, Eds. C.M.Ferreira and M.Moisan (Plenum Press, New York, 1993) p. 117.
4. Furukawa M., Kawamura K.and Shindo H. Jpn. J. Appl. Phys. 1998, 37, L1005.

# LARGE MICROWAVE REACTOR FOR THIN FILMS DEPOSITION

## J.Marec, C.Boisse-Laporte, N.Benissad
Laboratoire de Physique des Gaz et des Plasmas*, B.212, Université Paris-Sud, 91405 Orsay, France

**Abstract.** A large microwave plasma reactor (2.45 GHz) based on surface wave principle is operated in HMDSO / $O_2$ mixtures for the deposition of $SiO_2$ –like films and plasma polymerized films. Plasma composition and film structure are investigated as functions of the mixture composition, the microwave power and the pressure via EOS, FTIR and mass spectrometry. Relative concentration of HMDSO is a key parameter for the process yielding vitreous films at low values and plasma-polymerized ones at high values. Importance of the substrate bias is clearly shown. It both increases the deposition rate and improves the quality of vitreous films (less carbonated and voids).

## 1. INTRODUCTION

There are increasing needs of large volume and surface plasma material processing, e.g., surface treatment, hard coating for abrasion and UV degradation protection, cleaning ... Microwave discharges are good candidates to solve these problems as they provide high power density, thus high ionization degree ($10^{-5} - 10^{-3}$), hence high production of active species.

We below report the performances of a large microwave plasma reactor which has been already described elsewhere [1] in HMDSO / $O_2$ mixtures with the main objective of $SiO_2$ – like thin film deposition. Nevertheless, possibilities of polymerized-plasma coating are discussed. First, characteristics of plasma reactor are presented, secondly the film characteristics and the influence of plasma and process parameters are discussed. Finally, two coating applications briefly reported.

## 2. CHARACTERISTICS OF THE PLASMA REACTOR

### 2.1. Plasma reactor

Plasma reactor device is shown in Fig.1. It is based on surfaguide excitation [2]. Microwave power is coupled by two surfaguides supplied by two magnetron sources (at 2.45 GHz) needed to maintain a discharge filling the tube in molecular gases at medium pressures. In such a large structure ( 12 cm inside

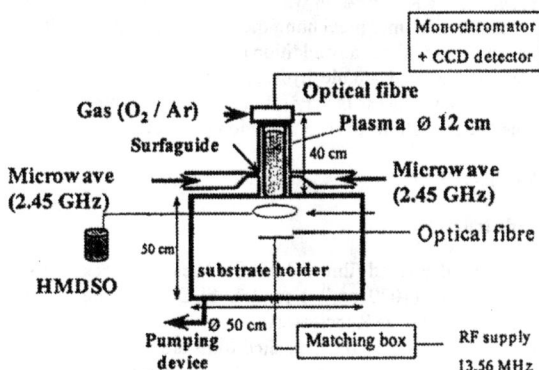

Figure 1. Plasma reactor design and diagnostics

diameter), it is expected 5 plasma modes are excited but only the dominant m = 3 azimuthal mode is excited. As shown in Fig.1, argon and oxygen gases are introduced by the top of the discharge tube whereas the monomer (HMDSO) is introduced just below the discharge end by a ring. Substrate holder can be RF biased. Active species are controlled by EOS from the discharge top or close to the substrate (providing information on the deposition mechanism).

## 2.2. Plasma homogeneity

It is of major importance for thin film deposition that the reactor is producing a plasma as homogeneous as possible. This structure has been tested from this point of view in an Ar / O$_2$ (2:98) mixture [3]. Therefore only the main features are recalled. From probes measurements, it is found that T$_e$ is homogeneous over the plasma section whereas n$_e$ is azimuthally quite constant and obviously radially decreasing from the axis to the wall (typical value is 1.5 x 10$^{12}$ cm$^{-3}$ at 100 mTorr with a 1200 W microwave power). Concerning the homogeneity of excited states, it strongly depends on the pressure and on the local electric field as they are created by the energetic electrons. Therefore it is rather good below 0.1 Torr. Actinometry shows the oxygen atom density is quite homogeneous whatever the density.

## 2.3. Plasma characteristics in HMDSO / O$_2$ mixtures

We first recall the structure of HMDSO (hexamethyldisiloxan): (CH$_3$)$_3$ – Si – O – Si – (CH$_3$)$_3$. EOS measurements show excited species and radicals are both found in the discharge and in the diffusion chamber (except for CO*). In the discharge, the HMDSO fragmentation is produced by electron impact and by oxygen species (O, O$_2$, O$_2^+$, O$_2^*$). In the diffusion chamber it is mainly produced by long life species as molecular metastable oxygen O$_2$(a$^1\Delta$) and atomic oxygen in ground state O$^3$P.This fragmentation yields excited atoms as Si*, C*, H* and radicals as SiO* and CH*. Recombination of the above species yields H$_2$* and C$_2$* while reactions of monomer fragments with O yields OH*. Species concentrations are found only depending on the monomer fraction R = [HMDSO] / [HMDSO + O$_2$].

The concentrations of species are studied via the emitted lines intensities (as shown in Fig.2, we plot the I$_{X^*}$ / I$_{Ar}$ ratios where I$_{Ar}$ is the 750 nm argon line, this choice is justified in section 4.3). In the case of oxygen atom, the actinometry ratios of the 776 and 844 nm oxygen lines, hence [O], are decreasing as R is increasing. Figure 2 shows three different regimes for the carbon containing species:

- Fragmentation regime, for R < 50% (production of C*, CH*, H*), owing to [O] and n$_e$ are high enough to dissociate the monomer.
- Middle regime, for 50% < R < 80%, in which the hydrocarbon radicals can recombine to produce C$_2$* (due to the lack of energy to further dissociate hydrocarbons).
- Weak fragmentation regime, for R > 80%, in which CH*, C$_2$*, slightly decrease. In that cm$^{-1}$case the [O] is to weak to oxidize the monomer preventing the recombination of hydrogenated and carbonated species, e.g., C quite disappears without any addition of O.

Concerning oxygenated species and atomic hydrogen, the concentrations reach a maximum value close to R = 20% as the dissociation process is very efficient due to the high amount of oxygen species in the discharge. Therefore the concentrations of recombination products as OH and CO increase.

## 3. FILM DEPOSITION PROCESS

### 3.1. Film composition and structure

Whatever the R value, SiO$_x$C$_y$H$_z$ deposited films exhibit Si-O features: rocking (450 cm$^{-1}$), bending (800 cm$^{-1}$) and asymmetrical stretching (1070 cm$^{-1}$ ) modes. Si-O stretching mode is dominant and its absorption and frequency both decrease as R increases.

At high R (typically R =80% as in Fig.3), the deposited films are the most carbon containing ones, they have a structure related to the initial monomer owing to the fragmentation is low. IR spectrum of Fig.3 exhibits CH$_x$ stretching modes (2877-90 cm$^{-1}$), CH$_2$ bending vibration (1460 cm$^{-1}$), CH$_3$ bending mode (1411 cm$^{-1}$), Si-CH$_2$ - Si bending band (1358 cm$^{-1}$), Si(CH$_3$)$_2$, Si(CH$_3$)$_3$ rocking bands (798 cm$^{-1}$ and 841

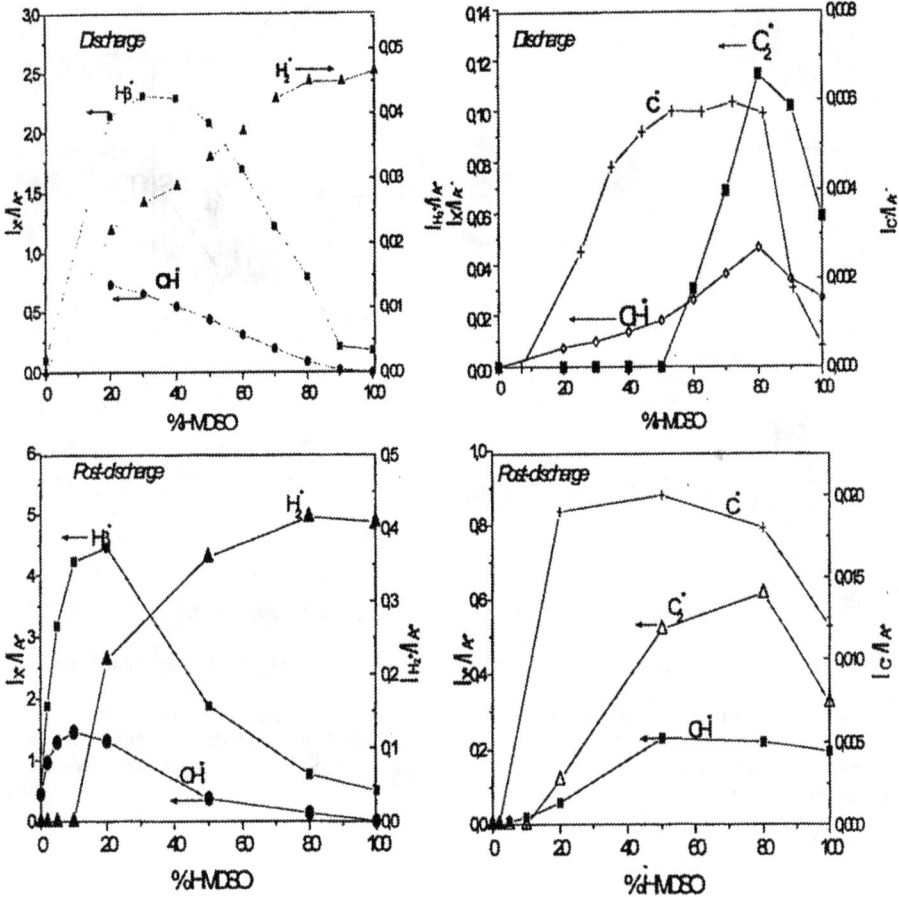

Figure 2. Line intensities of excited species normalized to 750 nm argon line both in and downstream the discharge

cm$^{-1}$) and SiH stretching vibration mode (2200 cm$^{-1}$). Presence of Si(CH$_3$)$_2$ and Si(CH$_3$)$_3$ groups shows the film chains are longer than in the monomer as HMDSO has been polymerized. Further, SiH bonds can explain the low OH* emission as they can prevent the OH absorption.

At low R (typically R = 20% as in Fig.3), deposited coatings result in vitreous silica-like structures thanks to the enhancement of the monomer dissociation in the gas phase and thanks to the oxidative reactions (due to the excess of oxygen). In that case, only Si-OH bonds are found besides Si-O ones.

## 3.2. Deposition rate

As shown in Fig.4, the deposition rate exhibits a maximum for R = 20% which corresponds to the range order of other microwave reactors and to a fast deposit compared to RF reactors. Such a behavior has also been observed in TEOS / O$_2$ mixtures [4]. The decrease of the deposition rate as R is increasing can be

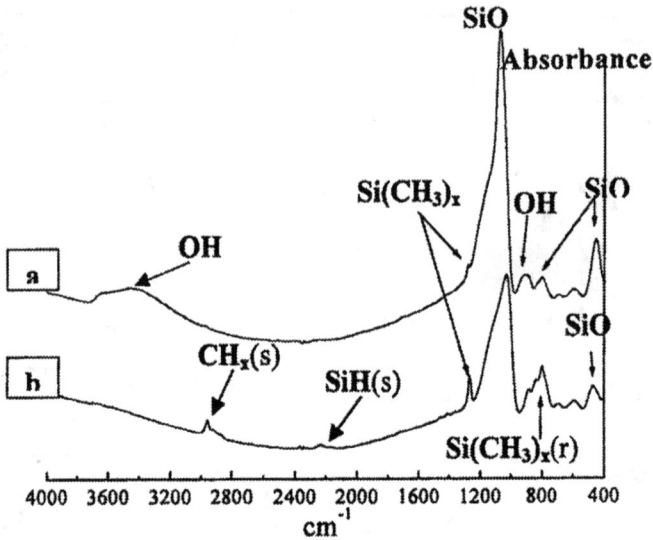

**Figure 3.** Infrared Spectra – $P_{MW}$ = 1200 W – p = 80 mTorr – (a) R = 20% - (b) R = 80%

explained by the decay of the oxygen concentration (cf. II .2) as the oxygen strongly contributes to the monomer dissociation in the gas phase.

Finally, we can say that the monomer is not fully dissociated without oxygen addition (or with only a few of oxygen) and the film is a "plasma-polymerized" type one. At the opposite, a strong dilution of the monomer in the oxygen favors the depletion of hydrocarbon depletion and high deposition rates but the deposit remains porous whatever the oxygen addition.

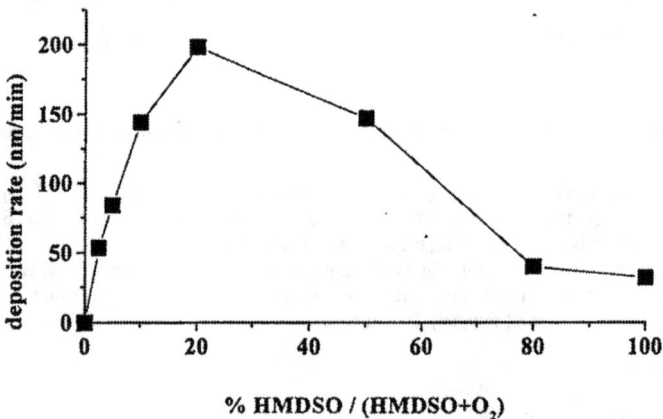

**Figure 4.** Deposition rate vs. HMDSO concentration – $P_{MW}$ = 1200 W – p = 80 mTorr

# 4. INFLUENCE OF PLASMA PARAMETERS ON THE DEPOSITION PROCESS

## 4.1. Influence of the microwave power

It has been already shown that increasing the microwave power in pure argon results in a higher electron density and a linear dependence of the 750 nm argon emission line which is thus caused by the increase of the electron density [5]. Hence, the intensity of this line can be regarded as a "gauge" of the electron density and is used for the normalization of the lines intensities of the other species. Moreover, for an excited state $X^*$ mainly populated by direct electron impact on the X ground state, the $I_{X^*} / I_{Ar^*}$ ratio is proportional to [X]. In Ar / $O_2$ mixtures, the increase of the $I_{O^*} / I_{Ar^*}$ with the power shows the higher the electron density, the higher the oxygen dissociation. A similar behavior is also observed in HMDSO / Ar mixtures. In HMDSO / $O_2$ mixtures, under the typical conditions R = 20%, 200 sccm at p = 80 mTorr, we find the argon line intensity $I_{Ar^*}$ quasi-linearly increases with the microwave power both in and downstream the discharge (Fig.5) while the $I_{O^*} / I_{Ar^*}$ ratio remains constant. It means the resulting excess of O is consumed for the monomer fragmentation and oxidation as it is shown by a higher production of oxygenated species OH and CO.

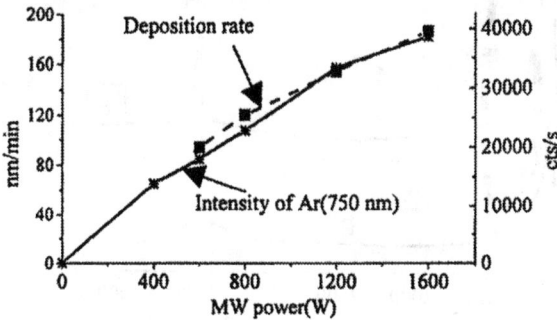

Figure 5. Intensity of 750 nm Argon line and Deposition rate vs. the microwave power – p = 80 mTorr – R = 20%

Concerning the film, the deposition rate increases with the microwave power (Fig.5) owing to the increase of the monomer dissociation thanks to the inelastic collisions with the electrons increasing the production of reactive fragments and radicals. Nevertheless, the FTIR spectra shown in Fig.6 show the film structure does not really depends on the microwave power. These spectra exhibit the three characteristic peaks of the Si-O-Si bond and the signature of the silanol groups between 3400 and 3600 $cm^{-1}$. Moreover, these films are carbonated as it is shown by the 1260 $cm^{-1}$ peak, signature of the symmetrical bending of $CH_3$ in $Si(CH_3)_x$ and by the stretching peak of $CH_x$ at 2900 $cm^{-1}$. Nevertheless, spectra of Fig.6 also show the absorbance of Si-O bond is increasing with the power, hence the density of siloxan bonds. Simultaneously, increasing the power decreases OH and carbon inclusion in the film. These results can be explained by two ways: increasing the power creates more CH radicals in the gas, probably meaning that the Si-CH₃ bonds of the monomer are broken with a corresponding higher density of the siloxan radicals depositing on the substrate – increasing the power increases the monomer fragmentation in the gas phase preventing the inclusion in the film both of the methyl groups linked to Si atoms and hydrocarbons $CH_x$. Heterogeneous reactions between oxygen and the surface (higher with the power increase) can also result in a decay of carbon inclusion. P-etch measures show the film is porous whatever the microwave power, only the grain size is larger at lower power.

We finally can conclude that the microwave increase improves the deposition rate but does not really change the film structure.

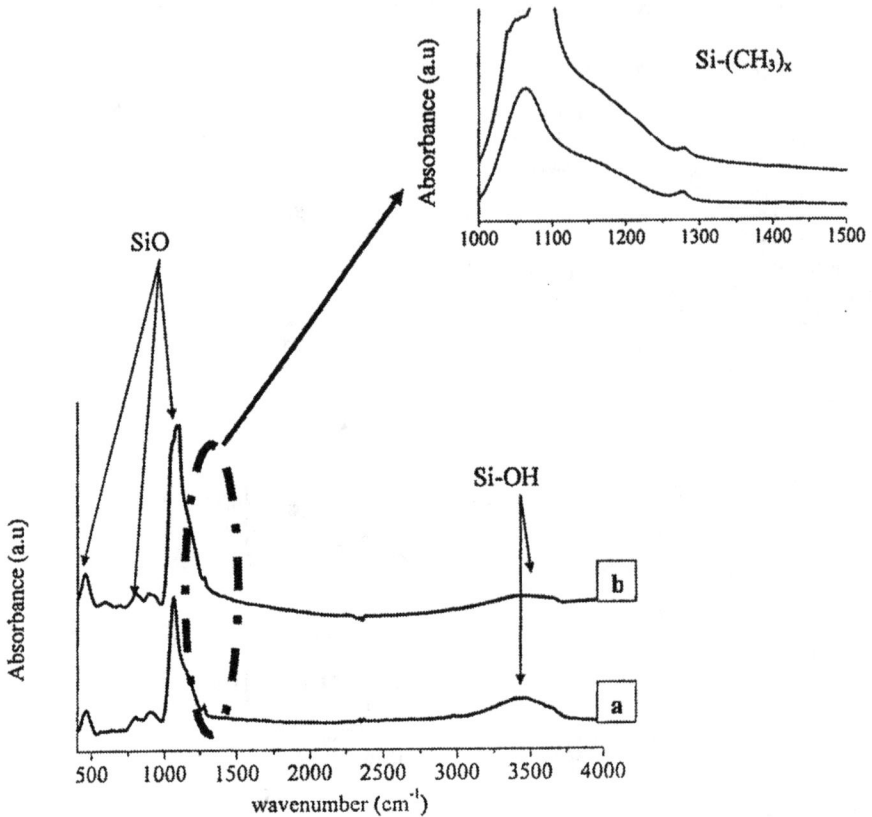

**Figure 6.** FTIR spectra for two microwave powers – (a) $P_{MW}$ = 600 W – (b) $P_{MW}$ = 1600 W – p = 80 mTorr – R = 20%

## 4. 2. Influence of the pressure

Pressure acts on electron temperature and density which are both decreasing as same as the mean free path for the electrons-neutrals collisions while the residence time is increasing. Mass spectrometry measurements show that the number of light ions is decreasing ($H^+$, $H_2^+$, $C^+$, $Si^+$, $CO^+$) whereas that of heavy ions is increasing (monomer fragments as $OSiCH_{2, 3}^+$, $Si(CH_3)_2^+$, $OSi(CH_3^+)$, $OSi_2(CH_3)_4^{++}$ and specially $(CH_3)_3 SiOSi(CH_3)_2^+$ whose the intensity increases by a factor 7 between 200 mTorr and 1 Torr – fragments resulting of recombination between HMDSO fragments as $HSiCH_3^+$, $H_2SiCH_3^+$, $SiOH^+$ or $HSi(CH_3)_2^+$). Formation of white powder is observed at pressures higher than 400 mTorr with R = 20% and at 1 Torr, the deposit is not adhesive. Such powder is due to the formation of free radicals requiring a small mean free path and a long residence time. Under such conditions, nucleation in the gas phase and film deposition processes are competing.

As expected, the deposition rate increases as the pressure increases but the film quality is worth (more carbonated and less dense). So it can be concluded that R = 20% and p = 80 – 100 mTorr are adapted conditions to have both a fast deposit and a relatively good quality of $SiO_2$–like films.

## 5. INFLUENCE OF THE RF BIAS

RF bias creates a plasma close to the substrate but does not change the microwave plasma itself as shown by the emission spectra whose line intensities remain the same whatever the RF power.

Concerning the film, spectra of Fig.7 show a strong decay of the absorption band of Si-OH (between 3400 and 3650 cm$^{-1}$). Contribution of bonded silanol groups markedly decreases, the dominant 3400 cm$^{-1}$ peak in the case without RF bias quite disappears at $P_{RF} = 80$ W. This peak is the signature of hydroxile groups, forming H bonds between them, which can be included after air exposure. Simultaneously a new absorption band appears for R > 50% which is attributed to the stretching vibration band of Si-H (2258 cm$^{-1}$) in $O_3SiH$ environment. Moreover, all layers are containing carbon as proved by the absorption peak of $Si(CH_3)_x$ at 1260 cm$^{-1}$.

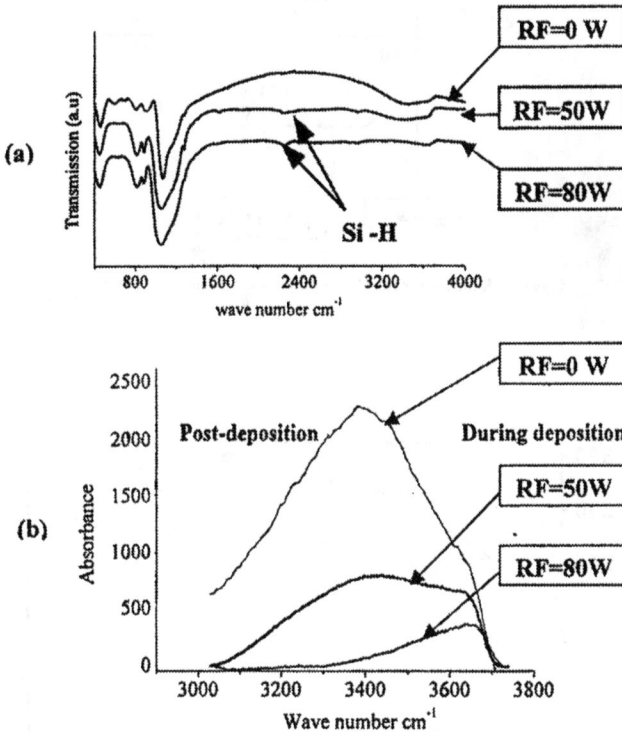

Figure 7. FTIR spectra – (a) Si-H absorption band appearing with RF power – (b) Decay of the Si-OH absorption band P = 80 mTorr – $P_{MW}$ = 1200 W

Refractive index also yields insights on the film quality. It is found close to 1.44 without RF much lower the index of the thermal silica whereas it strongly increases with the RF power, reaching 1.56 at 80 W (Fig.8). This increase can be due both to a densification of the layer and to SiH groups (index of a-Si:H material is 3.6). Etching rate of layers obtained with RF bias is about 2 – 3 times higher than in thermal silica, it is characteristic of weakly porous films whereas this rate was about 10 times higher without RF.

Deposition rate increases from 200 to 320 nm / mn from $P_{RF} = 0$ to 50 W with R = 20%, $P_{MW} = 1200$ W at p = 80 mTorr (Fig.9 – Recent results not plotted here show the deposition rate is continuously increasing up to 330 nm / mn). The film is generally more dense with RF bias, hence the coating are less thick for the same deposition time than without bias. Moreover, $O_2^+$ ions can extract $H_2O$ molecules and

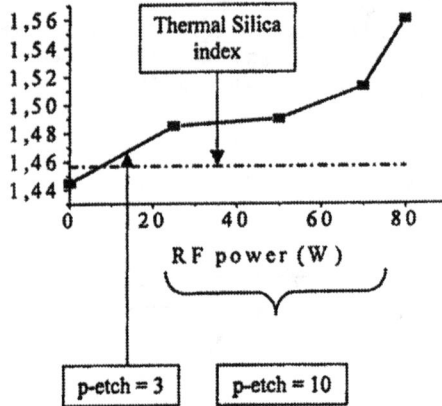

**Figure 8.** Refractive index vs. RF power – p = 80 mTorr – R = 20%

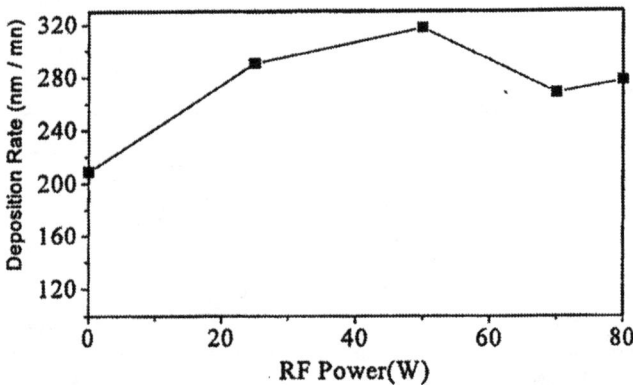

**Figure 9.** Deposition rate vs. RF power – $P_{MW} = 1200$ W – p = 80 mTorr – R = 20%

etch the organic part again decreasing the coating thickness. Such an etching process can also be performed by neutrals having got enough energy from ions as H atoms which are able to etch the silica.

259

However ion bombardment favors the creation of absorption sites and the RF plasma increases the monomer dissociation, hence it increases the density of radicals precursors of the deposition leading to an enhancement of the deposition rate. Therefore, we can say the volume reactions are the dominant processes in our reactor.

We can summarize the effects of the RF bias: less silanol groups and new bonds in the film, hence higher refractive index and more dense coating and especially a higher deposition rate which no more exhibits a peak at R = 20% but continuously increases with R (Fig.10).

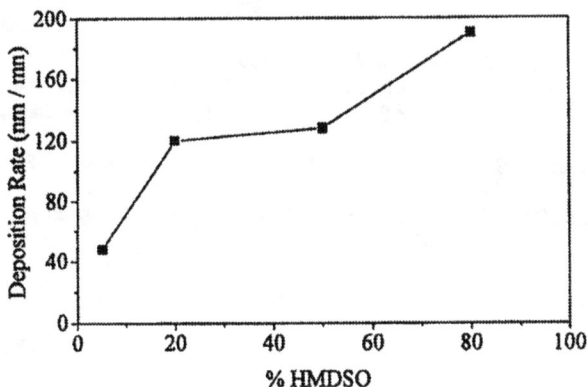

Figure 10. Deposition rate vs. HMDSO concentration with RF Power – p = 80 mTorr – $P_{RF}$ = 50 W – $P_{MW}$ = 1200 W

## 6. EXAMPLES OF APPLICATIONS

Accounting for their properties of hardness, transparency, resistance to abrasion, $SiO_2$ – like films have been used for:
- coating of polycarbonate car lenses (protection against abrasion and UV degradation) via a multi-steps process along which the layers are progressively going from plasma-polymerized one to $SiO_2$ – like.
- anti-reflection layers of glass lenses
- food packaging (anti-oxidation layers)

Recently, plasma-polymerized coating has been used for iron corrosion protection. Films have deposited at R = 80%, they exhibit a good corrosion resistance (> $10^8$ Ω) with 1-4 μm thickness at a high deposition rate ( ~ 100 nm/mn). The layers are strongly cross-linked, dense with only a few voids and well adhesive over metallic surfaces. A special application is the confinement of toxic compounds in iron containers.

## 7. CONCLUSION

We have used a surface wave type reactor accounting for the good homogeneity of active species and radicals produced both in and downstream the discharge. HMDSO has been chosen as a precursor instead of Silane as it does not requires so cautious operating conditions or TEOS whose the vaporization temperature is higher. It is found that all the monomer bonds are broken in the discharge unlike the Si-O bonds. Kinetics depends on R: a high fragmentation regime is observed at low [HMDSO] (O in excess) and a low dissociation one at high [HMDSO].

Obviously, the film nature depends on the [HMDSO]: inorganic ($SiO_{1,9}$) in rich oxygen mixtures and $SiO_xC_yH_z$ type in rich monomer mixtures. Moreover the deposition rate is high whatever the [HMDSO]. The deposition process can be controlled in situ by EOS close to the substrate (detection or not of CH*, C* and H* reveals if the coating is or not organic). The microwave power plays a quantitative role for the production of species in the plasma and by decreasing the OH bonds content (a strong porosity is observed below 600 W). Finally, in rich oxygen HMDSO / $O_2$ plasma, RF bias of the substrate holder enables high deposition rate ($\sim 300$ μm/mn) of dense and good homogeneity films. Therefore we think this reactor is a well adapted tool for coating applications, a typical set of parameters for silica-like film deposition being P = 1200 W, P = 50 W, p = 80 mTorr, R = 20%.

# References

1. Boisse-Laporte C., Benissad N., Bechu S. J.Phys IV, 1998, **8**, 187
2. Bechu S., Boisse-Laporte C., Leprince P., Marec J. J.Vac.Sc.Tech., 1997, **A15**, 668
3. Bechu S., Boisse-Laporte C., Leprince P., Marec J. 12th Int.Symp. on Plasma Chemistry, Minneapolis 1995, 303
4. Bogart K.H.A., Ramirez S.K., Gonzales L.A., Bogart G.R., Fisher E.R. J.Vac.Sc.Tech., 1998, **A16**, 3775
5. Bluem E., Bechu S., Boisse-Laporte C., Leprince P., Marec J. J.Phys.D:Appl.Phys., 1995, **28**, 1529

* Laboratory associated to CNRS – UMR 8578

# DEPOSITION OF CARBON MATERIALS IN THE ECR APPARATUS

**Osamu Matsumoto, Ken-ichi Itoh, Yukinori Takahashi**

Department of Chemistry, Aoyama Gakuin University, Tokyo 157-8572, Japan

**Abstract.** The effects of bias voltage, pressure, and addition of hydrogen to methane on the deposition of the carbon material films from methane plasma prepared in the ECR discharge were investigated. The transparent and semi-conducting carbon film was deposited on the substrate negatively biased at the voltage of -50 V, though the deposit was polyethylene-like without application of negative bias voltage. With increasing pressure, the deposit changed from the transparent and semi-conducting polyene-like film to polyethylene-like film. The film changed from transparent and semi-conducting to polyethylene-like film also with increasing mixing ratio of hydrogen. Since the plasma parameters depend on the discharge parameters, the dependence of plasma parameters on the deposition process was investigated.

## 1. INTRODUCTION

We have studied the preparation of carbon material film using ECR plasma apparatus with pure methane as carbon source at the pressure of $4 \times 10^{-2}$ Pa [1,2]. In this case, the difference between the plasma potential and the floating potential on the wall of the fused silica discharge tube inserted into the discharge chamber reached to -43 V at ECR resonant point and the wall was negatively self-biased. Transparent and semi-conducting polyene-like carbon films were deposited on the substrate placed on the wall. Small amounts of carbonaceous ions formed in the center of the plasma would drift to the wall by the potential gradient and the three dimensions cross-linked polymer film including graphite was deposited.

ECR plasma has the capability to operate at lower pressure than 1 Pa and to create higher plasma density, with corresponding higher degree of ionization [3]. In some cases, following data are given: an ionization density; $10^{12}$ cm$^{-3}$, a degree of ionization; 10 %. We have estimated by means of electrical probe method that the electron temperature (kTe) and the ion density ($n_i$) have the maximum values of around 15 eV and an order of magnitude of $10^{11}$ cm$^{-3}$, respectively, at the ECR resonant point in the center of the cavity [1]. The degree of ionization was estimated as few percent at the ECR resonant point in the center of the cavity [1]. The degree of ionization was also estimated as few percent at the pressure of the order of $10^{-2}$ Pa. CH$_x^+$ and C$_2$H$_x^+$ ions were taken in the plasma by means of quadrupolar mass analysis (QMA) [4,5]. The degree of ionization was also estimated as few percent by QMA. Therefore, the methane plasma prepared in the ECR plasma apparatus at the pressure of $4 \times 10^{-2}$ Pa would be decided as the ECR plasma.

The properties of the deposited films depend strongly on the deposition conditions. The parameters determining properties of deposited films are proposed as the ion density distribution, the bias voltage on the substrate, the gas pressure, and the feed gas compositions [6,7].

Diamond-like carbon films have been prepared in the ECR plasma of pure hydrocarbons at low pressure, with RF power supply to the substrate to apply the negative bias voltage [8-11]. The effect of negative bias voltage on the deposition rate and the structure of the deposits were examined. Moreover, the dependence of deposition process on the negative bias voltage and the gas pressure were investigated. Although the reported number is a little, some papers according to the gas pressure on the ECR plasma parameters have been reported [12,13]. The plasma parameters depended on the gas pressure [14].

In the diamond deposition in the ECR plasma, the carbon sources, such as hydrocarbons, were diluted with much amounts of hydrogen, though the pure hydrocarbons were used as carbon source in the deposition of diamond-like carbon. Thus the effect of addition of hydrogen to the carbon source on the carbon film deposition has also been investigated.

In the present paper, the effects of several discharge parameters, such as bias voltage on the substrate, gas pressure, and feed gas composition, on the plasma parameters and the properties of the deposited carbon films are investigated.

## 2. EXPERIMENTALS

The apparatus used for the deposition of the carbon films was the ECR apparatus as schematically shown in Fig. 1 [15]. The steel chamber was surrounded by two electromagnets and the magnetic flux intensity is set to meet the electron cyclotron resonance at 80 mm leaving from the fused silica window. The distribution of the magnetic flux intensity is also shown in Fig. 1. Substrates, the fused silica plate and the silicon wafer, were mounted on the stainless steel holder placed at several positions in the fused silica discharge tube inserted into the discharge chamber. The holder was coupled to the RF generator as shown in Fig. 1. The RF power was supplied between the grounded stainless steel discharge chamber and the substrate holder. The substrate on the holder was placed at 150 mm from the fused silica window. A negative bias voltage ($V_b$) from 0 V to $-100$ V vs. the chamber was applied to the substrate. The substrate was also placed on the wall of the fused silica discharge tube inserted into the chamber at the ECR resonant point, where the substrate was negatively self-biased, as reference. After the chamber was evacuated below $1 \times 10^{-3}$ Pa, $CH_4$ was introduced into the chamber and the pressure was maintained at $4 \times 10^{-2}$ Pa at first. Basically, $CH_4$ was selected as plasma gas and the pressure was kept at $4 \times 10^{-2}$ Pa. The pressure was varied from $4 \times 10^{-2}$ Pa to 1 Pa to examine the pressure effect on the deposition of carbon film. To examine the effect of mixing hydrogen into methane on the deposition of carbon film, methane was diluted with hydrogen from pure methane to methane/hydrogen mixture gas in desired ratio. The microwave power of 200 W at 2450 MHz was transferred to the chamber from the magnetron to the discharge gas in the chamber through the fused silica window. The deposition was carried out for the period of 1 h.

During the discharge, optical emission spectra from the plasma were recorded using a monochromator. The double probe technique was used to estimate the electron temperature and the ion density and the Langmuir probe technique was used to estimate the floating potential on the wall of the discharge tube and the plasma potential during the discharge. The ionized species in the plasma were analyzed by quadrupolar mass spectrometer set downstream the gas flow. After the deposition, the deposits on the substrates were characterized by means of several characterization methods, such as UV/vis absorption spectroscopy, ellipsometry, Raman spectroscopy, and IR spectroscopy. The fused silica substrate was used in UV/vis spectroscopy, ellipsometery, and Raman spectroscopy and the silicon wafer substrate was used in IR spectroscopy. The electric conductivity was measured using conventional four probe method.

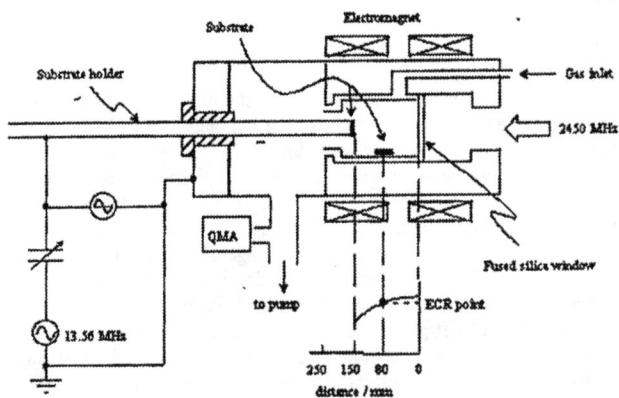

Figure 1. Experimental apparatus for the deposition of carbon materials with ECR.

# 3. RESULTS

## 2.1. Effect of bias voltage

### 3.1.1. Background

The transparent and semi-conducting polyene-like carbon films were deposited on the fused silica substrate that was placed on the wall of the discharge tube at the ECR resonant point. The substrate was negatively self-biased at -43 V. Thus we have carried out the deposition of the carbon film on the substrate, on which the negative bias voltage was applied by the supply of the RF power. The deposits were characterized and the results obtained are compared with those obtained on the self-biased substrate placed on the wall of the discharge tube and the effect of the application of negative bias voltage on the deposits are investigated.

### 3.1.2. Estimation of the setting position and the applied bias voltage on the substrate.

Based on plasma potential and the floating potential [1], the setting position of the substrate was decided to 150 mm downstream the gas flow from the fused silica window (70 mm from the ECR resonant point). At that point, from 0 V to -100 V of the bias voltage vs. the chamber was applied to the holder to compare the deposits to that on the wall of the discharge tube. To investigate the bias voltage effect only on the deposition, the substrate positions were selected in approximately the same temperature of 500±20 K.

### 3.1.3. Characterization of the deposits on the substrate mounted on the holder supplied negative bias voltage

The characters of deposit on the substrate mounted on the negatively biased holder was compared with those of the deposits on the self-biased substrate placed on the wall of the discharge tube at the ECR resonant point. Four types of deposits on the substrates, which were biased between 0 V and -100 V were selected as samples.

The absorption curves of the samples were measured for the wavelength between 800 nm and 200 nm. By the application of bias voltage of -50 V, the absorption curve changed from that of polyethylene, which was the deposit without bias voltage, to the curve that had the absorption maximum at 260 nm as shown in Fig.2. This is similar to that of the deposit on the wall and is assigned due to the $\pi-\pi^*$ transition of butadiene or 1,3,5 hexatriene. The absorption curve of the deposit at the bias voltage of -100 V did not show the maximum. The infrared absorption curves of the deposit on the silicon wafer substrate mounted on the negatively biased holder are shown in Fig. 3. The absorption due to the conjugated double bond was identified at around 1600 cm$^{-1}$ in the deposit biased at -50 V as like in the deposit on the wall, though the absorption intensity was weak. The deposit without bias voltage exhibited the curve due to polyethylene. Two strong peaks were observed in Raman spectra of the deposit on the substrate biased at -50 V as shown in Fig. 4 like graphite or ion impinged hydrocarbon polymer [16]. The spectra were similar to those observed in the deposit on the wall. The peaks were not observed in the deposit without application of negative bias voltage. As results, the deposits on the biased substrates at around -50 V were polyene-like polymers including graphite. On the other hand, the deposit without bias voltage was polyethylene-like polymer and that deposited on the substrate biased at -100 V was assigned as graphite.

The refractive index (n) and the film thickness (d) of the deposits are given in Table 1. The refractive index of the deposit on the substrate negatively biased at -50 V was 2.07 and it corresponded to that of the deposit on the wall (n=1.99). On the other hand, the refractive index of the deposit without negative bias voltage was 1.65. The former value corresponds with that of amorphous carbon formed by means of carbon ion beam method (n=2.02) [17,18]. The latter value corresponds to that of amorphous carbon prepared by means of glow discharge of hydrocarbons (n=1.74) [19]. Moreover, the refractive index of the deposit at the negative bias voltage of -100 V corresponds to the refractive index of graphite of 2.30. The deposits were divided into three types. The deposits on the negatively biased substrate at -50 V was approximately the same with that of the deposit on the wall, which corresponded with that of the deposit by means of carbon ion beam deposition. The deposit on the substrate without negative bias voltage corresponds with that deposited by means of glow discharge of hydrocarbons. Moreover, the deposit changed to graphite with increasing negative bias voltage.

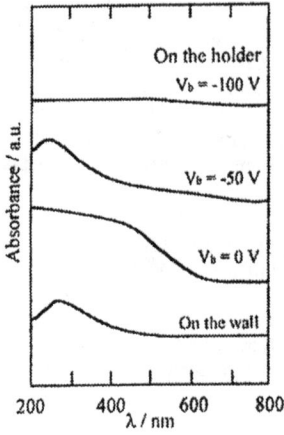

Figure 2. UV/vis spectra of the deposits on fused silica substrates varying bias voltage ($V_b$).

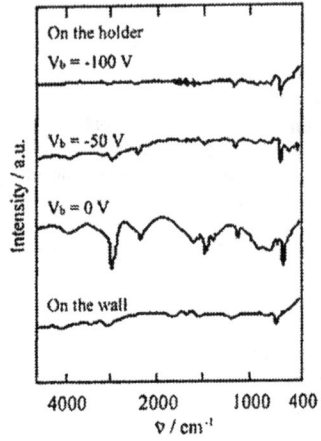

Figure 3. IR spectra of the deposits on silicon wafer substrates varying bias voltage ($V_b$).

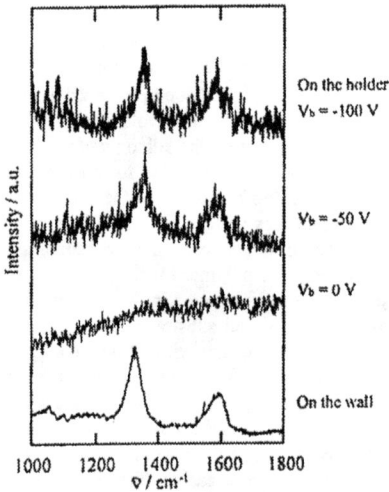

Figure 4. Raman spectra of the deposits on fused silica substrates varying bias voltage ($V_b$)

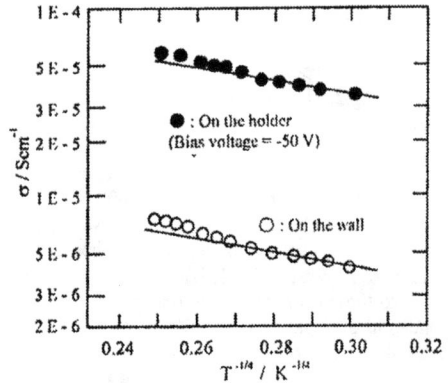

Figure 5. Electric conductivity ($\sigma$) vs. $T^{-1/4}$ for the deposits on fused silica substrates varying bias voltage ($V_b$).

### 3.1.4 Electric conductivity

The electric conductivity of the deposits at room temperature is given in Table 2. The deposit on the substrate without the application of the negative bias voltage was an insulator. When the bias voltage of -50 V was applied on the substrate, the deposit was changed to the semi-conductor that was similar to that deposited on the wall. In the temperature dependence of the electric conductivity, the conductivity decreased with decreasing temperature at low temperature. As the three dimensional variable hopping is dominant in the plot of the electric conductivity ($\sigma$) vs. $T^{-1/4}$ as shown in Fig. 5. Deposits may contain small amounts of conducting materials, such as graphite and be consisted of a three dimensions cross-linking network [20]. At higher negative bias voltage, the deposit changed to the conducting material such as graphite.

Table 1. Refractive index (n) and film thickness (d) of the deposits varying applied bias voltage

| Substrate position | Bias voltage / V | n | d / nm |
|---|---|---|---|
| On the holder | 0 | 1.65±0.01 | 125±9 |
| | -50 | 2.07±0.03 | 136±2 |
| | -100 | 2.30±0.01 | 148±5 |
| On the wall | -43 | 1.99±0.05 | 114±9 |

Table 2. Electric conductivity at room temperature of the deposits varying applied bias voltage ($V_b$)

| Substrate position | Bias voltage / V | Electric conductivity / Scm$^{-1}$ |
|---|---|---|
| On the holder | 0 | less than $1\times10^{-7}$ |
| | -50 | $7.4\times10^{-4}$ |
| | -100 | $1.2\times10^{-2}$ |
| On the wall | -43 | $6.9\times10^{-3}$ |

## 3.2. Effect of pressure [14]

### 3.2.1 Background

Although the reported number is a little, some papers according to the effect of pressure on ECR plasma are reported. Pelletier reported that the electron temperature and the potential decreased with increasing pressure [12]. Moreover, Asmussen found that, when gas pressure is increased, there is a transition from ECR heating to collisional heating of the electron gas in the investigation of advantage of the ECR plasma [13]. The effect of the pressure on the plasma parameters and the properties of the deposits in the ECR plasma are investigated.

### 3.2.2 Dependence of the deposits on the pressure

The plasma parameters were affected with pressure as mentioned above. On the other hand, we reported that unique transparent and semi-conducting carbon film was obtained only on the substrate, on which the bias voltage of –50 V was applied at the pressure of $10^{-2}$ Pa. Thus the dependence of properties of deposit on the pressure was investigated in the deposition of carbon film on the substrate applied the bias voltage. Two kinds of samples, which were mounted on the holder negatively biased at – 50 V and on the negatively self-biased wall of the discharge tube, were investigated.

The UV/vis absorption curves measured for two kinds of samples which were deposited on the holder and the wall in the plasma varying pressure are shown in Fig. 6. In both kinds of samples, approximately the same results were obtained. The deposits at the pressure of $10^{-2}$ Pa exhibited $\pi$–$\pi^*$ transition due to the conjugate double bond of polyenes at around 260 nm. With increasing pressure, the absorption at 260 nm disappeared and the curve changed to that due to polyethylene-like polymer. This tendency toward the change of deposits with increasing pressure was also identified by means of infrared absorption

266

spectroscopy. Raman spectra of the deposits changed from the spectra of polyene-like polymer including graphite to the spectra of polyethylene-like polymers with increasing pressure. The refractive index of both samples deposited varying pressure was measured with ellipsometry. The refractive index is given in Table 3 with the thickness of the deposits. The refractive index decreased with increasing pressure and reached to that of polyethylene-like polymer at the pressure of 1 Pa. The deposits at both positions at $10^{-2}$ Pa were the semi-conductors with the conductivity of around $10^{-5}$ Scm$^{-1}$ as given in Table 4. With increasing pressure, the deposit changed from semi-conductor to insulator.

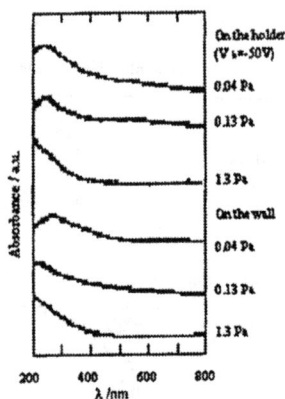

Figure 6. UV/vis spectra of the deposits on fused silica substrates varying bias voltage ($V_b$).

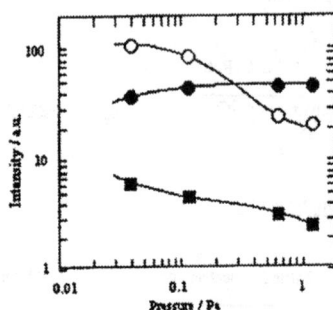

Figure 7. Normalized intensities of band head and line head in the plasma vs. pressure. (o : CH, • : H, ■ : CH$^+$)

Table 3. Refractive index (n) and film thickness (d) of the samples deposited varying pressure

| Substrate position | Pressure / Pa | n | d / nm |
|---|---|---|---|
| On the holder | 0.04 | 2.07±0.05 | 136±2 |
| | 0.13 | 1.99±0.01 | 122±3 |
| | 1.3 | 1.58±0.05 | 130±5 |
| On the wall | 0.04 | 1.99±0.05 | 114±9 |
| | 0.13 | 2.01±0.09 | 154±3 |
| | 1.3 | 1.56±0.01 | 159±9 |

Table 4. Electric conductivity of the samples deposited varying pressure

| Substrate position | Pressure / Pa | Electric conductivity / Scm$^{-1}$ |
|---|---|---|
| On the holder | 0.04 | $7.4\times10^{-4}$ |
| | 0.13 | less than $1\times10^{-7}$ |
| | 1.3 | less than $1\times10^{-7}$ |
| On the wall | 0.04 | $6.9\times10^{-5}$ |
| | 0.13 | less than $1\times10^{-7}$ |
| | 1.3 | less than $1\times10^{-7}$ |

### 3.2.3  Pressure effect on the plasma parameters

In optical emission spectra, several species due to dissociation of methane were identified. The dependence of the band and line head intensities on the pressure is shown in Fig. 7. CH radicals and CH$^+$ ions decreased with increasing pressure, while the intensity due to hydrogen atom was almost independent with pressure. Large amounts of methane molecules could dissociate and ionize at lower pressures. By means of electric double probe methods, the dependence of electron temperature (kTe) and the ion density ($n_i$) in the plasma on the pressure is studied. Results obtained are shown in Fig. 8. The kTe considerably decreased with increasing pressure. The $n_i$ decreased at around 1 Pa. As a result, the degree of ionization decreased with increasing pressure. The dependence of the plasma potential and the floating potential measured by means of electric single probe method on the pressure is shown in Fig. 9. The plasma potential decreased and the floating potential increased with increasing pressure. As a result, the difference between the plasma potential and the floating potential, which corresponds to the self-bias voltage applied on the samples placed on the wall of the discharge tube.

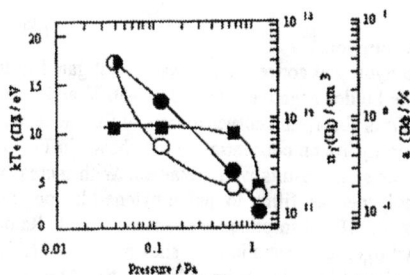

**Figure 8.**  Electron temperature (o), ion density (■), and and degree of ionization (●) vs. pressure.

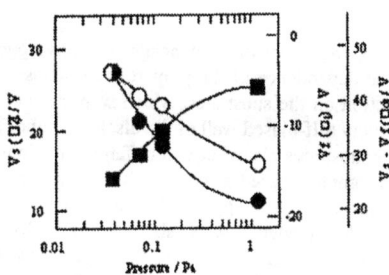

**Figure 9.**  Plasma potential (o), floating potential (■), and between plasma potential and floating potential (●) vs. pressure.

The mean power absorbed by an electron in the high frequency discharge, P, is given by

$$P = [ n_e e^2 E^2 / 2 m_e v_e ][ v_e^2 / (v_e^2 + \omega_e^2 )] \tag{1}$$

(where $n_e$ = electron density, e = charge on an electron, E = electric field strength, $m_e$ = mass of electron, $v_e$ = electron collision frequency, and $\omega_e$ = excitation frequency). At very low pressure, the mean free path of electron-neutral and electron-ion collisions becomes very low. Thus the collision frequency was considerably lower than the discharge frequency, $v_e \ll \omega_e$, and the equation (1) becomes

$$P = [ n_e e^2 E^2 / 2 m_e v_e ][ v_e^2 / \omega_e^2 ] \tag{2}$$

Since the collision frequency, $v_e$, is directly proportional to pressure, the lowering of pressure gives the increasing of the electric field strength, E, even the P of the same value is applied. With increasing pressure, $v_e$ increased and E decreased. Therefore, the high degree of ionization, that is the characteristic of ECR plasma, would be kept at high electric field strength at lower pressures.

With varying pressure, the plasma parameters were changed. The electron temperature and the degree of ionization decreased with increasing pressure. This exhibited that the collision between particles in the plasma increased with increasing pressure and positive ions were decreased by the recombination with electrons. Since the plasma potential and floating potential decreased with increasing pressure, the difference between the plasma potential and floating potential, which corresponded to the self-bias voltage, was decreased.

By comparison between the results in the plasma diagnostics and the properties of deposits in the methane plasma prepared using ECR apparatus, the impingement of larger amounts of ions on the negatively biased substrate would contribute to the deposition of the unique transparent and semi-conducting carbon films at lower pressures. Both plasma conditions, higher degree of ionization and self-bias voltage, were obtained in the ECR plasma prepared at the pressure as low as the order of magnitude of $10^{-2}$ Pa.

## 3.3. Effect of hydrogen [7]

### 3.3.1 Background

Hydrogen tends to exclude the $sp^2$ structure from the film and improve the $sp^3$ to $sp^2$ ratio, thereby enhancing the diamond-like behavior. In this case, hydrogen can participate chemically in the gas phase reaction as well as in the reaction with the film deposition. Especially, this is clearly identified in the diamond deposition from hydrocarbon/hydrogen plasmas with hydrocarbons as carbon sources. The effect of the hydrogen to methane ratio on the deposition of carbon films in the ECR plasma is briefly studied.

### 3.3.2 Dependence of the deposits on the hydrogen concentration

The dependence of the properties of deposits on the hydrogen concentration was investigated in the deposition on the substrates, which were mounted on the holder negatively biased at –50 V and on the negatively self-biased wall of the discharge tube. The UV/vis absorption curves measured for the deposits on the substrates placed at two different positions varying hydrogen concentration are shown in Fig. 10. In the samples placed at both positions, approximately the same results were obtained. With increasing hydrogen concentration, the deposits changed from polyene-like films to polyethylene-like polymer films. This tendency was also identified by means of IR absorption spectroscopy and Raman spectroscopy. Raman spectra of the deposits varying hydrogen concentration are shown in Fig. 11. The change of deposits from polyene-like polymer to polyethylene-like polymer was also identified by means of Raman spectroscopy. The refractive index measured by means of ellipsometry is given in Table 5 with thickness of the deposits. The refractive index decreased with increasing and reached to that of polyethylene-like polymer at the hydrogen concentration of 50 %. The change of deposits from semi-conducting carbon film to the insulator with increasing the hydrogen concentration in the plasma as given in Table 6.

Table 5. Refractive index (n) and film thickness (d) of the samples deposited varying hydrogen concentration.

| Substrate position | Mixing ratio | n | d / nm |
|---|---|---|---|
| On the holder | $CH_4(1.00)$ | 2.07±0.05 | 136±2 |
| | $CH_4(0.75)/H_2(0.25)$ | 1.99±0.01 | 122±3 |
| | $CH_4(0.50)/H_2(0.50)$ | 1.58±0.05 | 130±5 |
| On the wall | $CH_4(1.00)$ | 1.99±0.05 | 114±9 |
| | $CH_4(0.75)/H_2(0.25)$ | 2.01±0.09 | 154±3 |
| | $CH_4(0.50)/H_2(0.50)$ | 1.56±0.01 | 159±9 |

Table 6. Electric conductivity of the samples deposited varying hydrogen concentration.

| Substrate position | Mixing ratio | Electric conductivity / $Scm^{-1}$ |
|---|---|---|
| On the holder | $CH_4(1.00)$ | $7.4\times10^{-4}$ |
| | $CH_4(0.75)/H_2(0.25)$ | $7.0\times10^{-5}$ |
| | $CH_4(0.50)/H_2(0.50)$ | less than $1\times10^{-7}$ |
| On the wall | $CH_4(1.00)$ | $6.9\times10^{-6}$ |
| | $CH_4(0.75)/H_2(0.25)$ | less than $1\times10^{-7}$ |
| | $CH_4(0.50)/H_2(0.50)$ | less than $1\times10^{-7}$ |

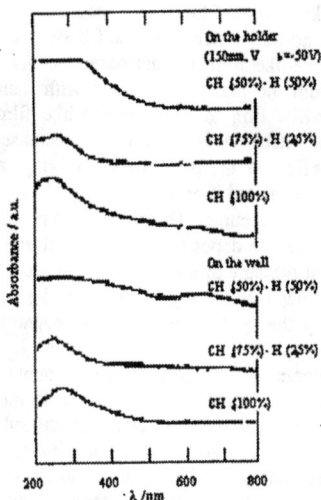

**Figure 10.** UV/vis spectra of samples deposited varying hydrogen concentration in the methane/hydrogen system.

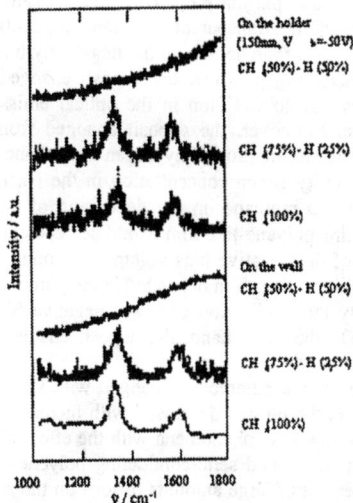

**Figure 11.** Raman spectra of samples deposited varying hydrogen concentration in the methane/hydrogen system.

### 3.3.3 Effect of addition of hydrogen on the methane plasma

In optical emission spectra, several species due to dissociation of methane were identified. The dependence of the band and line head intensities on the hydrogen concentration is shown in Fig. 12. CH radicals and $CH^+$ ions decreased with increasing hydrogen concentration. The CH radicals and $CH^+$ ions would be quenched by the hydrogen molecules and/or atoms added into the plasmas. Since the deposits changed from polyene-like films to polyethylene film with increasing hydrogen concentration, the proper amounts of $CH^+$ ions would be necessary to deposit the unique transparent and semi-conducting carbon films.

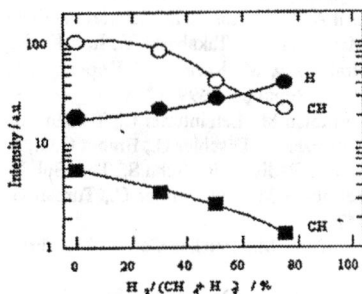

**Figure 12.** Normalized intensities of band head and head in the plasma vs. hydrogen concentration.

## 4. SUMMARIZING REMARKS

The dependence of properties of deposits on some discharge parameters of ECR plasma, such as pressure, hydrogen mixing ratio, and bias voltage on the substrate, were investigated. The results obtained are summarized and the effects of the discharge parameters on the deposition of carbon films in the ECR are discussed. As results, the deposition mechanism of carbon film is briefly discussed.

Both conditions, the higher degree of ionization and higher self-bias voltage, were obtained in the

ECR methane plasma prepared at the pressure as low as the order of magnitude of $10^{-2}$ Pa. At these conditions, the transparent and semi-conducting polyene-like carbon film was deposited with pure methane as carbon source on the negatively biased substrate and on the negatively self-biased substrate. With increasing pressure, the deposit changed to polyethylene-like film. In this case, the band head intensity due to $CH^+$ ion in the optical emission spectra from the plasma decreased with increasing pressure. Moreover, the deposit changed from the polyene-like film to polyethylene-like film with increasing mixing ratio of hydrogen to methane. The band head intensity of $CH^+$ ions also decreased with increasing hydrogen concentration in the plasma. From the effect of the pressure and mixing ratio of hydrogen to methane on the deposition of carbon film, it was found that the transparent and semi-conducting polyene-like film would be deposited with $CH^+$ ions as precursor. The application of optimum values of the negative bias voltage of around –50 V is necessary to deposit the transparent and semi-conducting carbon film deposited in the pure methane plasma at pressure of the order of magnitude of $10^{-2}$ Pa. By the application of lower negative bias voltage, the deposits were polyethylene-like polymer films. On the other hand, the deposit changed to graphite by the application of higher negative bias voltage.

The plasma parameters changed with varying discharge parameters. Especially, the amounts of ion species in the plasma decreased with increasing pressure and mixing amounts of hydrogen to methane. Considering these phenomena with the effect of the application of negative bias voltage on the substrate, the transparent and semi-conducting polyene-like carbon film was obtained by the contribution of the impingement of large amounts of ions on the negatively biased substrate. When both parameters, higher degree of ionization and optimum negative bias voltage, were attained, the transparent and semi-conducting polyene-like carbon film would be obtained in the pure methane ECR plasma prepared at the pressure as low as $10^{-2}$ Pa at the negative bias voltage of around – 50 V.

## References

1. Fujita T., Matsumoto O., J. Electrochem. Soc., 1989, 136, 2625.
2. Fujita T., Inagaki C., Uyama H., Matsumoto O., J. Electrochem. Soc., 1990, 137, 1645.
3. Gril A., Cold Plasma in Material Fabrication, IEEE Press, New York, 1993, 6.
4. Matsumoto O., Takahashi Y., Itoh K., Fujita T., Proc. 14th Intern. Symp. Plasma Chemistry, Ed. M. Hrabovsky, M. Konrad, V. Kopecky , Prague, Institute of Plasma Physics, Academy of Sciences, Czech Republic, 1999, 1735.
5. Zarrabian M., Leteinturier C., Turban G., Plasma Sources Sci. Technol., 1998, 7, 607.
6. Bubenzer A., Dischler G., Brandt G.G., Koidl P., J. Appl. Phys., 1983; 54, 4590.
7. Seth J., Padiyath R., Subu S., Thin Solid Films, 1992, 212, 251.
8. Zarrabian M., Leteinturier C., Turban G., Mahric C., Lancin M., Diamond Relat. Mater., 1997, 6, 542.
9. Zarrabian M., fouches-Coulon N., Turban G., Mahric C., Lancin M., Appl. Phys. Lett., 1997, 79, 2535.
10. Kawamoto K., Domoto Y., Hirano N., Kiyama S., Tada S., Appl. Surf. Sci., 1997, 113/114, 227.
11. Yoon S.F., Yang H., Ahn R.J., Zhang R., J. Electron Mater., 1998, 27, 44.
12. Pelliterier J., in Microwave Exited Plasma, Ed. M. Moisan, J. Pelletier, Elsevier, Amsterdum, 1992, 419.
13. Asmussen J., J. Vac. Sci., Technol., 1989, A7, 883.
14. Takahashi Y., Matsumoto O., Trans. Mater. Res. Soc. Japan, 2000, 25, 27.
15. Kiyooka H., Matsumoto O., Plasma Chem. Plasma Process., 1996, 16, 547.
16. Lee E.H., Henbree D.M., Jr., Rao G.R., Mansur L.K., Phys. Rev. 1993, B48, 15540.
17. Savvides N., J. Appl. Phys., 1986, 4133.
18. Savvides N., Thin Solid Films, 1988, 163, 13.
19. Smith S.W., J. Appl. Phys., 1984, 55, 764.
20. Mott N.F., Conduction in Non-Crystalline Materials, Oxford Press, Claredon, 1987, 27.

# MICROWAVE PLASMA ASSISTED CHEMICAL MICROPATTERNING OF POLYMERIC BIOMATERIALS

## A. Meyer-Plath, D. Keller, K. Schröder, G. Babucke, A. Ohl

Institut für Niedertemperatur-Plasmaphysik, D-17489 Greifswald, Germany

**Abstract.** Microwave plasmas exhibit specific advantages for the generation of surface functional groups. This is of relevance for a number of polymer surface modification techniques including biocompatibility improvement. In this paper the use of microwave plasmas for the new technique of plasma induced chemical micropatterning is discussed. The development of this technique was initiated by requirements of biomaterials research. The here exemplarily discussed technique is based on a sequence of plasma chemical surface modification steps. In a first step the whole polymer surface is modified by a plasma process to get a specific functionalization. Then, another plasma process is applied to regions of the polymer surface which were selected by an adequate micromasking technology. To get a sufficiently high chemical contrast both of the processing steps require well defined plasma conditions including specially adapted plasma sources. A newly-designed planar disk microwave plasma source is used for the first processing step. Optional it can be used to generate flat, disk-like or toroidal active plasma regions with diameters of about 8". For the second processing step a specially prepared downstream microwave plasma source was used.

## 1. INTRODUCTION

Plasma assisted surface modification is frequently used for biocompatibility improvement of polymeric biomaterials. It exhibits a number of advantages. It is an actual surface treatment and thermal substrate loads can be reduced to a negligible level. Therefore, bulk material properties remain unchanged during treatment. Plasma processing is a safe procedure, i.e. it includes inherently sterilizing effects and generation of hazardous leachable substances can be avoided. Further, specimen exhibiting complex geometry can be treated. The most important advantage, however, is the possibility to generate active surface sites useful for covalent bonding. This way high performance surface biocompatibility can be obtained via bonding of biomolecules. To date, a set of plasma methods is known which allow precise and reproducable control of cell adherence.

This was the starting point for the development of plasma assisted techniques for chemical micropatterning. Chemical micropatterns exhibit chemical but no morphological contrasts. They can be used to induce patterned growth of cell cultures and may serve to reproduce the microscopic structure of living tissue in an artificial environment. This is of interest for a number of applications, e.g. biosensor improvement or scaffolds of bioarticial organs.

Various techniques for chemical micropatterning have been developed [1]. Following the background given above it is not surprising that in many cases these techniques include plasma surface modification steps and that even completely plasma based techniques were developed. A first mainly plasma based technique was reported in the early nineties [2]. A rf-discharge in a $H_2$-$CH_3OH$ gas mixture was used to generate OH-functionalities on the surface of a fluorinated co-polymer. In a subsequent wet chemical process these functionalities reacted with aminosiloxanes. The chemical micropattern was obtained by covering the substrate surface with a TEM grid mask during plasma treatment. In this way a contrast between highly nonadherent, fluorinated surface regions and adherent, amino-functionalized regions could be obtained. Later on this technique was used for a number of cell biological investigations. Another mainly plasma based technique makes use of a rf-discharge in oxygen. A photolithographic masking procedure followed by a wet chemical treatment with surfactants led to chemical patterns on polystyrene [3]. A completely plasma-based technique was reported by some of the authors of the present paper for polystyrene and poly (ether ether ketone) [1, 4, 5]. The aim of the present paper is to describe and discuss this technique with respect to the use of microwave plasmas.

## 2. PLASMA ASSISTED CHEMICAL MICROPATTERNING

### 2.1. Processing sequences of plasma assisted micropatterning

In principle, the processing sequences of chemical micropatterning are similar to those of other micropatterning techniques. Substantial differences exist due to the exclusively chemical nature of the modifications. No morphologically but chemically contrasted surfaces are desired. This makes the process sensitive against contamination. Further, sufficiently plane substrates and appropriate masking procedure are necessary.

A model procedure for plasma assisted chemical micropatterning can be described as follows. First a plasma treatment is applied to the whole substrate surface which generates a well defined chemical functionalization. Optional, this treatment can be completed by a further biofunctionalization. Then, another plasma process leading to the desired contrast is applied to regions of the polymer surface which were selected by an adequate micromasking technology. For micromasking two different variants can be chosen. One variant makes use of a thin grid mask by which the substrate surface is covered. The other variant makes use of standard microlithographic techniques. The latter variant includes some intermediate steps. First the substrate is covered with a polymeric resist film, e.g. by spin casting a resist solution. Then this resist film is irradiated with patterned ultraviolet light or electron beams. After this, the pattern will be developed, i.e. the resist will be partially removed leaving a micropatterned resist mask on the surface. Now the substrate is prepared for the second plasma treatment which can be the same as for the other masking variant. After mask removal the chemical micropattern is ready for use.

While mask removal is very simple for the grid mask it is often difficult for standard microlithographic technique. It must be expected that the solvents used for removal destroy the chemical surface pattern. A similar problem may exist for spin casting. Indeed, it was observed that resist procedure and surface modification are conflicting. Special means have to be taken to solve this problem [3, 6].

Compared to the resist procedures the grid mask procedure is simple. Here, only the mechanical and heat sensitivity of the mask may cause problems. Heating of the mask by the plasma has to be low. Otherwise it may be destroyed. Another risk is that the thin mask could be crinkled by heating and, therefore, become useless to sufficiently cover the substrate. In this respect the microwave plasma afterglow exhibits considerable advantages. It is a mild plasma with very low mean electron energy,

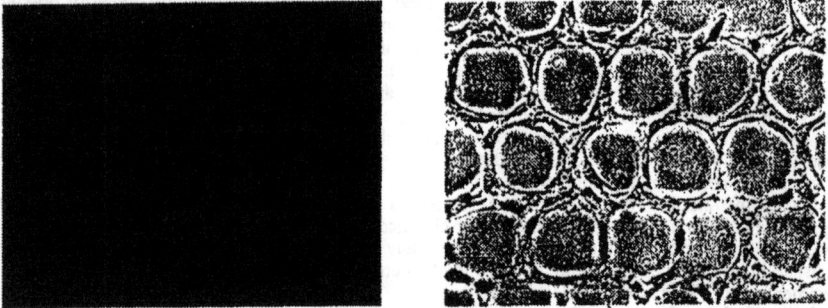

Figure 1. Microphotographs of chemical micropatterns on polystyrene surfaces. Left: Visualization of the chemical contrast by a fluorescence marker which indicates the presence of amino functional groups (bright areas). Right: Vizualisation of the protein adherence contrast by a cell culture. The width of the pattern bars is 30 μm. The length of the pattern period is 100 μm.

comparatively low charge carrier densities but still high enough chemical reactivity due to active neutrals. In general, microwave plasmas seem to be particularly useful for plasma assisted chemical micropatterning. Due to low mean electron energies the effect of microwave plasmas is limited to the topmost surface layers of the substrate, i.e. actual surface processing is improved. Further, the mechanism of microwave field absorption in plasmas offers good possibilities to generate localized plasmas. At the same time microwave plasmas are highly reactive, i.e. the density of charged particles and reactive neutrals is comparatively high. These are excellent conditions for control of thermal substrate load and chemical surface modification by downstream processing. Also, low substrate damage can be achieved, i.e. topological modifications can be avoided. Another reason for the use of microwave plasmas is the sensitivity of chemical micropatterning procedures against contamination. Microwave plasmas in conjunction with discharge vessels which ensure very low base pressure are known to be a good choice for the generation of lowly contaminated plasmas.

## 2.2. Example of a plasma assisted micropatterning procedure

Plasma treatment can be used to generate a number of different surface functional groups. Therefore the above described patterning procedure, in principle, may be used to generate many different chemical micropatterns. A certain limitation of available contrasts is given by the procedure itself, since the second plasma treatment has to be compatible to the first one. Another limitation is given by the materials used. Not every polymer can be treated by a plasma in the same way. Much depends on the desired contrast and application.

For polymeric biomaterials chemical contrasts are mostly desired which are useful for control of protein attachment. They can be classified into two types, protein-protein contrasts and adherent-nonadherent contrasts. The present paper refers to a patterning procedures which generates a universal adherent-nonadherent contrast [1, 4, 5]. It was developed for cell culturing applications, especially for the commonly used substrate material polystyrene. For this special application a lateral resolution in the micron scale is sufficient. Such a resolution can be easily obtained by using micromasks. In the present case these micromasks were laser cut metal masks.

For the first plasma step a $H_2/NH_3$ microwave plasma (p = 30 – 80 Pa, P = 700 W) was used. Following XPS-measurements, the resulting surfaces contained up to 4 % nitrogen, compared to the surface carbon content. The portion of $NH_2$-groups was about 50% of the overall nitrogen surface densities. Such surface properties are specially useful for culturing of adherent animal cells, i.e. the proteins strongly attach to these surfaces.

In the second plasma step using the micromask these surfaces were further modified to get low protein attachment. A downstream $Ar/H_2$ microwave plasma (p = 30 Pa, P = 70 W) was used to remove functionalities. This corresponds to a passivation of the surface. The resulting low energetic surface is characterized by low level of remaining functional groups, mainly oxygen functional groups and amino groups. Further, it is highly cross-linked. The low protein attachment is believed to be due to a predominance of methyl terminal groups. Visualizations of chemical patterns obtained by this technique are given in Fig.1. They clearly demonstrate that a lateral resolution in the micron scale is sufficient for cell culture applications.

## 3. EXPERIMENTAL: MICROWAVE PLASMA EQUIPMENT

### 3.1. Planar disk microwave plasma source for plasma functionalization

For the plasma modification with $H_2/NH_3$ microwave plasma best result were obtained for substrates located close to the active plasma zone. This, however, is critical concerning thermal load of the polymeric substrates. It becomes more critical if extended planar substrates with additional three dimensional structures like wells, rims or grooves have to be treated. Exactly this is often the case for biomaterials modification. For example the here described investigations were performed using different polystyrene cell culture dishes. Processing of such substrates is a challenging task for microwave plasma engineering. It requires planar plasmas with large lateral extension and good homogeneity. The here used so called "planar disk microwave plasma source" meets these requirement in a special way. It is a newly-designed, home made plasma source. A schematic cross section of this source is given in Fig. 2. The actual plasma excitation structure is a conical line. This conical line is terminated by a structure which consists of the stainless steel top plate of the discharge vessel, a round quartz window inside of the line and of the plasma itself. Microwave power is coupled to this line from a standard WR 340 rectangular waveguide via a rigid coaxial line.

The design of this source follows two ideas. First, the source makes use

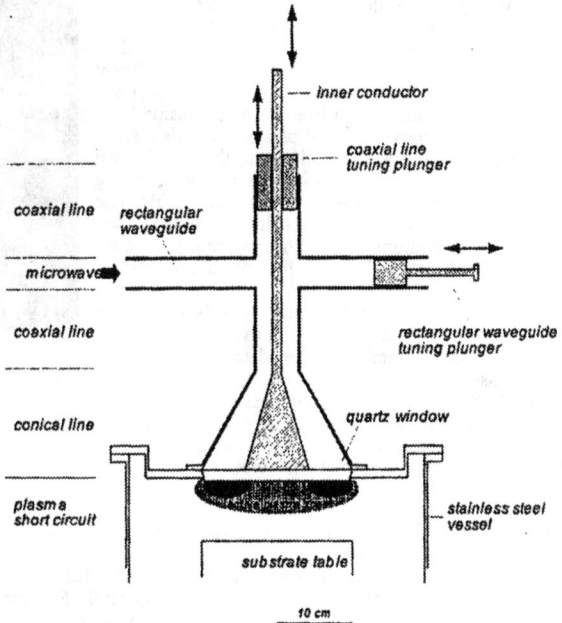

Figure 2. Schematic cross section of a planar disk microwave plasma source.

274

of the character of microwave propagation in a conical line. This wave propagation is comparable to a spherical wave with its origin in the cusp of the line. This means that the wavefront propagation direction and the quartz window are in an oblique angel. Accordingly, the termination of the conical line which consists of the inner and outer wall of the line and of the plasma itself has similarity with a tapered absorber. This design was chosen to improve the tuning behaviour and operation range of the source.

The second design idea for this source was the generation of a toroidal plasma excitation region. Spatial distribution of active species, especially active neutrals in low pressure microwave plasmas is controlled by diffusional transport and wall recombination. For cylindrical discharge vessels and homogeneous excitation this leads to inhomogeneous radial species distributions with maximum concentrations along the axis of the discharge. A toroidal excitation region would allow to generate radial species distributions with improved homogeneity.

The source includes two tuning plungers: One plunger which terminates the rectangular waveguide and another, coaxial plunger which terminates the coaxial line. In addition, the inner cone of the conical line can be lifted. Positioning of the inner cone was intended to control spreading of microwave power to the centre of the discharge, i.e. to control the extension of the plasma torus. Practical experience showed that this additional tuning element had little influence. It is assumed that there is already noticeable spreading of microwave power to the centre of the discharge due to a relatively thick quartz microwave window.

Generally, tuning depends on the applied microwave power. For low microwave power different ball-like plasma configurations could be obtained including free burning ball-like plasmas, plasma balls touching the window and oval plasmas. In this respect this source shows certain similarity to the well known microwave plasma ball reactors. For example, tuning also depends on the position of the subtrate table. The pronounced difference consists in the high power plasma configurations. Here, the generation of both toroidal or planar, disk-like active plasma regions is possible. Disk-like plasmas are useful for processes which require the glow treatment of substrates. The toroidal configuration is advantageous for homogeneous treatment under moderate down stream conditions.

It should be mentioned, that besides these stable plasma configurations also a number of unstable plasma configurations were seen. Often these unstable plasma were rotating structures with a fixed n-fold symmetry (n = 3,4,...).

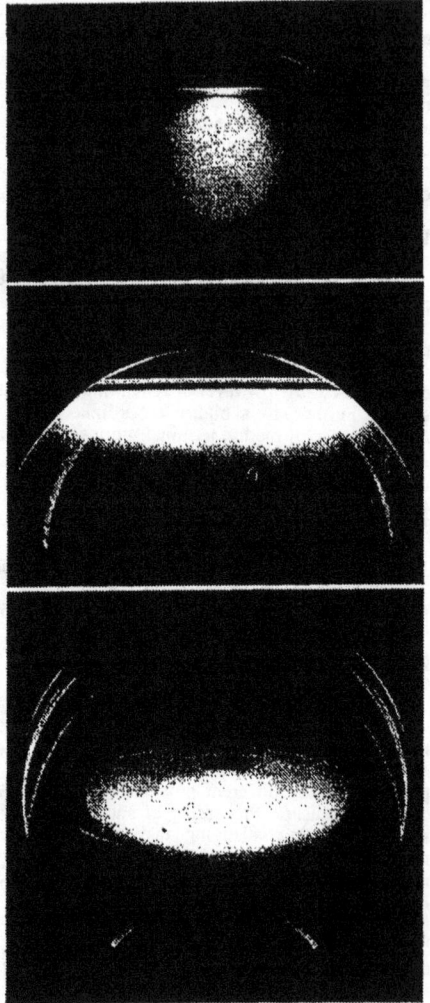

Figure 3. Photographs of different plasma configurations of a planar disk microwave plasma source (Ar/H2 = 4 : 1, total pressure 0.5 mbar, top: ball-like , centre: planar disk, bottom: torus). Different viewing directions were chosen to visualize the structures.

## 3.1. Down stream microwave plasma source for plasma passivation

For the above mentioned reasons mild plasmas and mild processing conditions should be preferred for plasma assisted chemical micropatterning in general. For the plasma passivation step of the here discussed procedure mild conditions are even crucial. Therefore, a specially-prepared standard downstream microwave plasma source is used for this purpose. The configuration of this source is given in Fig. 3. It consists of a quartz discharge tube which is inserted into a rectangular waveguide. The tube is fitted to a large diameter quartz vessel which contains the substrate table. This has the purpose to improve mild plasma conditions. In parallel it allows treatment of large diameter substrates. Special care is taken to avoid any contamination. This includes clean gas supply, differential pumping of o-ring sealings and a careful selection of wall materials. By this means a good reproducibility of the passivation could be obtained.

Passivation is the key step of the patterning procedure described here. Therefore, a characterization of the plasma properties in front of the substrates was performed. Properties of charged carrier components were determined using a cylindrical Langmuir-probe and a retarding field analyzer which was specially designed for use in microwave plasma at pressures in the mbar-range [7]. Results of retarding field analyzer measurements are given in Fig. 4. According to these results the mean ion energy for plasma conditions typical for plasma passivation is about 1 eV. As a results of probe measurements densities of about $10^8$ cm$^{-3}$ were found. This demonstrates that on substrate level mild plasma conditions exist. The average energy available for chemical reactions of hydrogen ions is not much higher than of hydrogen atoms.

For the present plasma conditions it can be assumed that atomic hydrogen is available in much higher concentrations than ionic hydrogen. Assuming a 10% degree of dissociation of molecular hydrogen in the plasma activation region a conservative estimation of hydrogen flow to the substrate

**Figure 4.** Schematic cross section of a down stream microwave plasma source which was used for plasma passivation.

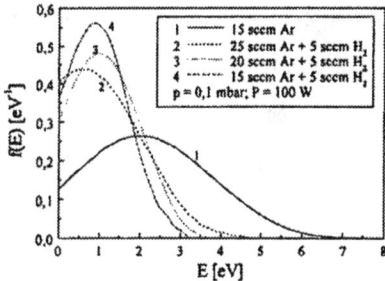

surface was performed. Following this the flux density of atomic hydrogen is in the order of $10^{15} - 10^{16}$ mm$^{-2}$s$^{-1}$. A rough estimation for the ions results in an ion flux density in the order of $10^{11}$ mm$^{-2}$s$^{-1}$. This suggests that neutral atomic hydrogen is a main component for surface modification reactions. Another reactive component of microwave plasmas is ultraviolet radiation. Especially for hydrogen plasmas it is

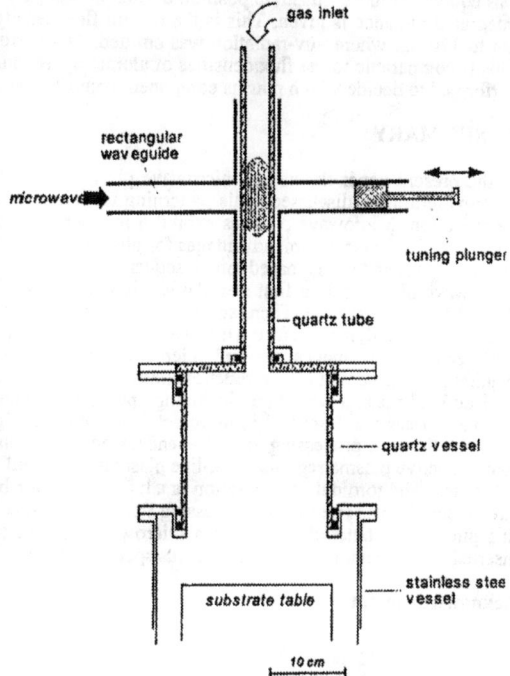

**Figure 5.** Ion energy distribution near a substrate in down stream location similar to that given in Figure 4, for different plasma conditions similar to plasma passivation conditions

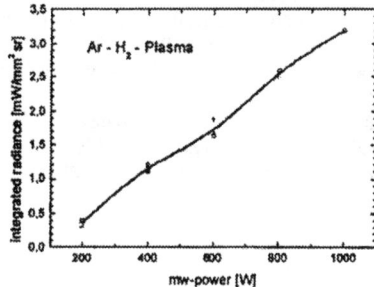

**Figure 6.** Microwave power dependence of the integrated radiance of the vuv-radiation which is irradiating the subtrate during plasma passivation.

276

known that ultraviolet radiation with extreme short wavelengths (vuv) plays a significant roll during polymer surface modification [8]. For this reason the discharge part of the plasma source was transferred to a vuv measuring station. Measurements of radiation flux density were performed for a position which was equivalent to the substrate position during plasma passivation. In Fig. 5 the power dependence of the integrated radiance is given. This is the radiant flux density integrated over the spectral range from 100 nm to 170 nm where vuv-radiation was emitted. The corresponding photon flux is $10^{13} - 10^{14}$ mm$^{-2}$s$^{-1}$. This is comparable to the flux densities of atomic hydrogen. As a result, further investigations have to be performed to decide which plasma component controls plasma passivation.

## 4. SUMMARY

In the present paper the use of microwave plasmas for the new technique of plasma induced chemical micropatterning is discussed. This patterning technique is of special interest for polymeric biomaterials modification. Microwave plasmas exhibit a number of advantages for chemical surface modification of polymers. They also exhibit advantages for plasma induced chemical micropatterning. The here described patterning procedure is based on a sequence of plasma chemical surface modification steps using microwave plasmas. In a first step the whole polymer surface is modified by a plasma process to get a desired functionalization. Then, another plasma process leading to another functionalization is applied to regions of the polymer surface which were selected by an adequate micromasking technology. Especially this second step makes use of microwave plasma advantages. For chemical micropatterns the requirements to be met by the functionalized surfaces are high. As a result, both processing steps require well defined plasma conditions including specially adapted plasma sources. The respective microwave plasma sources are described in more detail. A newly-designed planar disk microwave plasma source is used for the first processing step. Dependent on tuning and power it generates flat, disk-like active or toroidal active plasma regions. Disk-like plasmas are useful for processes which require glow treatment of substrates. The toroidal configuration is advantageous for homogeneous treatment under moderate down stream conditions. The second processing step requires specifically clean and mild plasma conditions. For this purpose a standard downstream microwave plasma source consisting of a quartz discharge tube inserted into a rectangular waveguide was specially prepared.

### Acknowledgements

The authors would like to thank G. Friedrichs and U. Kellner for technical assistance and H. Lange and P. Holtz for performing the VUV measurements.

### References

1. Ohl A., Schröder K., Surface and Coatings Technol., 1999, 116-119, 820.
2. Vargo T.G., Thompson P.M., Gerenser R.F., Valentini R.F., Aebischer P., Hook D.J., Gardella J.A. Langmuir, 1992, 8, 130
3. Lhoest J.B., Detrait E., Dewez J.L., Van den Bosch de Aguilar P., Bertrand P., J. Biomat. Sci., Polym. Edn. 1996, 7, 1039
4. Ohl A., Schröder K., Keller D., Husen B. Proc. 13th Intern. Symp. Plasma Chem., Ed. C.K. Wu, Beijing: Peking University Press, 1997, 1259
5. Ohl A., Schröder K., Keller D., Meyer-Plath A., Bienert H., Husen B., Rune G.M. J. Mat. Sci.: Mat. in Medicine, 1999, 10, 747
6. Flounders A.W., Brandon D.L., Bates A.H. Biosens. Bioelectron., 1997, 12, 447
7. Guo X.M., Ohl A. J. Phys. D: Appl. Phys., 1998, 31, 2018
8. Wilken R., Holländer A., Behnisch J. Plasmas and Polymers, 1998, 3, 165

# MICROWAVE PLASMA JET FOR SENSITIVE SPECTROSCOPY

## K.F.Sergeichev, I.A.Sychov, D.V.Vlasov*

General Physics Institute, 38 Vavilov St., Moscow, Russia, 117942
*Natural Science Investigations Center of the GPI, 38 Vavilov St., Moscow, Russia, 117942

Abstract. In this work the coaxial microwave torch is used for spectrum emulation. Such torches were used in the 60's both for plasma chemical reactions and for welding small parts of metals under reducing atmosphere and are now revived with respect to melting of refractory metals. Data of both the jet microwave discharge parameters and the spectroscopy technique tests are presented.

## 1. INTRODUCTION

It is known that plasmatrons working at high pressures allow to get high temperatures of plasmas much more those which could be achieved by means of chemical ways in result of exothermic reactions. Owing to high plasma temperature and high brightness the plasma jets are wide applied in a spectroscopy last time because the torches increase sensitivity of spectral technique by orders. This circumstance is very important for measuring of chemical element traces in different probes. It is naturally that HF and microwave jet discharges are most pure for the spectroscopy because both don't need electrodes.

The history of similar plasmatrons creation ascends to 60-s. HF electrodless discharge in a stream of gas which is flowing out in atmosphere, or the inductive plasma torch (ICP - torch), as one usually names this device, was realized by Reed [1] in 1960. Kononov and Yakushin [2] had obtained the temperature of plasma exhausting in atmosphere of order 10000 K in a similar device with power up to 40 KW in a stream of an air (argon) at atmospheric pressure, in quartz tube of the diameter 6 см (gas flow rate up to 1 l/s). Now spectral analyzers applying ICP, are produced by well known companies (Hewlett-Packard, Perkin-Elmer).

The rise of frequency in electrodless discharges promotes not only reduction of plasmatron size proportionally to a working wavelength reduction, but also rise the intensity of power entered in a discharge, that is important for deriving high temperatures of plasma at rather smaller dimensions and cost of devices.

The progress in a microwave (MW) electronics engineering in the 60-s years resulting in creation of serial magnetrons of continuous high power operation, has allowed even in those years to create microwave plasmatrons with open "flame" as the kind of coaxial plasma torch [4]. Originally they tried to use such plasmatrons with a power from one up to several kW for plasma chemistry reactions at high temperatures, for welding of small details from refractory metals in reducing atmosphere. However last years MW plasmatrons are even more often used as radiation sources of spectra excitation for high-temperature spectroscopy. For example short "flame" in argon was obtained in [5] with help of coaxial resonator on the edge of a coaxial line supplied by MW source of 100÷300 W power on frequency 2,45 ГГц. The plasma was exhausted from the resonator in atmosphere, with electron temperature $\approx 5500$ K, electron density $\sim 10^{14}$ см$^{-3}$ and gas temperature $\sim 4000$ K.

## 2. DEVICE

Basic elements of the torch are coaxial metal tubes of diameters 2a and 2b, divided by insulating quartz tube, preventing of arc ignition between conductors (Fig.1). Interior tube in which the working gas moves has a tip with a narrow orifice - nozzle forming a plasma jet.

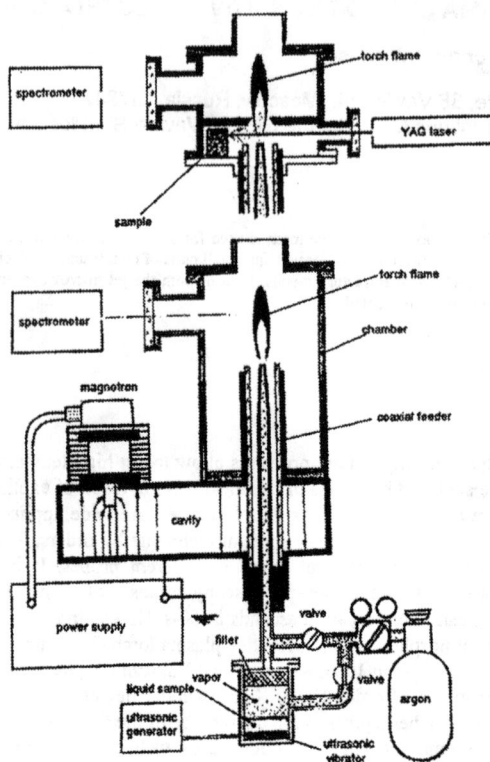

**Figure 1. Facility scheme.**

Thus the torch represents a waveguide coaxial junction with a transformation coefficient and VSWR close to 1. Central tube of the torch coaxial feeder passes through a rectangular waveguide, then through an adjusted coaxial stub and is finished by a nippel for junction with a balloon of working gas given through the reduction gearbox. Quartz tube is pasted between broad walls of the rectangular waveguide for vacuum isolation of the coaxial. The nozzle on the coaxial edge is made from copper.

The creation of such torches has become possible today due to widely spread and cheap magnetrons for household microwave ovens with MW power about 1 kW. In addition the modern engineering allows to create compact and inexpensive sources of the stabilized high voltage power supply for magnetrons.

An equivalent circuit of the plasma torch represents ohmic and capacity resistance connected in parallel . The capacity reactivity is compensated by tuning of the unit therefore its ohmic resistance R plays the principal role in torch operation. For optimum energy transmission to a load (flame) it's necessary to execute a condition of equality of coaxial line and plasma jet impedances:

$$Z_c = 60 \, \varepsilon^{-1/2} \ln(a/b) = R \, [\text{Ohm}] , \qquad (1)$$

where $\varepsilon$ - permitivity of dielectric filling the coaxial line.

Gas flow rate is appeared to has a small influence on reactive component of the load, that excludes need of additional unit matching at variation of gas flow rate in the jet.

As experience showed plasma jet is a good matched load for magnetron, that allows to exclude a circulator from the system. The magnetron operates in continuous generation mode.

It's necessary to study a number of questions for use of such plasma jet as spectral source. First, what temperature could be achieved by plasma electrons, what is the gas temperature, is the discharge equilibrium? Second, what is a role of nozzle-plasma contact and has it a reflection in spectra?

## 3. FEATURES

Fig.2 represents the photo of coaxial MW jet operating with argon. The gas outflow rate is $(0.5 \div 3) \cdot 10^3$ SCCM. Continuous magnetron power is ~ 850 W. Spectral investigations had shown that copper lines in green spectral range are absent. That allowed to conclude that copper impurities in discharge plasma inserted by the nozzle are negligible. When thin copper wire inserted in plasma jet it was melt and characteristic green luminescence of the copper appeared.

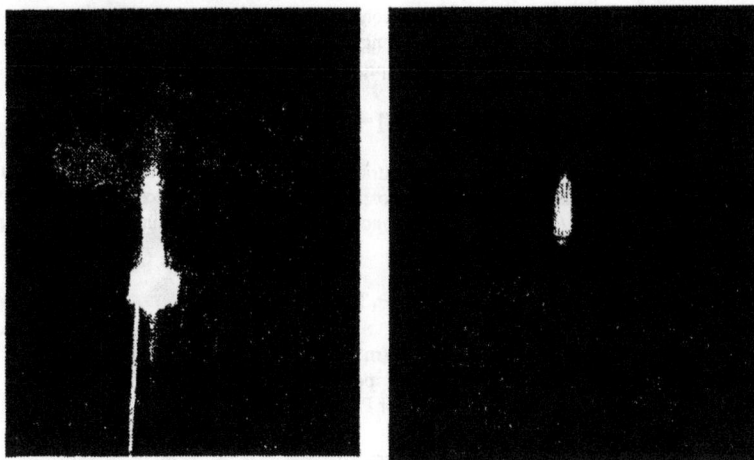

**Figure 2.** Photo of MW jet in operation: left – ordinary exposure, right – exposure through UV2 filter.

The radius and length of plasma jet kernel are $\rho \sim 0.1$ cm and $L \sim 1$ cm correspondingly. Variation of argon outflow rate had showed that under small outflow velocities << 10 m/s plasma jet transfers to interelectrode arc discharge or dies away while under large outflow velocities $\geq$ 10 m/s continue to burn a little changing in size.

An insignificant heating of the unit during a long operation time points on the absence of noticeable heat transfer from the plasma jet towards the torch. One needs to suppose that the regime of slow burning is realized in the jet when thermal conductivity wave moving towards the nozzle is continuously carried out in opposite direction by the gas flow [3].

MW radiation scattered by the plasma jet to environment (leakage) is respectively small because the intensity of MW background at the distance from the jet of order few tens centimeters not exceeds allowable sanitary standard. It means that MW power absorption coefficient in the plasma jet is closed to 1, that gives a possibility to evaluate the plasma parameters in the jet.

## 3. EVALUATIONS

The scheme of realization of approximate calculation looks as follows: proceeding from the condition

(1)of impedance matching and measured jet geometry ($\rho$,L) one can define average specific conductivity of the plasma

$$\sigma = L/(R\pi\rho^2) \quad [\text{Ohm·m}]^{-1}, \qquad (2)$$

which connected with electron density and electron collision rate in the gas (plasma) $\nu$ via well known expression

$$\sigma = e^2 n_e/m\nu, \qquad (3)$$

where e and m are charge and mass of electron. The approximation of electrostatics is used in this which is valid when electron collision rate exceeds significantly the angular frequency of electromagnetic field $\nu \gg \omega = 2\pi f$ while the plasma jet length is smaller than electromagnetic wavelength $l \le \lambda/2\pi = c/\omega$.

For calculation of average density $n_e$ one needs to know the electron-atom collision rate $\nu$ which depends on pressure (i.e. neutral molecule or atom concentration) of the gas surrounding the plasma jet and on electron temperature $T_e$. For argon in the range of electron temperatures $0.5 < T_e < 5$ eV and discharge burning in normal atmosphere the linear approximation is possible [6] for dependence $\nu(T_e)$ as:

$$\nu = 3.3 \cdot 10^9 \, p \, [\text{Torr}] \cdot T_e[\text{eV}] = 6 \cdot 10^{10} \, T_e/T_g, \, [s^{-1}], \qquad (4)$$

where $p = N_g k T_g$ – gas pressure, $N_g$ – atom concentration, $T_g$ – gas temperature in the plasma jet. As plasma jet is in equilibrium with surrounding atmosphere in transverse direction then $N_0 T_0 = N_g T_g$, where $N_0$ and $T_0$ are normal (room) gas atom concentration and temperature. In high temperature discharges the main energy losses are radiation losses:

$$P_{rad} = \varepsilon(T_g)\sigma_0 T_g^4 S, \qquad (5)$$

where $\varepsilon(T_g)$ – emissivity factor, $\sigma_0 = 5,67 \cdot 10^{-8} \, [\text{W}/(\text{m}^2 \cdot \text{K}^4)]$ – Stefan-Boltzmann constant, $S = 2\pi\rho L$ – radiation surface of the body. Assuming that MW power fed to discharge and absorbed in it $P_0$ is transferred mainly to radiation of black body of power $P_{rad}$ one can obtain $T_g$:

$$T_g = [P_0/(\varepsilon\sigma_0 S)]^{1/4}. \qquad (6)$$

Taking into account that at high temperature $\varepsilon$ could be in the range $0.5 \div 1$ one obtains $T_g \approx (4000 \div 4800)$ K.

From the other hand the plasma conductivity at high ionization degree can be define by electron-ion collisions

$$\nu_{ei} = 2.9 \cdot 10^{-6} Z N_i \cdot \ln\Lambda \cdot T_e^{-3/2}, \qquad (7)$$

where $n_i$ [см$^{-3}$] – plasma ion density, Z – ion charge order, $T_e$[eV], $\ln\Lambda = 5 \div 20$ – Coulomb logarithm. It's more probable that conductivity in high pressure discharges is defined by collision rate $\nu$, however one should take into account possible competition of the collision rate $\nu_{ei}$.

Set of expressions (2) - (6) allows to define the dependence $n_e(T_e)$:

$$n_e = 2 \cdot 10^{18} \sigma \, T_e/T_g = 2 \cdot 10^{18} \, (T_e/T_g) \cdot L/(Z_c \pi \rho^2), \, [\text{м}^{-3}]. \qquad (8)$$

For finding of $n_e$ и $T_e$ one should have once more equation connecting them. For not highly accurate evaluations the Saha equation [6] can serve for that which is valid for thermodynamically equilibrium electron density under not high ionization degree $n_e \ll N_g$. At $Z=1$:

$$n_e^2 = 6,06 \cdot 10^{21} (g_+/g_a) N_g T^{3/2} \exp\{-(I/kT)\}, \, [\text{м}^{-6}], \qquad (9)$$

here T [eV] is assumed to be equilibrium medium temperature: $T_e=T_i=T_g$; for noble gases the factor $(g_+/g_a) = 6$; I – ionization potential (for argon I=15,8 eV).

Evaluations performed according to this scheme had given $T_e \approx 6000$ K which is closed to $T_g$. At this time the electron density which should provide optimum plasma jet conductivity is $n_e \approx 8.5\cdot10^{19}$ [m$^{-3}$]. Calculations of electron collision rates according to formulas (4) и (7) have shown the assumption on domination of electron-neutral collisions to be correct: $v \approx 10^{11}$, c$^{-1}$, a $v_{ei} \leq 10^{10}$, c$^{-1}$.

As the result of evaluations performed above the electron temperature appeared to be higher than gas temperature evaluated from power absorption, as it points out that it's necessary to do additional speculations in validity of which one could doubt. Therefore it would be better to have experimental data. Because of we did not measure gas temperature and electron density in our experiments let's refer the works investigated the discharges in quartz tube passing through waveguide at the direction of pump electric field for various sorts of gases [7,8]. In [7] the discharge studied at wavelength λ=12 cm in a waveguide with cross-section 7.2×3.4 cm$^2$. The quartz tube radius was 1 cm and discharge radius ~0.5 cm. The power of 1÷2 kW was introduced in plasma. At these conditions the gas temperature of the air at atmospheric pressure was ~4000 K while gas temperature of nitrogen – 5000 K. Matching the discharge with power source one could introduce to plasma up to 80÷90% of power. In [8] the waveguide of the same cross-section and quartz tube of diameter 0.8 cm were used. The nitrogen at atmospheric pressure was studied. Measurements of vibrational and rotational temperatures together with electron density showed that plasma state is closed to equilibrium. The temperature ~6000 K was weakly depending on MW power. Otherwise for MW discharge in argon studied in [9] at the same conditions the electron temperature ~7000 K noticeably exceeded the atom temperature ~4500 K. It was explain by the fact that there is no such electron energy losses channel as impact excitation of molecule vibrations.

In this report we tried to follow the simplicity of our calculations as much as possible. More strict consideration of processes of temperature and electron density establishment, local thermodynamic equilibrium in high pressure discharges one can find in literature [10-15].

At last for justification of previous evaluations it's necessary to show a possibility of full absorption of MW wave in the plasma jet. For this let's use the propagation characteristics of MW wave in the plasma with parameters evaluated. In [6] formulas are presented for calculation of propagation constant $\beta=2\pi/\lambda_{pl}$ and attenuation constant $\alpha$ as well as skin depth $\delta$ for the case of plane wave incidence on a homogeneous plasma layer. Here $\lambda_{pl}$ to be MW wavelength in the plasma. The case under consideration satisfies to two conditions: $n_e \gg n_c v/\omega$ and $v \gg \omega$, where $n_c=m\omega^2/(\varepsilon_0 e^2)$ – critical plasma density for given frequency $\omega=2\pi f$, $\varepsilon_0$ – vacuum permitivity. Thus formulas for $\alpha$ and $\beta$ are simplified:

$$\beta \approx \alpha \approx (\omega/c)[n_c\omega/(2n_c v)]^{1/2}, \qquad (10)$$
$$\delta = 1/\alpha. \qquad (11)$$

Substitution in these formulas of our plasma parameters gives $\beta \approx \alpha \approx 5$ cm$^{-1}$ followed by the wavelength in the plasma $\lambda_{pl}$=1.3 cm, power absorption at this length is in $e^2 \approx 7.4$ times, skin depth $\delta \approx 0.2$ cm. These values are good coincided with observed sizes of plasma jet which we used in previous calculations.

Let's evaluate the energy convection losses from discharge by gas jet. Linear specific energy of the plasma jet is $W_l=\pi\rho^2 N_g \kappa T_g$ so the power flowed out by the get is

$$P_{conv}=W_l\cdot v, [W], \qquad (12)$$

where v to be gas flow velocity. In our velocity range 5÷20 m/s convection losses are no more than 10 W i.e. no more than 1.5% of forwarded power. This also confirms the assumption about radiation character of power losses from the plasma jet.

Finally consider the conditions of discharge ignition. Maximum (breakdown) MW power transferred along the coaxial waveguide without dielectric filling can be calculated according to

$$P_{br} = 8,3\cdot10^{-3} E_{br}^2 b^2 \ln(a/b), [W], \qquad (13)$$

Where a,b have dimension [cm] and $E_{br}$[V/cm] to be breakdown electric field near central conductor where it has maximum value. $E_{br}$ for air at atmospheric pressure to be 29 kV/cm. Corresponding power $P_{br} \approx 600$ kW. For argon at atmospheric pressure $E_{br} \approx 17$ kV/cm and $P_{br} \approx 200$ kW that exceeds our operating power by orders. Actual field stress on the central conductor of our coaxial line at MW power 850 W is ~0.54 kV/cm. No doubt a cusp at the nozzle end plays some positive role in breakdown power reducing. However this is not the only circumstance; it's necessary take into account the resonance in a waveguide at subbreakdown stage. Before a breakdown the rectangular waveguide with a wave of the lowest type $H_{10}$ having waveguide length $\lambda_g = 16.7$ cm and loaded on a matched coaxial line represents a resonator of high Q-factor in which electric field approaches breakdown value due to amplification in tens times at the edge of coaxial waveguide. It is possible, if the rectangular waveguide has length L to be multiple integer of half-waves $L = n\lambda_g/2 = n \cdot 8,35$ cm. At casual sizes of a waveguide the plasma jet had been ignited artificially creating a spark on the coaxial torch edge. However by selection of the waveguide length without ferrite circulator it's possible to realize a self-breakdown of the torch and steady combustion of a discharge in continuous generation mode.

## 4. APPLICATION

Let's remind that our main goal is the application of MW plasma jet designed by ours for analytical spectroscopy.

Classical analytical spectroscopy methods (including resonance fluorescence) for many years are known as most reliable, precise and sensitive tools for wide range of industrial and scientific applications. Usually laser beam is focused at the sample and generates a hot plasma spark. Plasma flux in this laser spark formed by explosion-like expanding of heated (to a temperature up to keV) of atoms of the sample. These ionized sample atoms emit specified and well defined characteristic spectral lines similar to those that are obtained in other types of plasma. High initial temperatures of laser plasma are quite enough to ionize any known element and this is why all elements of Mendeleev's table can be detected. This emission light are collected by mirrors and dispersed by diffraction grating of high resolution spectrometer with one or several CCD arrays on its output. Spectral intensity distribution of emitted light is detected by CCD arrays, amplified, filtered and cleared from all types of noises and transferred to computer nearly immediately (for the time less then millisecond).

Up to some time this technique having the resolution ppm (part per million) satisfied needs of the researchers and manufacturers of materials. However recently owing to growing role of purification of new materials technology and integrated chips there was a need for sensitivity magnification of analytical methods. The basic disadvantage of laser method of a microprobe research limiting its sensitivity is that lifetime of laser plasma creating an excited states of atoms is short < 1 μs. The magnification of sensitivity can be reached by a location of microprobes in continuous plasma resulting in time increase of atomic luminescence by few orders so that the higher gas temperature, the more probability of atomic transitions.

This idea was realized in above mentioned ICP-devices which allowed to magnify the analytic methods sensitivity up to ~ppb (part per billion). However ICP-devices have the following disadvantages:
- first, the device is very expensive due to high cost of plasma source and complexity of its operation;
- second, microprobe in ICP device is spread over large plasma volume ~few tens $cm^3$ that does not allow to magnify the sensitivity by focusing of plasma image on spectrum analyzer input.

As it is seen from our previous consideration the unit based on microwave plasma jet is free from those disadvantages. Fig.3 represents the photo of the experimental model of the facility (which we shall name by analogy MWCP), available in General Physics Institute of the Russian Academy of Sciences. It consists of three units (not counting the laser): a microwave head, spectral analyzer (monochromator) and computer. Its most expensive part is CCD matrix serving for light quanta registration (~1÷2 k$). The circumstance that the flame of the torch has a small size allows to focus its image on a monochromator slot. Thus it is possible to expect that the sensitivity of MWCP method will be higher than ICP. The invariance of parameters of a discharge at a changing over a wide range of a gas flow velocity allows to doze a microprobe content in the discharge. Great advantage of this method is the possibility to research liquid microprobes without use of laser sputtering. In addition the device can be made portable.

283

Figure 3. View of the experimental facility (General Physics Institute, 2000).

As demonstration of this method possibilities we studied impurities of various elements in water for instance content of Na, K and other atoms in it (see Fig.4,5,6). We began investigations from usual water in which a lot of elements were observed. Then water was distilled one, two and three times. It was proved by other way (mass spectrometer analysis) that content of impurities in one time distilled water no more than $10^{-7}$. But even in three times distilled water we could distinct the Na line on the wavelength (Fig.6). Evaluation of Na atoms content in this three times distilled water shows that it is at a level of few ppb.

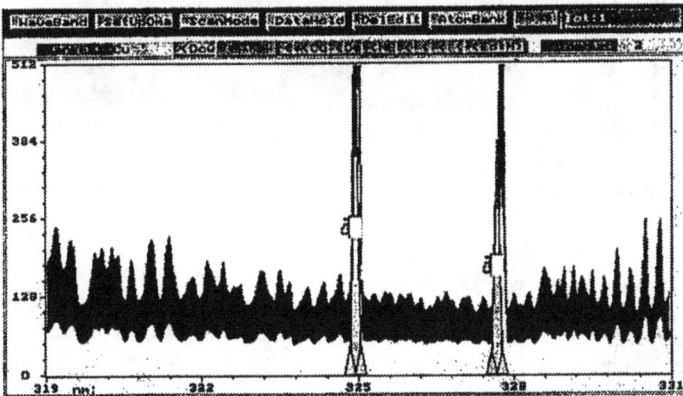

Figure 4. Detection of Cu impurities in unprocessed water (the Cu-lines intensity resulting from the copper nozzle is negligible respectively to these intensities).

284

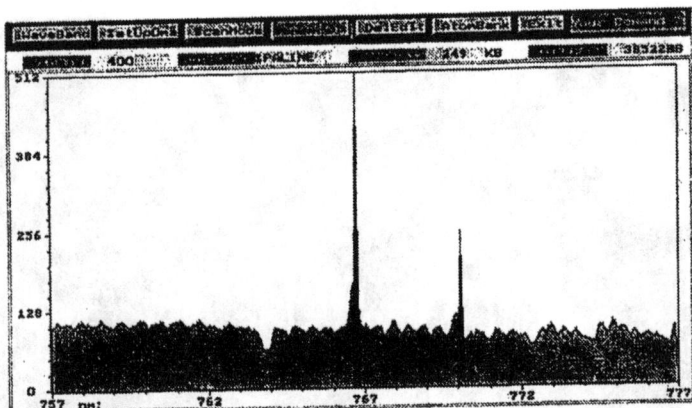

Figure 5. Detection of K impurities in the water obtained by combustion of hydrogen in oxygen which were obtained by electrolysis of one time distilled water.

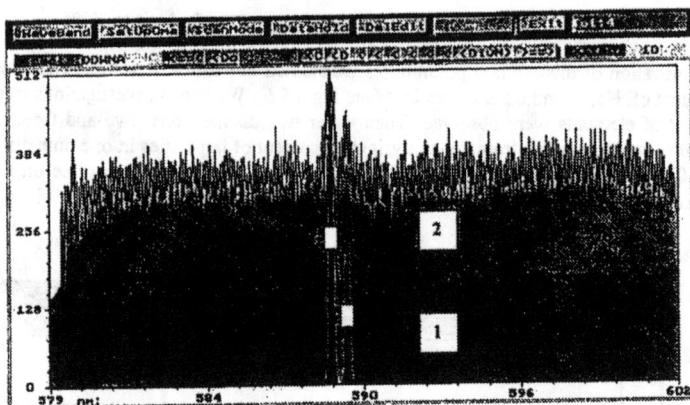

Figure 6. Detection of Na impurities: 1 – in one time distilled water; 2 – in three times distilled water.

References

1. Reed T.V. J. Appl. Phys., 1961, **32**, 821.
2. Kononov S.V., Yakushin M.I. J. prikladnoi mekhaniki i tekhnicheskoi fiziki,1966, №6,67 (in Russian).
3. Raizer Yu.P. J. prikladnoi mekhaniki i tekhnicheskoi fiziki , 1968, №3, p.3 (in Russian).
4. Pushner H. Heating with microwave. Fundamental component and circuit technique. 1966.
5. Wong S.K., UTIAS, Tech. Note, Can # 225.
6. Golant V.E. SVCh metody issledovaniya plazmy, M. Nauka, 1968 (in Russian).
7. Blinov L.M., Volod'ko V.V.et al. In "Generatory nizkotemperaturnoi plazmy", M. Energiya, 1969 (in Russian).
8. Baltin L.M., Batenin V.M.et al. Teplofizika vysokikh temperatur, 1971, **9**, 1105.
9. Baltin L.M., Batenin V.M.et al. In " Generatory nizkotemperaturnoi plazmy", M. Energiya, 1969 (in Russian).

285

10. Filkenburg V., Mekker P. Electric arcs and thermal plasma, 1961.
11. Plasma diagnostic technique. Ed. by R.Huddlestone and S.Leonard, N.Y., 1965.
12. Kolesnikov V.N.in "Fizicheskaya optika" M. Nauka, 1971 (in Russian).
13. Newman V. Beitr. aus der Plasmaphysik, 1971, **Bd11**, #3, 248.
14. Pichler G., Vujnovic V. Phys, Lett., 1972, **A40**, #5, 397.
15. Shumaker. J.B., Popenoe C.H. J. of Research National Bureau of Standards, 1972, **A76**, #2, 71.

# ACTIVITIES FOR DIAMOND DEPOSITION USING DIFFERENT PULSED DISCHARGES

A.L. Vikharev, A.M. Gorbachev, V.A. Koldanov

Institute of Applied Physics, 46 Ulyanov str., Nizhny Novgorod, 603600, Russia

**Abstract.** The paper reviews works on deposition of diamond films by using microwave discharges ignited in the pulse-periodic regime. Plasma parameters, kinetic processes, growth rate and quality of the diamond films in microwave reactors operating in the continuous wave (CW) and pulsed regimes are compared. The results of numerically modeling the reactor based on the cylindrical cavity excited in the CW and pulsed regimes at the $TM_{013}$ mode are compared. Basing on the results of experimental and theoretical studies the possibility to apply pulse-periodic microwave discharges for diamond film deposition is estimated.

## 1. INTRODUCTION

The interest in deposition of diamond films is due to their unique properties and a wide range of practical application [1]. Diamond films are synthesized by plasma-assisted chemical vapor deposition (CVD) method. Different types of gas discharges are used for creation of plasma. However, microwave discharges ignited by radiation of 2.45 GHz are used wider due to the following advantages: absence of electrodes, high specific power contribution, high densities of excited and charged particles, relatively large area and high homogeneity of the film. As a result, microwave plasma-assisted CVD reactor (MPACVD reactor) provides deposition of high-quality diamond films at the growth rate of 1-2 μm/h.

Currently one can state that diamond films have made a sure step from a fundamental research into the market. We will list here only some of the products, which use diamond films and are currently available in the market. They include laser diode heat sinks, thin film cutting tool inserts and diamond coated tools. The application range of diamond films is rapidly extending.

Let us look closer at one of the diamond film applications, which we have been recently working with. It is known that additional plasma heating under the ECR conditions in the next step fusion machines (e.g. ITER tokamak) requires microwave generators operating in the CW regime at the power of 1 MW and frequency of 170 GHz. Required generators (gyrotrons) have been already created [2,3]. They can use only diamond output windows, which have high thermal conductivity and good dielectric parameters [3,4]. A diamond disk of 1.5-2 mm thickness and 80-100 mm in diameter with loss tangent of $10^{-5}$ is needed for the window of the 1 MW CW gyrotron. Currently such disks are grown by De Beers (UK) and Digaskron (Russia) by the MPA CVD method [4,5]. Such disks grow during 5-6 weeks at continuous operation of the MPA CVD reactor. This time is too long, and therefore diamond plates are prohibitively expensive.

For this and other applications of diamond films that have been mentioned it is very important to improve the diamond growth process to get higher deposition rates, while retaining acceptable purity and crystalline quality. And this is a task of the fundamental character.

One of the ways to improve the growth rate of diamond films may be the use of the pulsed regime of reactor operation instead of the continuous wave (CW) regime. This idea is not altogether new; however, the pulsed regime has not become an accepted practice. There are single works on operation of the CVD reactor in the pulsed regime. The aim of this paper is use a few known examples in order to analyze the possibility of application of pulse-periodic microwave discharge for diamond film deposition.

## 2. COMPARISON OF THE PULSED AND CONTINUOUS REGIMES OF CVD REACTOR OPERATION

Let us briefly analyze what the pulsed plasma in the reactor can introduce into the process of diamond film synthesis. There are a limited number of external parameters that can influence the deposition process, when the CVD reactor with CW discharge operates: input power, pressure and composition of the gas mixture, gas flow rate and substrate temperature, $T_s$, when it is cooled externally. In the reactor, where the gas mixture is activated by means of a continuous microwave discharge, the value of the reduced electric field, E/N, is not high (here E is amplitude of the electric field in plasma, N is density of molecules of the working gas). Hence, the processes of dissociation stay far away from the optimal regimes. At the same time calculations show that the main dissociation channel even at high gas temperatures ($T_g = 2000\text{-}3000^0C$) is dissociation with an electron impact [6]. The study described in [7] proved that concentrations of components in diamond film deposition do not correspond to the equilibrium ones. Based on the increase in growth rate possible with higher dissociation levels, a promising diamond deposition process improvement seems to be the use of powerful pulsed microwave discharges.

High values of the amplitude of the electromagnetic field in pulsed discharges will provide efficient dissociation of the gas in the mixture and the choice of repetition rate and duration of microwave pulses will make it possible to control or to change the gas temperature. By that it may turn out that the same deposition rate and final quality of diamond films as in a continuous wave discharge will be achieved at lower levels of mean microwave power or at lower values of substrate temperature. The latter is important for deposition of diamond films on different types of substrates. So the pulsed mode in comparison with the continuous mode of operation allows independent adjustment to increase gas dissociation (density of hydrogen atoms) by increasing microwave peak power while controlling substrate temperature by changing repetition rate and pulse duration.

Growth rate and quality of diamond films are determined by a complex set of bulk and surface reactions. However, the growth rate is essentially dependent on the density of atomic hydrogen at the substrate surface. According to results of D.G.Goodwin [8], the growth rate $G$ may be estimated by the following formula:

$$G \propto \frac{[CH_3]_{sur}[H]_{sur}}{3 \cdot 10^{15} cm^{-3} + [H]_{sur}} . \tag{1}$$

Relative defect density (the defect fraction in the film) $X_{def}$ that determines film quality can be estimated as

$$X_{def} \propto \frac{G}{[H]_{sur}^2} . \tag{2}$$

Here $[H]_{sur}$ is the atomic hydrogen concentration and $[CH_3]_{sur}$ the methyl concentration at the surface. Often the gas-phase reaction,

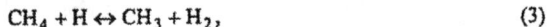

$$CH_4 + H \leftrightarrow CH_3 + H_2, \tag{3}$$

is rapid and near partial equilibrium [8]. The methyl concentration is coupled chemically to atomic hydrogen. Thus, the diamond film quality and growth rate are determined by density of atomic hydrogen at the substrate surface.

## 3. EXPERIMENTS ON DIAMOND DEPOSITION IN PULSED REGIME BY USING DIFFERENT TYPES OF MICROWAVE DISCHARGE

Let us consider the known works on deposition of diamond films in the pulsed regime. Having compared these works one can conclude that the design of the reactor is one of the important factors that affect density of atomic hydrogen, radicals and their distribution in the reactor. Advantages of the pulsed regime manifested differently in different reactors.

## 3.1. Diamond deposition using the standing wave discharge in a waveguide

One of the first experiments with the pulsed microwave discharge of this type [9,10] was performed at the gas pressure p = 30 Tor. Figure 1 shows the schematic diagram of this experimental set-up. The experimental conditions were as follows: gas composition $H_2 + CH_4 = 100$ sccm + 0.5 sccm, peak power $P_k = 400$ and 800W, average power P = 200 and 400W, pulse repetition rate F = 200-5000Hz, CW regime power $P_{CW} = 200$ and 400W. It was found that plasma synthesis of diamond films was successful only at low peak powers, of the order of 800 W, which is only twice as high as the average power. As the peak power exceeded 800 W, the microwave discharge became spatially inhomogeneous, and that led to a significantly inhomogeneous temperature profile across the substrate [9]. In this experiment the rate of film growth and its quality were determined by the average microwave power, similar to the case of using continuous microwave discharges. As the result, no improvement in growth rate or diamond film quality compared with CW was observed.

Reference [11] describes an experiment performed on a similar setup at a higher gas pressure, p = 60 Torr, and the same level of peak power, $P_k = 800W$. The experimental conditions were as follows: gas composition $H_2 + CH_4 = 50$ sccm + 0.3 sccm, peak power $P_k = 800W$, average power P = 400W, pulse repetition rate F = 500Hz, CW regime power $P_{CW} = 400W$, substrate temperature $T_s = 900^0C$. That experiment demonstrated better quality of the film as compared with the CW regime.

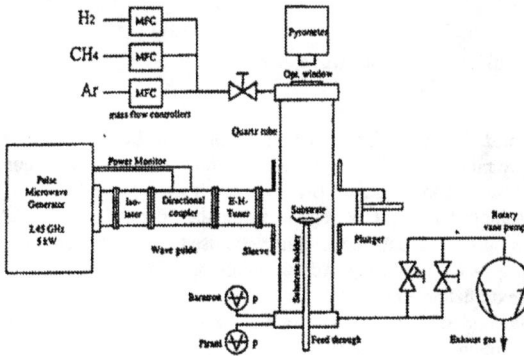

Figure 1. Schematic diagram of the experimental setup used for pulsed microwave plasma-assisted diamond deposition [9].

## 3.2. Diamond deposition using the microwave ECR discharge

Figure 2 shows two schematic diagrams of experimental setups for diamond film synthesis by the pulsed ECR plasma taken from Ref. [12] and [13]. The experimental conditions, e.g. in [13], were as follows: gas composition $H_2 + CH_4 + O_2 = 100$ sccm + 10 sccm + 9 sccm; gas pressure p = 3 Torr, peak power $P_k = 3kW$, average power P = 300W, pulse repetition rate F = 120 Hz, CW regime power $P_{CW} = 300W$, substrate temperature $T_s = 600^0C$. These experiments showed that the quality of the film is better, and the growth rate is at least an order of magnitude higher in the pulsed regime as compared to the continuous one at the same average microwave power [12, 13]. These results were obtained due to more efficient gas dissociation, which resulted in higher density of atomic hydrogen at pulse modulation of the incident microwave power. The experiments described in [12, 13] were performed at low pressures of the gas mixture (p = 0.1-3 Torr), and with a constant magnetic field, in order to improve gas dissociation. In the general case, the growth rate becomes higher at a higher pressure of the gas mixture, and in the majority of cases diamond films are deposed under higher pressures in the absence of the magnetic field.

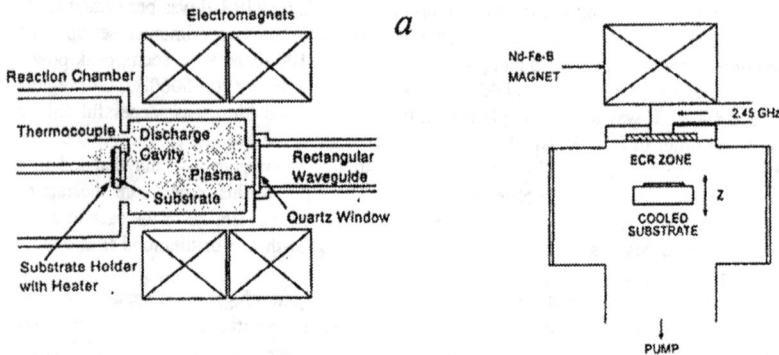

Figure 2. Experimental setups (*a* and *b*) for diamond film deposition by the pulsed ECR plasma taken from [12] and [13] respectively.

### 3.3. Diamond deposition using the surface wave discharge

One of the promising reactors for deposition of diamond films in the pulsed regime is the reactor based on the use of the surface wave microwave discharge. Figure 3 shows schematic configurations of the reactors operating in the CW regime at 2.45 GHz and 915 MHz taken from paper [14]. Large-area deposition of the diamond film was achieved in the CW regime using the 76 mm substrate, at the growth rate of $0.15 \square$m/hour. The following experimental conditions were used: gas composition $H_2 + CH_4 = 100$ sccm + 0.7 sccm, gas pressure p = 12–30 Torr, power of continuous wave radiation $P_{CW} = 4.5$ kW, substrate temperature $T_s = 800^0$C (external heating).

We think that the diamond film growth rate in such a reactor can be made sufficiently higher by passing over to the pulsed regime of operation. Such deductions are supported by the results of the works mentioned below.

In [15] it was stated that the main mechanism for propagation of the pulsed discharge on the surface wave is the breakdown wave in an inhomogeneous electric field at the front edge of the ionization front. By that, the velocity of the breakdown wave is significantly influenced by the effect of field strengthening in the region of plasma resonance at the discharge front [16]. As seen from the results of numerical calculations for the pulsed discharge (Fig. 4), field strengthening can be appreciable. Therefore, along with ionization, intense dissociation of the gas occurs at the ionization front in a pulsed surface wave discharge.

Besides, it was stated in [17] that the degree of hydrogen dissociation grows in the pulsed regime as compared with the continuous one due to a lower microwave power density, which leads to lesser heating of the discharge tube wall and hence, the losses of H-atoms by wall recombination.

Thus, in the pulsed microwave surface wave discharge the H-atom density can be significantly higher than in the CW discharge due to the more intense character of dissociation of hydrogen molecules and reduction of H-atom losses.

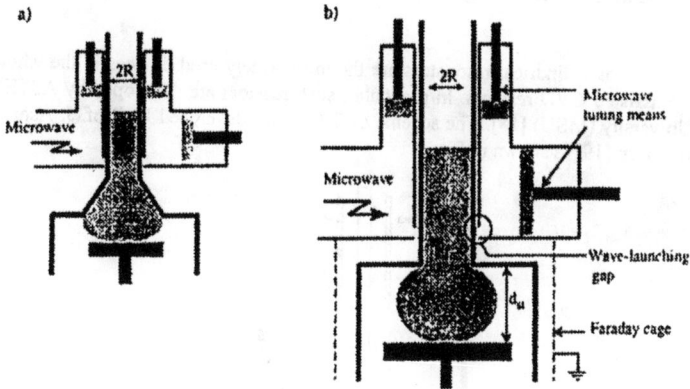

Figure 3. Schematic diagrams of the diamond deposition reactors with excitation of surface wave discharges at the wave frequencies: a- 2.45 GHz and b- 915 MHz. Illustration from [14].

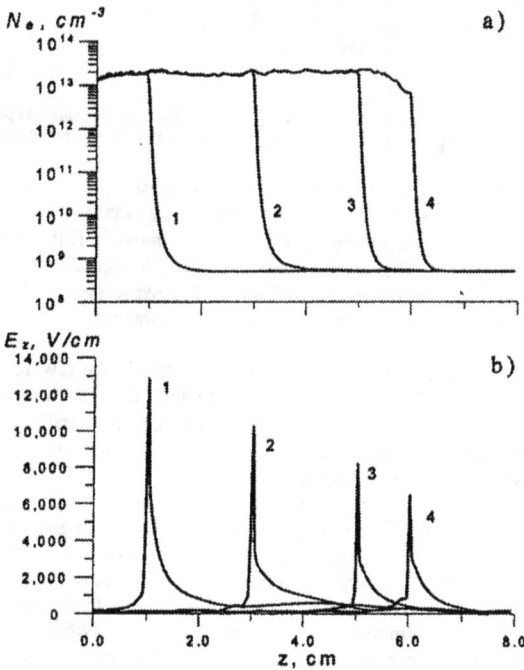

Figure 4. Longitudinal profiles of (a) the electron density and (b) $E_z$ component of the electric field at the instants (1) 32, (2) 70, (3) 100 and (4) 130ns for the pulsed surface wave discharge at the wave frequency 10 GHz, peak power $P_k = 50$kW, air pressure p= 1.5 Torr and tube diameter 8mm. Illustration from [16].

## 3.4. Diamond deposition using the cavity-type discharge

Currently reactors made as cylindrical resonators are the most widely used devices in the whole class of microwave plasma-assisted CVD reactors. In particular, such reactors are developed by ASTEX and the Michigan State University (MSU) [18]. The scheme of the reactor developed by Prof. J. Asmussen from the MSU and taken from [19] is shown in Fig. 5.

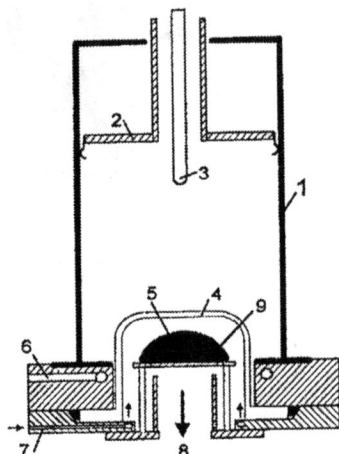

Figure 5. Schematic diagram of the MSU diamond deposition reactor: 1- cavity, 2- sliding short, 3- power coupling probe, 4- quartz dome, 5- plasma, 6- cooling water tunnel, 7- gas input tunnel, 8- pump, 9- substrate. Illustration from [19].

This reactor is excited at the $TM_{013}$ mode by means of a coaxial waveguide connected to the resonator at the center of the end wall. At the other end wall of the reactor a quartz dome is situated, which limits the discharge volume. The upper wall of the resonator can be moved. This provides tuning of the resonator to the resonance frequency. This reactor operates in the CW regime at the frequency 2.45 GHz at moderate and high pressures in the range 20-140 Torr at microwave power 4-5 kW. It provides the growth of diamond films with an area of several tenths of square centimeters at a rate of several micrometers per hour [19].

As seen from the above, such a reactor has good characteristics in the CW regime. Therefore, the study of the MSU reactor operation in the pulsed regime (at which the growth rate must become even higher) seem to be rather promising. Preliminary investigation of reactor operation in the pulsed regime were performed by the teams of the Michigan State University and the Institute of Applied Physics in a joint experiment in the framework of the CRDF grant (REI-352). The experimental conditions were as follows: gas composition $H_2 + CH_4 = 100$ sccm + 1.5 sccm, gas pressure $p = 15$ and 20 Torr, peak power $P_k = 2$kW, average power $P = 1000$W, pulse repetition rate $F = 60$Hz, CW regime power $P_{CW} = 1000$W, substrate temperature $T_s = 600^0$C. As it was found out, at the same average power the plasma volume in the resonator in the pulsed is significantly larger than that in the CW regime due to higher peak powers. As the result, at the same gas pressure the specific power contribution into the plasma became lower in the pulsed regime. Hence, it was determined that for a reactor with the cavity-type discharge the pulsed and CW regime of operation should be compared at different gas pressures, at which specific power into plasma are equal in the both cases. Currently these experiments are going on.

293

# 4. NUMERICAL CALCULATIONS OF THE IMPORTANT PARAMETERS OF A MICROWAVE CVD REACTOR

The microwave MSU reactor operating in the CW and pulsed regimes was studied also by numerical simulations [20]. The distribution of the electromagnetic field in the resonator was found from the Maxwell equation by the finite difference time domain (FDTD) method (Fig. 6) [21]. The microwave electric field interactions with the plasma discharge are described using a finite-difference solution of the equation for electron current [22]. The scheme of the calculation sequence is shown in Fig. 7. As the result, H-atom concentration at the surface of the substrate was determined, as well as specific power and plasma volume in the resonator at CW and pulsed regimes.

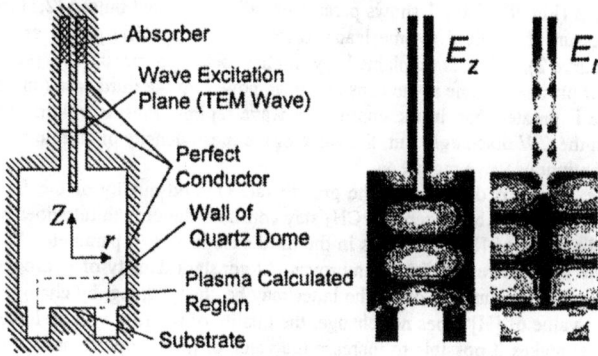

Figure 6. Microwave cavity plasma reactor configuration used for numerical simulation (a) and spatial distributions of electric field components in the resonator (b).

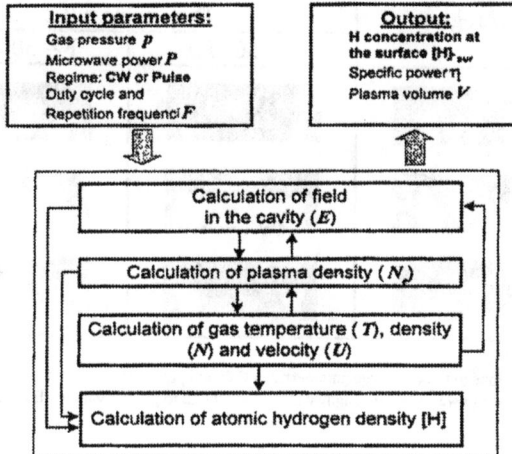

Figure 7. The scheme of the calculation sequence.

At high pressures the CW discharge plasma becomes significantly inhomogeneous when the power reaches some threshold, Fig.8. The discharge plasma stratifies due to development of the ionization-heating instability [23]. The discharge passes over to the inhomogeneous state in a certain time. Therefore, the use of the pulse-periodic regime of discharge maintenance with pulse duration shorter than the time of inhomogeneous discharge formation makes it possible to obtain a large-size homogeneous discharge. Numerical modeling showed that the use of pulses with pulse repetition rate (F) of several kHz makes it possible to obtain large-volume homogeneous plasma at a high level of power (Fig.8). In this case the atomic hydrogen density modulation amplitude is low. The average value of atomic hydrogen density in the pulsed regime approximately equals the density in the continuous wave regime at the same values of average specific power.

The use of the pulsed regime at a relatively low pulse repetition rate may be also useful for improving the quality of the films (Fig. 9). Table 1 shows parameters of CW (1) and pulsed (2,3) regimes. It is seen that at the same pressure the pulsed regime leads to lower mean specific power and mean density of atomic hydrogen (curve 2). This is explained by higher peak power in the pulsed regime, and, consequently, a larger plasma volume at the constant mean power. In the third case (curve 3), the pressure in the pulsed regime is greater than in the continuous wave regime. Plasma volume and mean specific power are same as in the CW discharge. But, the value of the mean density of atomic hydrogen squared is greater than in the continuous wave regime.

Analysis of the expressions determining the growth rate (1) and quality of the films (2) shows the following. If densities of atomic hydrogen and $CH_3$ stay constant, the growth rate does not change either. However, if at that the value of $[H]^2$ grows, as in the third case in Fig. 9, parameter $X_{def}$ becomes lower, and the quality of the film, correspondingly, improves. At constant density of hydrogen atoms the film growth rate increases as $CH_3$ density grows. The latter may be changed, e.g. by changing the contents of $CH_4$. However, if the value of $[H]^2$ does not change, the quality of the film will be lower. The use of the pulse regime (curve 3) makes it possible to increase the value of $[H]^2$ proportionally, thus increasing the growth rate while maintaining the quality of the film.

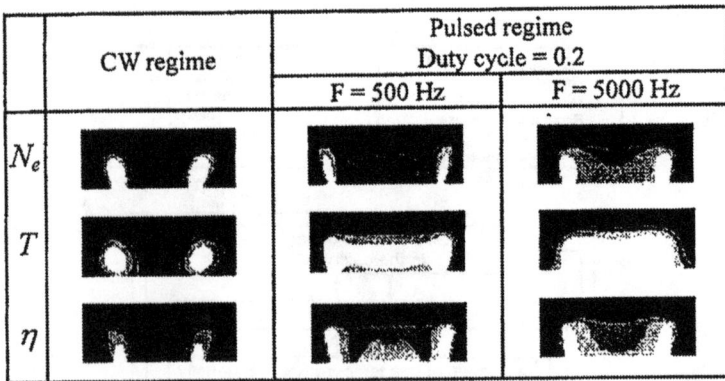

| | CW regime | Pulsed regime Duty cycle = 0.2 | |
|---|---|---|---|
| | | F = 500 Hz | F = 5000 Hz |
| $N_e$ | | | |
| $T$ | | | |
| $\eta$ | | | |

Figure 8. Spatial distributions of the discharge parameters in CW and pulsed regimes at a gas pressure of 75 Torr and an average microwave power of 4 kW. Lighter tones correspond to greater values. Illustration from [20].

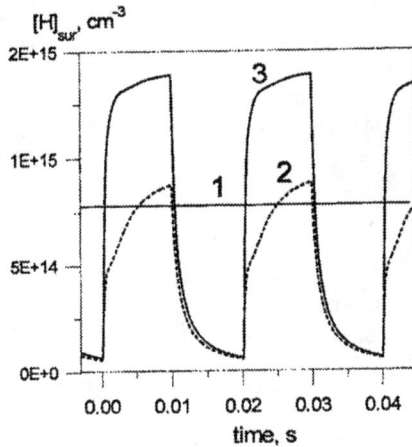

Figure 9. Time dependence of the atomic hydrogen density at the surface in CW (1) and pulsed regimes (2,3) at repetition frequency F = 50 Hz.

Table 1. Parameters of CW and pulsed regimes of operation of CVD reactor

| Curve | Power $<W>$, W | Repetition frequency F, Hz | Duty cycle | Pressure $p$, Torr | Average specific power $\eta$, W/cm$^3$ | $\langle[H]_{sur}\rangle$ cm$^{-3}$ | $\langle[H]^2_{sur}\rangle$ cm$^{-6}$ |
|---|---|---|---|---|---|---|---|
| 1 | 2000 | CW regime | | 100 | 79 | $7.8\cdot10^{14}$ | $6.1\cdot10^{29}$ |
| 2 | 2000 | 50 | 0.5 | 100 | 47 | $4.5\cdot10^{14}$ | $3.0\cdot10^{29}$ |
| 3 | 2000 | 50 | 0.5 | 150 | 78 | $7.5\cdot10^{14}$ | $8.8\cdot10^{29}$ |

Thus, the numerical analysis performed for the regimes of microwave discharge maintenance in the CVD reactor revealed the basic regularities of its operation. It was shown that the pulsed regime of discharge may be used for generation of homogeneous plasma at higher pressures than the CW regime and for improvement of the quality and growth rate of diamond films.

## 5. CONCLUSION

This paper analyzes the possibility to use the pulse-periodic microwave discharge for deposition of diamond films in different CVD reactors. Basing on Goodwin's model [8] we compared the CW and pulsed regimes of operation of CVD reactors taking into consideration the bulk processes only. Though it is known that the growth rate and quality of diamond films are determined by a complex set of bulk and surface reactors, simplified consideration makes it possible to conclude that the pulsed regime of operation of CVD reactor is attractive for application.

At low pressure diamond quality improved and growth rates are higher in the pulsed mode compared to continuous wave operation with the same average input microwave power.

At moderate and high pressures when the power reaches a certain threshold, the discharge becomes inhomogeneous. The use of the pulsed regime of discharge maintenance at high pulse repetition rate (a few kHz) makes it possible to obtain homogeneous plasma at a high level of microwave power and to increase the value of specific power. The use of microwave pulses with a not-too-high pulse repetition rate (tens of Hz) makes it possible to improve quality of the diamond film.

## Acknowledgements

This work was supported by CRDF under grant RE1-352, NWO under grant 047.011.00001, Russian Foundation for Basic Research (RFBR) under grant 00-02-16413 and Russian Ministry of Industry, Science and Technologies. The authors would like to express their gratitude to Professors J.Asmussen, T.A.Grotjohn, D.K.Reinhard (Michigan State University) for collaboration. Thanks are also due to Drs. K.F.Sergeichev and I.A.Sychov (General Physics Institute) for their helpful discussions.

## References

1.  Handbook of Industrial Diamonds and Diamond Films, Ed. by M.A.Prelas, G.Popovici and L.K.Bigelow, , New York-Basel-Hong Kong: Marcel Dekker, 1998.
2.  Litvak A.G. et al., Proc. 17th IAEA Fusion Energy Conference. Yokohama, Japan, 19-24 October 1998, IAEA-FI-CN-69/FTP/24.
3.  Sakamoto K. et al., Proc. 23rd Intern. Conf. on Infrared and Millimeter Waves, Ed. by T.J.Parker. S.R.P.Smith, University of Essex, UK, 1998.
4.  Kasugai A. et al., Rev. Scientific Instruments. 1998, **69**, 2160.
5.  Ralchenko V.G. et al., Diamond and Related Materials, 1997, **6**, 417.
6.  Markelevich Yu.A., Rakhimov A.T., Suetin N.B., Fizika Plazmy, 1995, **21**, 921 (in Russian).
7.  Weimer W.A., Cerio F.M., Johnson C.E., J. Mater. Res., 1991, **6**, 2134.
8.  Goodwin D.G., J. Appl. Phys. 1993, **74**, 6888 and 6895.
9.  Laimer J., Shimokawa M., Matsumoto S., Diamond and Related Materials, 1994, **3**, 231.
10. Laimer J., Matsumoto S., Plasma Chemistry and Plasma Processing, 1994, **14**, 117.
11. Poucques L., Bougdira J, Hugon R. et al., Proc. 24th Intern. Conf. on Phenomena in Ionized Gases, Warsaw, Poland, 1999, **1**, 203.
12. Hatta A., Kadota K., Mori Y. et al., Appl. Phys. Lett., 1995, **66**, 1602.
13. Ring Z., Mantei T.D., Tlali S., Jackson H., Appl. Phys. Lett., 1995, **66**, 3380.
14. Schelz S., Campillo C., Moisan M., Diamond and Related Materials, 1998, 7, 1675
15. Ivanov O.A. et al., J. Phys. IV France, 1998, **8**, Pr7-317.
16. Ivanov O.A., Koldanov V.A. Plasma Physics Reports, 2000, **26**, 902.
17. Rousseau A., Tomasini L., Gousset G. et al., J. Phys. D: Appl. Phys. 1994, **27**, 2439.
18. J. Asmussen, in High density plasma sources, Ed. by O.A.Popov, Park Ridge, NJ: Noyes, 1995, 251.
19. Kuo-Ping K., Asmussen J. Diamond and Related Materials, 1997, **6**, 1097.
20. Gorbachev A.M., Koldanov V.A., Vikharev A.L., Proc. Inern. Workshop "Strong Microwaves in Plasmas", Ed. A.G.Litvak, Nizhny Novgorod: IAP, 2000, **1**, 329.
21. Yee K.S., IEEE Trans. Antennas Propagat., 1966, **AP-14**, 302.
22. Tan W., Grotjohn T.A., Diamond and Related Materials, 1995, **4**, 1145.
23. Vikharev A.L., Gorbachev A.M., Ivanov O.A., Kolisko A.L. JETP, 1994, **79**, 94.